H. SCHEBITZ / H. WILKENS

Atlas of Radiographic Anatomy of the Dog and Cat

4th Edition

Atlas der Röntgenanatomie von Hund und Katze

4. Auflage

ATLAS
DER RÖNTGENANATOMIE
VON HUND UND KATZE

Dr. Dr. h. c. H. SCHEBITZ †

o. ö. Professor, ehem. Vorstand der Chirurgischen Tierklinik
Ludwig-Maximilians-Universität
München

Dr. H. WILKENS

Professor, Vorsitzender des Vorstands des Anatomischen Instituts
Tierärztliche Hochschule
Hannover

Vierte, neubearbeitete erweiterte Auflage

Mit 127 Röntgenbildern, 147 Röntgenskizzen und 79 Lagerungsskizzen sowie 4 Tabellen

1986

VERLAG PAUL PAREY · BERLIN UND HAMBURG
W. B. SAUNDERS COMPANY · PHILADELPHIA UND TORONTO

ATLAS OF RADIOGRAPHIC ANATOMY OF THE DOG AND CAT

Dr. Dr. h. c. H. SCHEBITZ †
Professor, former Head of the Clinic of Veterinary Surgery
Ludwig-Maximilians-Universität
Munich

Dr. H. WILKENS
Professor, Head of the Department of Anatomy
Tierärztliche Hochschule
Hanover

Fourth, revised enlarged edition

With 127 Radiographs, 147 Radiographic Sketches, 79 Positioning-Drawings, and 4 Tables

1986

PAUL PAREY SCIENTIFIC PUBLISHERS · BERLIN AND HAMBURG

W. B. SAUNDERS COMPANY · PHILADELPHIA AND TORONTO

Bilingual published in German and English by Paul Parey Scientific Publishers, Berlin and Hamburg, Germany.
Addresses: Lindenstrasse 44–47, D-1000 Berlin 61; Spitalerstrasse 12, D-2000 Hamburg 1.
Sales Rights for United States of America, Canada, Latin America, Australia and New Zealand: W. B. Saunders Company, Philadelphia and Toronto.

The 1st edition was published in 1967 under the title
"Atlas of Radiographic Anatomy of Dog and Horse"
ISBN 3-489-74816-6;

the 2nd edition was published in 1973 under the title
"Atlas of Radiographic Anatomy of Dog and Horse"
ISBN 3-489-75116-7.

(Beginning with the 3rd edition, there is a new arrangement: This volume deals with the "Radiographic Anatomy of the Dog and Cat" and an additional volume covers the "Radiographic Anatomy of the Horse".)

3rd edition 1977
ISBN 3-489-75416-6.

8887774

© 1986 by Paul Parey Scientific Publishers, Berlin and Hamburg, Germany
Addresses: Lindenstrasse 44 – 47, D-1000 Berlin 61; Spitalerstrasse 12, D-2000 Hamburg 1
Printed in Germany by Felgentreff & Goebel, D-1000 Berlin 61, Germany
Binding by Lüderitz & Bauer, D-1000 Berlin 61, Germany

ISBN 3-489-56416-2 Paul Parey Scientific Publishers, Berlin and Hamburg
ISBN 0-03-012122-1 W. B. Saunders Company, Philadelphia and Toronto · RAMAC 612122-2

Vorwort zur vierten Auflage

Durch die Aufnahme der Katze in den Atlas der Röntgenanatomie war bei der Vorbereitung der 3. Auflage eine Teilung nach Tierarten, und zwar Hund und Katze einerseits und Pferd andererseits, vorgenommen worden, die den Wünschen der praktizierenden Tierärzte entgegenkam. Der Verkauf dieser Auflagen innerhalb von acht Jahren im deutschen und englischen Sprachraum sowie die japanische Übersetzung in drei Tierartenausgaben haben Verlag und Herausgeber von der Richtigkeit der getroffenen Entscheidung überzeugt. Bereits seit 1982 haben sich die Herausgeber mit der Vorbereitung der 4. Auflage befaßt, die durch die Erkrankung von HORST SCHEBITZ zeitlich beeinträchtigt wurde. Dennoch hat er 24 neue Röntgenbilder von Hund und Katze, insbesondere des Gliedmaßenbereichs, erstellt und die dazu erforderlichen Lagerungen mit 11 weiteren Abbildungen bearbeitet, bis der Tod seinem arbeitsreichen Leben im Frühjahr 1985, kurz vor der Vollendung seines 65. Lebensjahres, das unerwartete Ende setzte. Dieser Verlust hat uns alle, aber mich in besonderem Maße schmerzlich getroffen. Um so mehr war es für mich eine Verpflichtung, seine Ideen umzusetzen und die Vorbereitung der überarbeiteten und erweiterten Neuauflage, die nunmehr auch 20 Röntgenskizzen der postnatalen Entwicklung des Gliedmaßenskeletts enthält, zum Abschluß zu bringen.

Mehr als zwei Jahrzehnte haben mich bei der Arbeit am Atlas der Röntgenanatomie mit HORST SCHEBITZ nicht nur kollegial, sondern auch freundschaftlich verbunden. Für die englischen Übersetzungen durfte ich die Hilfe von Herrn Dr. Dr. V. SOKOLOVSKY, Chikago, in Anspruch nehmen, der schon HORST SCHEBITZ uneigennützig und kollegial bei früheren Auflagen unterstützt hatte. Dafür danke ich Herrn Dr. Dr. SOKOLOVSKY besonders herzlich, auch wenn ich nicht alle wohlgemeinten Ratschläge für diese Neuauflage einarbeiten konnte.

Mein Dank gilt ebenso den Mitarbeitern in der Chirurgischen Universitäts-Tierklinik München für die neuen Röntgenaufnahmen sowie Herrn Maler und Graphiker E. THEEL für die Reinzeichnungen der Röntgenskizzen und Lagerungsabbildungen. Besonders danke ich Frau M.-L. MEINECKE für ihre unermüdliche, gewissenhafte Unterstützung bei allen vorbereitenden Arbeiten zur Drucklegung.

Herrn DDr. h. c. F. GEORGI und seiner Mitarbeiterin Frau H. LIESE gilt mein Dank für die altbewährte hervorragende verlegerische Betreuung und Gestaltung der Neuauflage im Verlag Paul Parey. Sie haben sich unseren Wünschen stets aufgeschlossen und entgegenkommend gezeigt und waren um die bestmögliche Wiedergabe der logetronisierten Röntgenbilder bei der modernen Drucktechnik rührend bemüht.

Ich hoffe, daß der Atlas der Röntgenanatomie von Hund und Katze, in dem die Arbeit von HORST SCHEBITZ fortlebt, eine gute Aufnahme findet.

Hannover, im Winter 1985/86

HELMUT WILKENS

Preface to the fourth edition

The cat was included for the first time in the previous edition of the Atlas of Radiographic Anatomy, thus making it convenient to divide the subject matter into two volumes, one dealing with the dog and cat and the other with the horse. This arrangement also served the wishes of veterinary practitioners and the demand shown during the past eight years for the German / English edition as well as for the Japanese translation of the three-species work, has confirmed the publishers in their judgement.

The 4th edition has been in preparation since 1982 but its publication was delayed because of the illness of HORST SCHEBITZ. Despite this, however, Dr. SCHEBITZ was able to provide 24 new radiographs, mainly on the limb regions of the dog and cat, and 11 illustrations of the positioning related to them. Dr. SCHEBITZ died in his 65th year in the spring of 1985. This loss has grieved us all, myself in particular, and I am cognizant of my obligation to incorporate his ideas in the new edition which was then being revised and enlarged. It now also contains 20 radiological sketches of the postnatal development of the appendicular skeleton. Through our work on the Atlas I have been associated with HORST SCHEBITZ, as colleague and friend, for more than two decades. Dr. Dr. SOKOLOVSKY of Chicago, who had already unselfishly aided HORST SCHEBITZ in previous editions, also helped me in the preparation of the English translation. For this I express my sincere thanks, even though it was not possible to incorporate all his well-intentioned suggestions in the new edition.

My thanks go to my collaborators in the Veterinary Surgical Clinic of Munich University for the new radiographs, and to the artist Mr. E. THEEL for the final drawings of the X-ray sketches and for illustrating the positioning of the animals. I have special thanks for Mrs. M.-L. MEINECKE for her tireless and conscientious support in the preparation of the material for printing.

I express my gratitude to DDr. F. GEORGI and Mrs. H. LIESE of the publishing house Paul Parey for the accustomed meticulous care with which they guided the new edition through all stages of publication. They were eager to comply with our wishes whenever possible, and were most concerned to ensure that the reproduction of the radiographs was to the highest standard that modern printing techniques can achieve.

I hope that the Atlas of Radiographic Anatomy of the Dog and Cat will find a ready acceptance, so ensuring that the work of HORST SCHEBITZ lives on.

Hanover, Winter 1985/86

HELMUT WILKENS

Vorwort zur ersten Auflage

Wenn man Röntgendiagnostik betreibt, ist es zweckmäßig, sich mit der Röntgenanatomie zu befassen. Auf diesem Spezialgebiet fehlten bislang Bücher, die dem Studierenden oder auch dem noch wenig mit der Röntgenologie vertrauten Tierarzt die normal-anatomischen Grundlagen für die Interpretation von Röntgenaufnahmen vermitteln. Da eine Einführung in dieses Fachgebiet nicht ohne vielseitiges Abbildungsmaterial erfolgen kann, haben Kliniker und Anatom gemeinsam einen Atlas der Röntgenanatomie von Hund und Pferd erstellt. Die Konzeption des vorliegenden Atlas ist auf die Belange der kurativen Praxis abgestimmt. Der Atlas enthält Hinweise für die Aufnahmetechnik, die bei gewisser Erfahrung und bei Verfügung über die notwendigen Geräte zu jederzeit reproduzierbaren Ergebnissen führen sollen.

Um einen möglichst großen Interessentenkreis zu gewinnen, erscheint der Atlas zweisprachig – deutsch und englisch.

Herrn Dr. Dr. V. SOKOLOVSKY, Chicago/USA, sei an dieser Stelle unser besonderer Dank für die Übersetzung der Legenden zu den Lagerungsskizzen und für die Beratung in allen weiteren Fragen bei der Übersetzungsarbeit ausgesprochen. Ebenso gilt unser Dank den Herren G. W. O. SPECKMANN und G. L. YOUNG für ihre tatkräftige Mithilfe bei der Übersetzung der Legenden zu den Röntgenskizzen, Herrn W. HEINEMANN für die Beratung bei der graphischen Gestaltung der Abbildungen, Herrn G. KAPITZKE für die Ausführung der Lagerungszeichnungen und Röntgenskizzen, Fräulein D. ABRAMOWSKI sowie Frau G. BÜRKLE für die Beschriftung der Röntgenskizzen, Herrn G. OBERST für die Vorlagen der Röntgenbilder, die er mit größter Sorgfalt angefertigt hat, und nicht zuletzt den Herren Drs. W. ZEDLER, M. SASCHEK, H. WISSDORF, K. NEURAND und Frau Dr. Chr. PAULICK sowie den med.-techn. Assistentinnen Fräulein M. L. MEINECKE, Fräulein B. GROSS und Fräulein A. VAN DER GROEBEN für ihre freundliche Unterstützung bei allen vorbereitenden Arbeiten zur Drucklegung.

Obwohl gegenüber der ersten Planung die Zahl der Röntgenaufnahmen von 60 auf 94 erhöht wurde, bleiben noch immer Wünsche nach Ergänzung einzelner Kapitel.

Dem Verlag Paul Parey gebührt für das Verständnis der erforderlich gewordenen Umfangsvermehrung des Atlas, insbesondere aber auch für die Ausstattung mit der gelungenen Wiedergabe der Röntgenaufnahmen, unser aufrichtiger Dank.

Möge der Atlas allen Interessenten ein brauchbarer Ratgeber sein.

München und Hannover, im Sommer 1967   HORST SCHEBITZ, HELMUT WILKENS

Preface to the first edition

To be proficient in the field of radiologic diagnosis, one must be familiar with the radiographic anatomy. Until recently, there has been no book providing the student or the insufficiently experienced veterinarian with the basic anatomy necessary for the interpretation of radiographs. As an introduction into this special field is not possible without abundant illustrations, clinician and anatomist have cooperated to compile an Atlas of Radiographic Anatomy of the Dog and Horse. This atlas was written for the practising veterinarian; consequently, it also contains instructions for radiographic technique which, with experience and the proper equipment, should produce consistently good radiographs.

In order to find access to a greater number of readers, the atlas is published in German and English.

Our special thanks go to Dr. Dr. V. SOKOLOVSKY, Chicago, U.S.A., for the translation of the legends to the positioning-drawings and consultation on all other aspects of the translation, and to Mr. G. W. O. SPECKMANN and Mr. G. L. YOUNG for their active participation in the translation of the legends to the drawings of the radiographs. We are indebted to Mr. W. HEINEMANN for consultation with respect to the illustrations, to Mr. G. KAPITZKE for the positioning-drawings and radiographic sketches, to Miss D. ABRAMOWSKI and Mrs. G. BUERKLE for the labelling of the drawings of the radiographs, to Mr. G. OBERST for the copies of the radiographs which were made with the greatest of care, and last but not least to Drs. W. ZEDLER, M. SASCHEK, H. WISSDORF, K. NEURAND, Chr. PAULICK, and also to the medical technicians Miss M. L. MEINECKE, Miss B. GROSS, and Miss A. VAN DER GROEBEN for their assistance in all preparatory work prior to printing.

Although the number of radiographs in this atlas has already been increased from 60 to 94, there are still individual chapters which call for improvement.

Our sincere thanks are due to the publishers, Paul Parey, for their understanding shown towards the inevitable increase in volume of the atlas, and for providing the fine reproductions of the radiographs. We hope that this atlas will be a useful addition to the veterinary medical literature.

Munich and Hanover, Summer of 1967   HORST SCHEBITZ, HELMUT WILKENS

Professor Dr. Dr. h. c. HORST SCHEBITZ
*20. April 1920   † 10. März 1985

Inhaltsverzeichnis – Contents

9

Einleitung

Die Gliederung des Atlas erfolgt nach Tierarten und innerhalb dieser bei den Aufnahmen von Skelett und Gelenken nach Körperregionen. Die Aufnahmen von den Organen der Körperhöhlen sind nach Organsystemen zusammengefaßt.

Die Röntgenaufnahmen wurden ausnahmslos an lebenden Hunden bzw. Katzen unter den für die Diagnostik zweckmäßigen Bedingungen angefertigt. Sie erscheinen im Atlas als Negativ reproduziert. Die Abbildung im Buch entspricht mithin der im durchfallenden Licht zu betrachtenden Röntgenaufnahme. Bei der Reproduktion von Röntgenaufnahmen müssen wegen der großen Schwärzungsdifferenzen geringe Verluste an Details in Kauf genommen werden. Da die Röntgenskizzen nach dem Original bei durchfallendem Licht gezeichnet wurden, sind in einigen Skizzen Details enthalten, die in den Röntgenabbildungen beim Druck verlorengegangen sind.

Jede Röntgenaufnahme wird durch eine Bildunterschrift erklärt: Es werden zunächst Körperteile und Organe genannt; darauf folgen die Lage des Tieres bzw. Körperteils und schließlich die Richtung des Strahlengangs.

Weiterhin werden technische Daten angegeben: Blende, Filmmaterial, Folien und Einstellung. Dabei gelten folgende Abkürzungen: FFA = Fokus-Film-Abstand, kV = Kilovolt; mAs = Milliampère-Sekunden-Produkt.

Jede Röntgenabbildung wird durch eine Röntgenskizze erläutert, die Abbildungshinweise enthält. Dabei erfolgt keine Beschränkung auf bestimmte Details. Dadurch erscheinen manche der Skizzen auf den ersten Blick mit Abbildungshinweisen überladen. Der Vorteil der Skizzen ist aber darin zu sehen, daß die Röntgenaufnahmen frei von erklärenden Hinweisen bleiben.

Folgende Hinweise sind bei den Röntgenskizzen zu beachten:

* Soweit differenzierbar und wenn in der Legende nicht besonders erwähnt, kennzeichnen Buchstaben und Zahlen mit hochgesetztem Strich (A'; 1″) filmferne Skelettteile.

** Bei Überlagerungen im Bereich der Gelenke sind die filmnahen Konturen durchgezogen, die filmfernen unterbrochen dargestellt.

Lagerungsskizzen veranschaulichen die Aufnahmetechnik mit der Fixierung des zu untersuchenden Tieres durch Hilfspersonen. Der Zentralstrahl ist durch eine gestrichelte Linie, seine Auftreffstelle am Tierkörper durch einen Punkt gekennzeichnet. Keiner Erörterung bedarf es, daß bei den Röntgenaufnahmen die Vorschriften über den Strahlenschutz zu berücksichtigen sind.

Im deutschen wie im englischen Sprachgebrauch werden ausschließlich Termini technici der Nomina Anatomica Veterinaria (1983) verwendet.

Richtungsbezeichnungen für den Strahlengang sind auf die Nomina Anatomica Veterinaria abgestimmt. Dabei sei auf die Publikation von HABEL et al. (1963) verwiesen.

Als Abkürzungen gelten: A. = Arteria, V. = Vena, M. = Musculus, N. = Nervus.

Die Bibliographie berücksichtigt nur die Bücher der Veterinär-Anatomie bzw. Röntgenologie sowie Publikationen, die zur Erstellung des Atlas herangezogen wurden oder bei vorbereitenden Untersuchungen Berücksichtigung fanden.

Auf ein Sachregister wurde bewußt verzichtet, da dieses nur eine Wiederholung der Termini darstellen würde und andererseits zur Benutzung des Atlas für bestimmte röntgenologische Fragestellungen anatomische Grundkenntnisse vorgesetzt werden dürfen.

Introduction

The atlas is arranged according to species, and for each species the radiographs of the skeleton and joints are arranged according to the body regions. The radiography of the internal organs are classified according to the organ systems.

All radiographs were taken of live dogs and cats under conditions best suitable for diagnostic purposes. They are reproduced as negatives. The illustrations, therefore, correspond to radiographs viewed under transillumination. Because of considerable differences in blacking, some detail is lost in the process of reproduction. Since the sketches were drawn from the original radiographs, trans-illuminated, some of them show details which were lost during reproduction of the radiographs.

Each radiograph is provided with a legend: skeletal parts and organs are mentioned first; this is followed by the positioning of the animal and the region for the body respectively and finally by the direction of the beam. Furthermore, technical data are provided: diaphragm, film, screen and setting.

The following abbreviations are used: FFD = Focus Film Distance, kV = kilovoltage, mAs = Milliampère second.

A detailed sketch with references accompanies each radiograph. At first sight many of the sketches appear to be overcrowded with references, however, it has the advantage that radiographs remain free of inscriptions.

The following directions should be observed in connection with the X-ray sketches:

* As far as skeletal parts can be identified and if not specifically mentioned in the legends, prime letters and numerals with stroke (A′, 1′) refer to parts next to the tube.

** In the case of superimposition of contours in the joint regions, those next to the film are indicated by continuous lines, those next to the tube by interrupted lines.

The drawings of the positioning illustrate the radiographic technique and the immobilisation of the animal. The central X-ray beam is shown by an interrupted line and the point of its impact on the body by a dot. Protective measures against radiation need hardly be mentioned.

The nomenclature used in this atlas is based on Nomina Anatomica Veterinaria (NAV 1983).

Directional terms for the X-ray beam are based on Nomina Anatomica Veterinaria. In this respect the reader is referred to the publication of HABEL et al. (1963).

Abbreviations: A. = Artery, V. = Vein, M. = Muscle, N. = Nerve.

In legends the official terms are not translated. Explanatory notes in German are translated.

The bibliography includes only those books and publications on veterinary anatomy and radiology which have direct bearing upon the compilation of this atlas.

We intentionally have forgone the index. It only would be a repetition of the termini technici. On the other hand, basic anatomic knowledge is a prerequisite in using the atlas for specific radiologic problems.

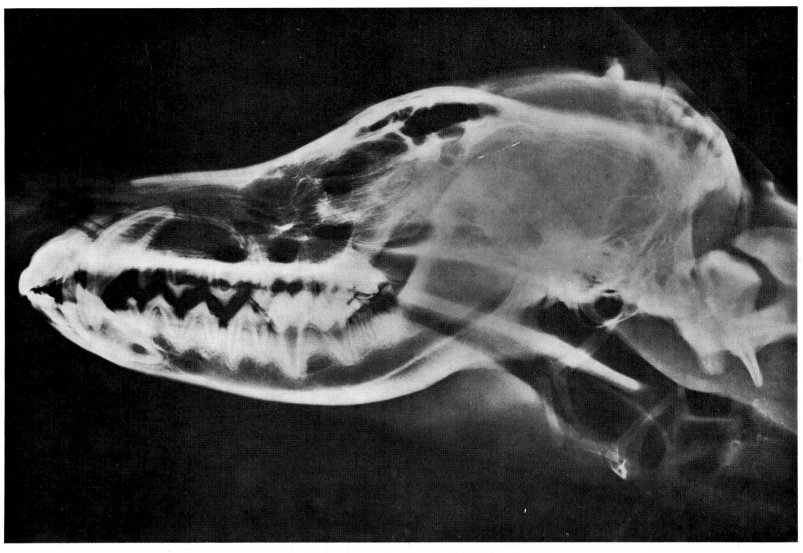

Abb. 1   Kopf. Latero-lateral. Deutscher Schäferhund, 5 Jahre.
Bucky-Blende — Feinzeichnende Folie — FFA 120 cm — 70 kV — 25mAs
Verkleinerung von 24 × 30 cm
Lagerung Abb. 7

Fig. 1   Head. Laterolateral. German shepherd dog, 5 years old.
Bucky diaphragm — High definition screens — FFD 120 cm — 70 kV — 25 mAs
Diminution of 24 × 30 cm
Positioning fig. 7

A   Ossa faciei;
B   Ossa cranii;
C   Mandibula;
D   Os hyoideum;
E   Atlas;
F   Axis;
G   Larynx;
H   Maxillare Dentes canini — Maxillary canine teeth;
J   Mandibulare Dentes canini — Mandibular canine teeth;
K   Maxillare Dentes praemolares IV — 4th maxillary praemolar teeth;
L   Mandibulare Dentes molares I —1st mandibular molar teeth;

Am Cranium - On the cranium:

1   Os incisivum;
2   Os nasale;
3   Os frontale;
4   Concha nasalis dorsalis;
5   Concha nasalis ventralis;
6   Maxilla, Crista conchalis;
7   Meatus nasi dorsalis;
8   Meatus nasi ventralis;
9   Canalis lacrimalis, rostrale Öffnung — Canalis lacrimalis, rostral opening;
10   Palatum durum;
11   Recessus maxillaris;

12   Processus uncinatus;
13   Foramen infraorbitale;
14   Canalis infraorbitalis;
15   Foramen maxillare;
16   Canalis palatinus major;
17   Foramen palatinum caudale;
18   Foramen sphenopalatinum;
19   Fossa sacci lacrimalis;
20   Orbita;
21   Os zygomaticum, Processus temporalis;
22   Os temporale, Processus zygomaticus;
23   Naht zwischen 21 und 22 — Suture between 21 and 22;
24   Sinus frontalis;
25   Labyrinthus ethmoidalis;
26   Fossa ethmoidalis;
27, 28   Jugum cerebrale:
27   Kaudale Begrenzung der Fossa ethmoidalis — Caudal border of the ethmoidal fossa,
28   Am Dach des Cavum cranii — At the roof of the cranial cavity;
29   Crista sagittalis externa;
30   Verschattung, die sich aus der Wölbung des Schädeldachs ergibt — Shadow formed by the curvature of the cranial roof;
31   Protuberantia occipitalis externa;
32   Tentorium cerebelli osseum;
33   Canalis sinus transversi;
34   Felsenbeinpyramide — Petrous temporal bone;

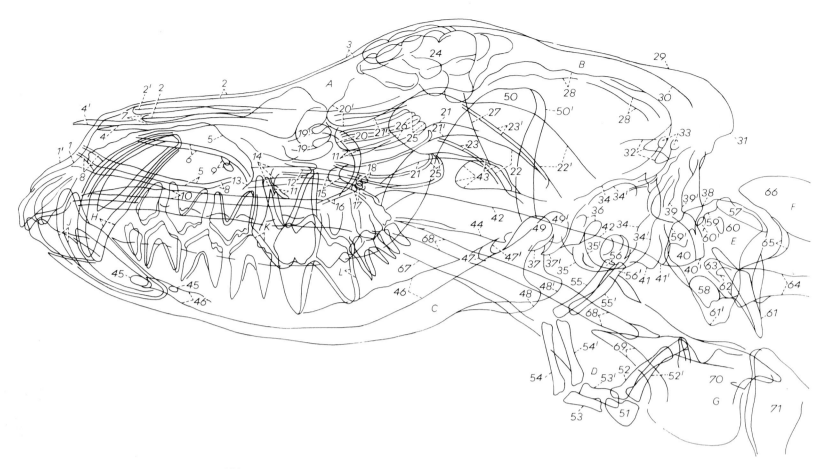

Abb. 2*   Röntgenskizze zu Abb. 1   Fig. 2*   X-ray sketch to fig. 1

35   Bulla tympanica;
36   Porus acusticus externus;
37   Processus retroarticularis;
38   Tuberculum nuchale;
39   Fossa condylaris dorsalis;
40   Condylus occipitalis;
41   Processus paracondylaris;
42   Basis cranii externa;
43   Os praesphenoidale;
44   Processus pterygoideus et Hamulus pterygoideus;
45   Foramina mentalia;

An der Mandibula — On the mandible:

46   Canalis mandibulae;
47   Foramen mandibulae;
48   Processus angularis;
49   Processus condylaris;
50   Processus coronoideus;

Am Os hyoideum — On the hyoid bone:

51   Basihyoideum;
52   Thyreohyoideum;
53   Ceratohyoideum;

54   Epihyoideum;
55   Stylohyoideum;
56   Tympanohyoideum;

Am Atlas — On the atlas:

57   Arcus dorsalis;
58   Arcus ventralis;
59   Fovea articularis cranialis;
60   Foramen vertebrale laterale;
61   Ala atlantis;
62   Foramen transversarium;

Am Axis — On the axis:

63   Dens;
64   Corpus;
65   Incisura vertebralis cranialis;
66   Processus spinosus;

67   Radix linguae;
68   Velum palatinum;
69   Epiglottis;
70   Cartilago thyreoidea;
71   Cartilago cricoidea.

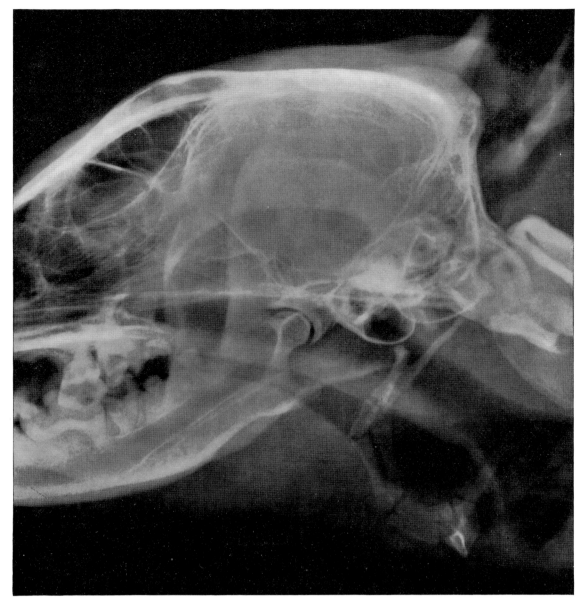

Abb. 3   Kiefergelenk. Latero-lateral. Deutscher Schäferhund, 7 Jahre (unvollständiges Gebiß).
Aufnahme aus der Durchleuchtung. Bucky-Blende — Mittelverstärkende Folie — FFA 100 cm —
60 kV — 12 mAs
Originalgröße (Ausschnitt aus 24 × 30 cm)

Fig. 3   Temporomandibular joint. Laterolateral. German shepherd dog, 7 years old (incomplete set of teeth).
Radiograph of fluoroscopic image. Bucky diaphragm — Standard screens — FFD 100 cm — 60 kV — 12 mAs
Original size (section of 24 × 30 cm)

A   Ossa faciei;
B   Ossa cranii;
C   Mandibula;
D   Os hyoideum;
E   Atlas;

a   Articulatio temporomandibularis;
b   Articulatio atlantooccipitalis;

Am Cranium — On the cranium:

1   Orbita;
2   Sinus frontalis;
3   Ethmoturbinale II im Eingang zum Sinus frontalis medialis — 2nd ethmoturbinate in the opening of the medial frontal sinus;
4   Meatus nasopharyngeus, Dach — Meatus nasopharyngeus, roof;
5   Palatum durum;
6   Canalis infraorbitalis, ventrale Begrenzung — Canalis infraorbitalis, ventral border;
7   Vomer;
8   Os zygomaticum, Processus temporalis;

9   Os temporale, Processus zygomaticus;
10   Naht zwischen 8 und 9 (Sutura temporozygomatica) — Suture between 8 and 9;
11   Processus retroarticularis;
12   Labyrinthus ethmoidalis;
13   Fossa ethmoidalis;
14, 15   Jugum cerebrale:
14   Kaudale Begrenzung der Fossa ethmoidalis - Caudal border of the ethmoidal fossa,
15   Am Dach des Cavum cranii — At the roof of the cranial cavity;
16   Verschattung, die sich aus der Wölbung des Schädeldachs ergibt — Shadow formed by the curvature of the cranium;
17   Crista sagittalis externa;
18   Crista nuchae;
19   Tentorium cerebelli osseum, Basis — Tentorium cerebelli osseum, base;
20   Canalis sinus transversi;
21   Meatus temporalis;
22   Foramen retroarticulare;
23   Protuberantia occipitalis externa;
24   Tuberculum nuchale;
25   Condylus occipitalis;

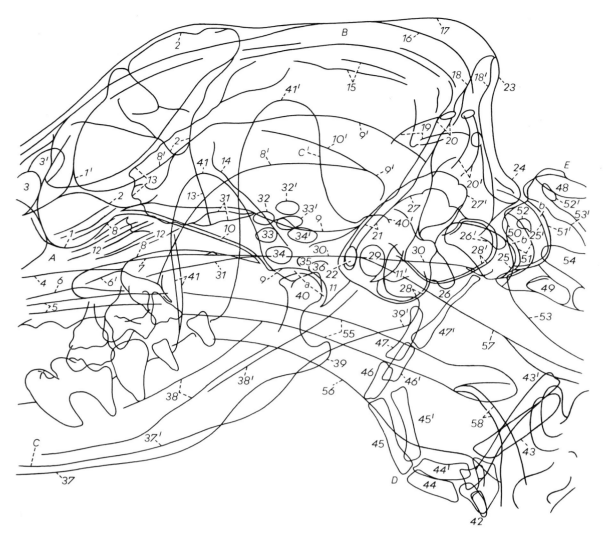

Abb. 4*   Röntgenskizze zu Abb. 3   Fig. 4*   X-ray sketch to fig. 3

26   Processus paracondylaris;
27   Felsenbeinpyramide — Petrous temporal bone;
28   Bulla tympanica;
29   Porus acusticus externus;
30   Basis cranii externa;
31   Os praesphenoidale;
32   Foramen opticum;
33   Fissura orbitalis;
34   Foramen alare rostrale;
35   Foramen alare caudale;
36   Foramen ovale;

An der Mandibula — On the mandible:

37   Corpus mandibulae;
38   Canalis mandibulae;
39   Processus angularis;
40   Processus condylaris;
41   Processus coronoideus;

Am Os hyoideum — On the hyoid bone:

42   Basihyoideum;

43   Thyreohyoideum;
44   Ceratohyoideum;
45   Epihyoideum;
46, 47   Stylohyoideum:
46   Distaler Abschnitt — Distal part,
47   Proximaler Abschnitt — Proximal part;

Am Atlas — On the atlas

48   Arcus dorsalis;
49   Arcus ventralis;
50   Fovea articularis cranialis, kranialer Rand — Fovea articularis cranialis, cranial border;
51   Verschattung, die sich aus der Konkavität von 50 ergibt — Shadow formed by the concavity of 50;
52   Foramen vertebrale laterale;
53   Ala atlantis;

54   Dens axis;
55   Velum palatinum;
56   Radix linguae;
57   Pharynx, dorsale Wand — Pharynx, dorsal wall;
58   Epiglottis.

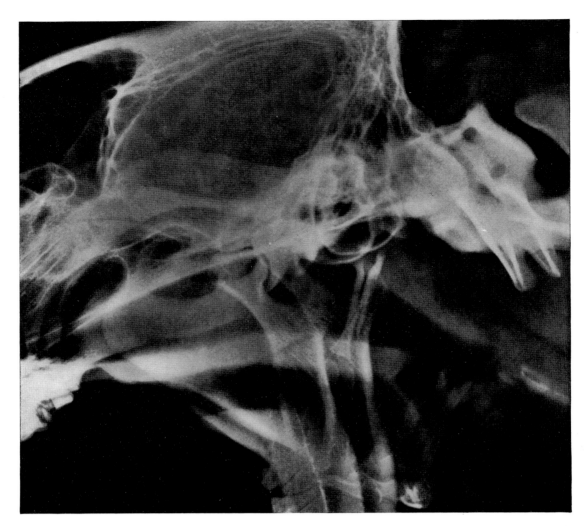

Abb. 5 Kiefergelenk bei geöffneter Mundhöhle. Latero-lateral. Deutscher Schäferhund, 5 Jahre.
Bucky-Blende — Feinzeichnende Folie — FFA 120 cm — 65 kV — 45 mAs
Originalgröße (Ausschnitt aus 24 × 30 cm)
Lagerung Abb. 8

Fig. 5 Temporomandibular joint with open mouth. Laterolateral. German shepherd dog, 5 years old.
Bucky diaphragm — High definition screens — FFD 120 cm — 65 kV — 45 mAs
Original size (section of 24 × 30 cm)
Positioning fig. 8

A Ossa faciei;
B Mandibula;
C Os hyoideum;
D Atlas;
E Axis;
M_2 Maxillarer Dens molaris II — 2nd maxillary molar tooth;
M_3 Mandibularer Dens molaris III - 3rd mandibular molar tooth;

a Articulatio temporomandibularis;
b Articulatio atlantooccipitalis;
c Articulatio atlantoaxialis;

Am Cranium — On the cranium:

1 Sinus frontalis;
2 Labyrinthus ethmoidalis;
3 Fossa ethmoidalis, rostrale Begrenzung — Fossa ethmoidalis, rostral border;
4, 5 Jugum cerebrale:
4 Kaudale Begrenzung der Fossa ethmoidalis — Caudal border of the ethmoidal fossa,
5 Am Dach des Cavum cranii — At the roof of the cranial cavity;

6 Crista nuchae;
7 Protuberantia occipitalis externa;
8 Tuberculum nuchale;
9 Tentorium cerebelli osseum, Basis — Tentorium cerebelli osseum, base;
10 Foramen mastoideum;
11 Condylus occipitalis;
12 Processus paracondylaris;
13 Basis cranii externa;
14 Felsenbeinpyramide — Petrous temporal bone;
15 Porus acusticus externus;
16 Processus retroarticularis;
17 Os praesphenoidale;
18 Fissura orbitalis;
19 Canalis opticus;
20 Arcus zygomaticus;
21 Naht zwischen Processus zygomaticus des Os temporale, kaudal, und Processus temporalis des Os zygomaticum, rostral, (Sutura temporozygomatica) – Suture between the zygomatic process of the temporal bone caudally and the temporal process of the zygomatic bone rostrally;

Abb. 6* Röntgenskizze zu Abb. 5   Fig. 6* X-ray sketch to fig. 5

22   Processus pterygoideus et Hamulus pterygoideus;
23   Foramen maxillare;
24   Palatum durum;

An der Mandibula — On the mandible:

25   Processus angularis;
26   Processus condylaris;
27   Processus coronoideus;

Am Os hyoideum — On the hyoid bone:

28   Basihyoideum;
29   Thyreohyoideum;
30   Ceratohyoideum;
31   Epihyoideum;
32   Stylohyoideum;
33   Tympanohyoideum;

Am Atlas — On the atlas:

34   Arcus dorsalis;
35   Arcus ventralis;

36   Fovea articularis cranialis, kranialer Rand – Fovea articularis cranialis, cranial border;
37   Verschattung, die sich aus der Konkavität von 36 ergibt – Shadow formed by the concavity of 36;
38   Foramen vertebrale laterale;
39   Ala atlantis;
40   Foramen transversarium;

Am Axis – On the axis:

41   Dens;
42   Processus articularis cranialis;
43   Incisura vertebralis cranialis;
44   Processus spinosus;

An Pharynx und Larynx – On the pharynx and larynx:

45   Velum palatinum;
46   Corpus linguae;
47   Epiglottis;
48   Cartilago thyreoidea;
49   Cartilago cricoidea.

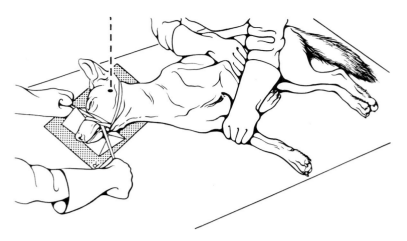

Abb. 7   Lagerung zur Aufnahme des Kopfes. Latero-lateral.

Der Kopf ist auf der Kassette (Bucky-Blende) so zu lagern, daß die Medianebene parallel zur Kassette und damit die beiden Unterkieferknochen senkrecht übereinander liegen. Mit Hilfe eines unter den rostralen Kopfbereich gelegten Schaumgummikeils kann der Kopf korrekt gelagert und fixiert werden.
Der Zentralstrahl sollte den Kopf in der Mitte zwischen Ohrgrund und temporalem Lidwinkel treffen und im rechten Winkel auf die Kassette einfallen.

Fig. 7   Positioning of head. Laterolateral.

The head should be aligned in such a manner, that the median plane runs parallel to the cassette (Bucky diaphragm) and both halves of the mandible lie perpendicularly upon each other.
A wedge-shaped foam rubber pad placed under the rostral part of the head will facilitate correct positioning and fixation.
The central beam should strike the head midway between the base of the ear and the lateral commissure of the eyelids and fall at the right angle on the cassette.

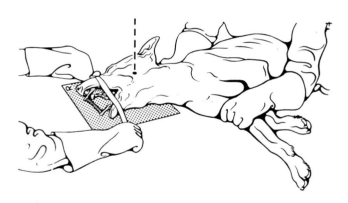

Abb. 8   Lagerung zur Aufnahme des Kiefergelenks bei geöffneter Mundhöhle. Latero-lateral.

Die Lagerung zur Aufnahme des Kiefergelenks bei geöffneter Mundspalte kann auf zweierlei Weise erfolgen. Zur Darstellung des filmfernen Kiefergelenks ist der Kopf so zu lagern, daß der Ober- und Unterkiefer der Kassette (Bucky-Blende) anliegen (Abb. 8). Der Winkel, den die palpierbare falsche Naht zwischen den Nasenbeinen mit der Kassette bildet, beträgt 15° rostral.
Zur Darstellung des filmnahen Kiefergelenks ist der Kopf so zu lagern, daß der Arcus zygomaticus der Kassette (Bucky-Blende) aufliegt. Die Kassette und die palpierbare falsche Naht zwischen den Nasenbeinen bilden nunmehr kaudal einen Winkel von 15° (Abb. 5). Diese Lagerung läßt sich durch Unterlegen eines Schaumgummikeils unter den geöffneten Fang erleichtern.
In beiden Fällen wird das zu untersuchende Gelenk ohne nennenswerte Überlagerungen vor dem der anderen Seite abgebildet.
Der Zentralstrahl sollte das zu untersuchende Kiefergelenk treffen und im rechten Winkel auf die Kassette einfallen.

Fig. 8   Positioning of temporomandibular joint with open mouth. Laterolateral.

The positioning for radiography of the temporomandibular joint with open mouth can be accomplished in two different ways. In order to demonstrate the temporomandibular joint next to the tube, the head must be placed with the upper and lower jaws resting on the cassette (Bucky diaphragm) (fig. 8). The angle between the palpable internasal suture and the cassette should be 15° rostrally.
For radiography of the temporomandibular joint next to film, the head should be placed with the zygomatic arch adjacent to the cassette (Bucky diaphragm). The angle between the cassette and the palpable internasal suture should be 15° caudally (fig. 5). This positioning can be facilitated by placing a foam rubber pad under the open fang.
•In both cases the joint to be examined can be portrayed without much superimposure of the opposite side.
The central beam should strike the joint and fall at the right angle on the cassette.

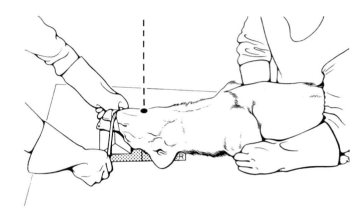

Abb. 9   Lagerung zur Aufnahme des Kopfes, Nase nicht anliegend. Ventro-dorsal.

Voraussetzung für eine sichere Orientierung ist eine gleichmäßige Abbildung der Kopfhälften. Da schon eine geringe Verkantung erhebliche Verzeichnungen und Überlagerungen verursacht, sollte in den Fällen, in denen sich der Patient aus Angst oder wegen Schmerzen gegen die Lagerung sträubt, ein Sedativum appliziert werden.
Die Lagerung eignet sich zur Anfertigung einer Übersichtsaufnahme. Es ist darauf zu achten, daß die Kehlränder des Unterkiefers in gleicher Höhe und der harte Gaumen parallel zur Kassette (Bucky-Blende) liegen. Durch Unterlegen eines Schaumgummikeils läßt sich die Lagerung erleichtern.
Der Zentralstrahl sollte den Kopf in der Mittellinie in Höhe des letzten Backenzahns treffen und im rechten Winkel auf die Kassette einfallen.

Fig. 9   Positioning of head with nose not resting on cassette. Ventrodorsal.

A symmetrical reproduction of both halves of the head is essential for proper radiographic evaluation. Even a minor tilting of the head causes considerable distortions and superimposures. Because of this, a restless patient (fear, pain) should be sedated.
This positioning is suitable for obtaining a routine radiograph. Both borders of the mandible must lie at the same level and the hard palate should be aligned parallel to the cassette (Bucky diaphragm). The positioning can be facilitated by using foam rubber pads.
The central beam should strike the head in the midline on the plane of the last molar and fall at the right angle on the cassette.

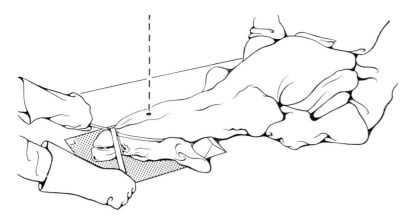

Abb. 10   Lagerung zur Aufnahme des Kopfes, Nasenrücken anliegend.
Ventro-dorsal.

Bei dieser Lagerung mit aufliegendem Nasenrücken sind die seitlichen
Buchten der Stirnhöhle und der rostrale Bereich der Schädelhöhle übersichtlicher dargestellt. Die Lagerung des Nasenrückens auf der Kassette
(Bucky-Blende) läßt sich durch Unterlegen eines Schaumgummikeils unter
die ersten Halswirbel erleichtern.
Der Zentralstrahl sollte den Kopf in der Mittellinie in Höhe des Angulus
mandibulae treffen und im rechten Winkel auf die Kassette einfallen.

Fig. 10   Positioning of head with dorsum nasi resting on cassette.
Ventrodorsal.

In this position with the dorsum nasi resting on the cassette (Bucky diaphragm), the lateral recesses of the frontal sinuses and the rostral part of
the cranial cavity are more clearly defined. Positioning of the dorsum nasi
can be facilitated by placing a wedge-shaped foam rubber pad under the
cranial cervical vertebrae.
The central beam should strike the midline of the head on a level of the
mandibular angle and fall at the right angle on the cassette.

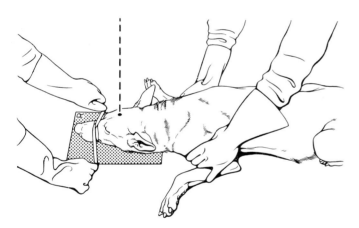

Abb. 11   Lagerung zur Aufnahme des Kopfes. Dorso-ventral.

Die Lagerung des Kopfes in Brust-Bauchlage ist für den Patienten beque-
mer als die Rückenlagerung. Bis auf wenige Ausnahmen kann bei dieser
Lagerung auf eine Sedierung verzichtet werden.
Die Kehlränder sollen der Kassette (Bucky-Blende) anliegen, so daß die
Medianebene senkrecht zur Kassette steht.
Der Zentralstrahl sollte den Kopf in der Mittellinie in Höhe des tempora-
len Augenwinkels treffen und im rechten Winkel auf die Kassette einfal-
len.

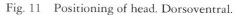

Fig. 11   Positioning of head. Dorsoventral.

Abdominal recumbency is more comfortable for the patient than lying on
its back. With few exceptions, this positioning does not require sedation
of the patient.
The borders of the mandible must rest on the cassette (Bucky diaphragm)
in order to align the median plane perpendicularly to the cassette.
The central beam should strike the midline of the head on a level of the
lateral commissure of the eyelids and fall at the right angle on the cassette.

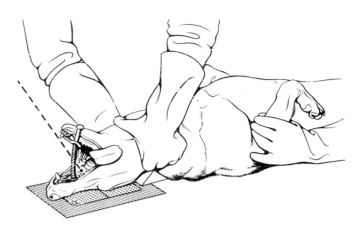

Abb. 12   Lagerung zur Aufnahme des Oberkiefers bei geöffneter
Mundspalte. Ventro-dorsal, Schrägprojektion.

Der Kopf ist mit der Stirn und dem Nasenrücken auf dem folienlosen
Film (Bleiunterlage unter dem Film nicht vergessen) zu lagern. Es ist darauf zu achten, daß die Medianebene senkrecht zum Film steht. Mit einem
unter die ersten Halswirbel gelegten Schaumgummikeil läßt sich die Lagerung erleichtern.
Der Zentralstrahl sollte in einem Winkel von 45 ° in die weit geöffnete
Mundhöhle einfallen und den harten Gaumen in der Mittellinie in Höhe
des 3. Prämolaren treffen.

Fig. 12   Positioning of upper jaw with open mouth. Obliquely
ventrodorsal.

The head must be placed with the frontal bones and the dorsum nasi upon
the non-screen film (lead blockers underneath the film not to be forgotten). Care must be exercised to align the median plane perpendicularly to
the film. The alignment can be facilitated by placing a wedge-shaped foam
rubber pad under the cranial cervical vertebrae.
The central beam should enter the widely opened oral cavity at an angle of
45 ° and strike the hard palate in the midline, on the plane of the 3rd
premolar.

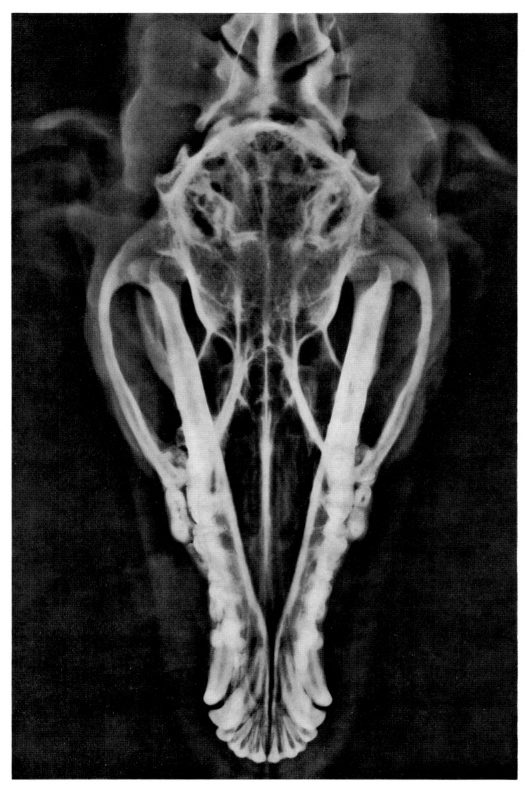

Abb. 13   Kopf. Nase nicht anliegend. Ventro-dorsal. Deutscher Schäferhund, 3 Jahre.
Bucky-Blende – Mittelverstärkende Folie – FFA 120 cm – 55 kV – 24 mAs
Verkleinerung von 24 × 30 cm
Lagerung Abb. 9

Fig. 13   Head. Nose not resting on cassette. Ventrodorsal. German shepherd dog,
3 years old.
Bucky diaphragm – Standard screens – FFD 120 cm – 55 kV – 24 mAs
Diminution of 24 × 30 cm
Positioning fig. 9

A	Axis;	
B	Atlas;	
C	Ossa cranii;	
D	Ossa faciei;	
E	Mandibula;	
F	Maxillarer Dens praemolaris IV – 4th maxillary premolar tooth;	
G	Mandibularer Dens molaris I – 1st mandibular molar tooth;	
H	Maxillarer Dens caninus – Maxillary canine tooth;	
J	Mandibularer Dens caninus – Mandibular canine tooth;	

Anteile des Os hyoideum und der Cartilago thyreoidea sind punktiert eingezeichnet – Parts of the hyoid bone and thyroid cartilage are shown by dotted lines;

a   Articulatio atlantoaxialis;
b   Articulatio atlantooccipitalis;
c   Articulatio temporomandibularis;

Am Axis – On the axis:

1 Dens;
2 Processus articularis cranialis;
3 Processus spinosus;
4 Incisura vertebralis cranialis;
5 Pediculus arcus vertebrae, nach medial zugleich seitliche Begrenzung des Foramen vertebrale - Pediculus arcus vertebrae medially forming the lateral border of the vertebral foramen;

Am Atlas – On the atlas:

6, 7 Arcus dorsalis:
6 Kranialer Rand – Cranial border,
7 Kaudaler Rand – Caudal border;
8, 9 Arcus ventralis:
8 Kranialer Rand – Cranial border,
9 Kaudaler Rand – Caudal border;
10 Fovea articularis cranialis;
11 Verschattung, die sich aus der Konkavität von 10 ergibt und in die seitliche Begrenzung des Foramen vertebrale übergeht – Shadow formed by the concavity of 10 and merging into the lateral margin of the vertebral foramen;
12 Fovea articularis caudalis;
13 Ala atlantis;
14 Incisura alaris;
15 Foramen transversarium;

Am Cranium – On the cranium;

16 Crista nuchae;
17 Squama occipitalis;
18 Tuberculum nuchale;
19 Condylus occipitalis;
20 Canalis condylaris;
21 Foramen magnum;
22 Kompaktaverstärkung an der Pars basilaris des Os occipitale – Basilar part of the occipital bone with thickened compacta;
23 Canalis nervi hypoglossi;
24 Foramen jugulare;
25 Processus paracondylaris;
26 Tentorium cerebelli osseum;
27 Canalis sinus transversi;
28 Öffnung und Übergang in den Sulcus sinus transversi – Opening leading into the groove for the transverse sinus;
29 Meatus temporalis;
30 Foramen retroarticulare;
31 Processus mastoideus;
32 Felsenbeinpyramide, mediale Begrenzung der Pars petrosa – Petrous temporal bone, medial border of the petrous part;
33 Crista partis petrosae;
34 Bulla tympanica;
35 Tuba auditiva;
36 Porus acusticus externus, in den Meatus acusticus externus übergehend – Porus acusticus externus merging into the external acoustic meatus;
37 Felsenbeinpyramide, äußere Begrenzung der Pars tympanica – Petrous temporal bone, lateral border of the tympanic part;
38 Foramen stylomastoideum;
39 Canalis nervi trigemini;
40 Canalis caroticus;
41 Foramen ovale;
42 Dorsum sellae;
43 Foramen rotundum;
44 Foramen alare caudale;
45 Canalis alaris;
46 Foramen alare rostrale;
47 Fissura orbitalis;
48 Canalis opticus;
49 Processus pterygoideus, nach kaudal in den Hamulus pterygoideus übergehend – Processus pterygoideus merging caudally into the hamulus of the pterygoid bone;
50 Crista sagittalis externa;
51 Cavum cranii, Wand – Cavum cranii, wall;
52 Fossa ethmoidalis;
53 Labyrinthus ethmoidalis;
54 Sinus frontalis;
55 Choanenrand mit Spina nasalis caudalis – Choanal border with caudal nasal spine;
56 Sutura palatina mediana;
57 Fissura palatina;

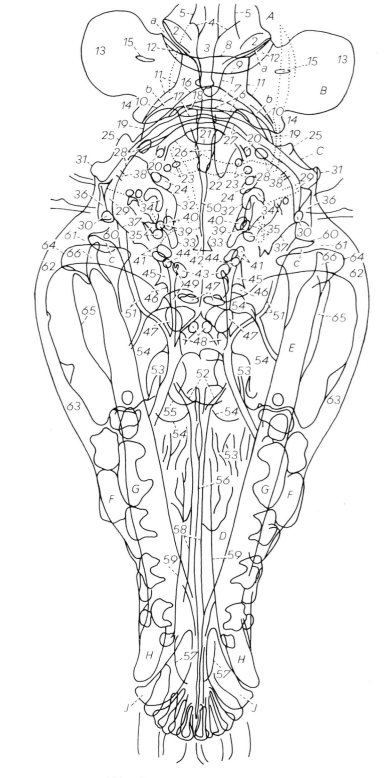

Abb. 14   Röntgenskizze zu Abb. 13

Fig. 14   X-ray sketch to fig. 13

58 Vomer;
59 Os nasale, lateraler Rand – Os nasale, lateral border;
60 Processus retroarticularis;
61 Fossa mandibularis;
62 Os temporale, Processus zygomaticus;
63 Os zygomaticum, Processus temporalis;

An der Mandibula – On the mandible:

64 Processus condylaris;
65 Processus coronoideus;
66 Processus angularis.

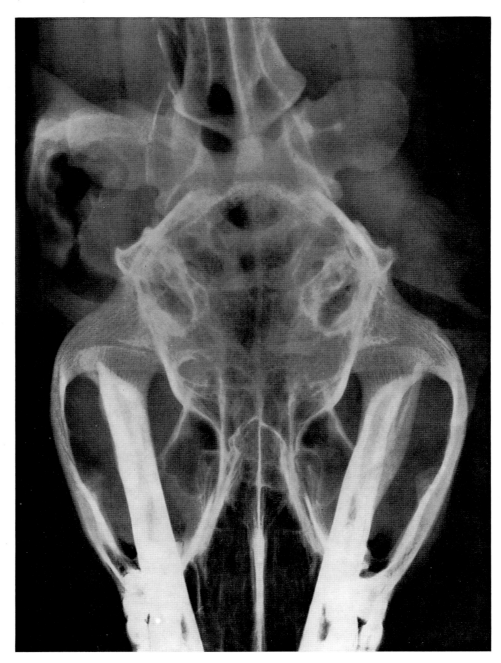

Abb. 15   Kopf. Nasenrücken anliegend. Ventro-dorsal. Deutscher Schäferhund, 4 Jahre.
Bucky-Blende – Feinzeichnende Folie – FFA 120 cm – 55 kV – 22 mAs
Verkleinerung von 24 × 30 cm
Lagerung Abb. 10

Fig. 15   Head. Dorsum nasi resting on cassette. Ventrodorsal. German shepherd dog,
4 years old.
Bucky diaphragm – High definition screens – FFD 120 cm – 55 kV – 22 mAs
Diminution of 24 × 30 cm
Positioning fig. 10

A   Axis;
B   Atlas;
C   Ossa cranii;
D   Ossa faciei;
E   Mandibula;
$M_1 - M_2$ Maxillare Dentes molares I-II – 1st and 2nd maxillary molar teeth;
$M_1' - M_3'$ Mandibulare Dentes molares I-III – 1st to 3rd mandibular molar teeth;
P_4 Maxillarer Dens praemolaris IV – 4th maxillary premolar tooth;

Anteile des Os hyoideum sind gestrichelt gezeichnet – Parts of the hyoid bone are indicated by broken lines;

a   Articulatio atlantoaxialis;
b   Articulatio atlantooccipitalis;
c   Articulatio temporomandibularis;

Am Axis – On the axis:

1   Dens;
2   Processus articularis cranialis;
3   Processus spinosus;
4   Incisura vertebralis cranialis;
5   Pediculus arcus vertebrae, nach medial zugleich seitliche Begrenzung des Foramen vertebrale – Pediculus arcus vertebrae medially forming the lateral border of the vertebral foramen;

Am Atlas – On the atlas:

6   Arcus dorsalis, kaudaler Rand – Arcus dorsalis, caudal border;
7   Arcus ventralis, kaudaler Rand – Arcus ventralis, caudal border;
8   Kranialer Rand des Arcus ventralis mit dem kranialen Rand des Arcus dorsalis ineinander projiziert – Cranial borders of the ventral and dorsal arches projected into one another;
9   Fovea articularis cranialis;

10 Verschattung, die sich aus der Konkavität von 9 ergibt und in die seitliche Begrenzung des Foramen vertebrale übergeht – Shadow formed by the concavity of 9 and merging into the lateral margin of the vertebral foramen;
11 Fovea articularis caudalis;
12 Ala atlantis;
13 Incisura alaris;
14 Foramen transversarium;

Am Cranium – On the cranium:

15 Crista nuchae;
16 Tuberculum nuchale, übergehend in die sich aus der Squama occipitalis ergebende Verschattung – Tuberculum nuchale merging into the shadow formed by the squamous part of the occipital bone;
17 Condylus occipitalis;
18, 19 Canalis condylaris:
18 Kaudale Öffnung – Caudal opening,
19 Rostrale Öffnung – Rostral opening;
20, 21 Foramen magnum:
20 Dorsaler Rand – Dorsal margin,
21 Ventraler Rand – Ventral margin;
22 Crista sagittalis externa;
23 Canalis nervi hypoglossi;
24 Foramen jugulare;
25 Processus paracondylaris;
26 Foramen retroarticulare;
26′ Meatus temporalis;
26″ Sulcus sinus transversi; Canalis sinus transversi nicht differenzierbar – Sulcus sinus transversi; canal of the transverse sinus not visible;
27 Processus mastoideus;
28 Felsenbeinpyramide – Petrous temporal bone;
29 Bulla tympanica;
30 Tuba auditiva;
31 Foramen stylomastoideum;
32 Porus acusticus externus;
33 Crista partis petrosae;
34 Foramen caroticum externum;
35 Foramen ovale;
36 Foramen alare caudale;
37 Canalis alaris;
38 Foramen alare rostrale;
39 Foramen rotundum;
40 Dorsum sellae;
41, 42 Fissura orbitalis:
41 Schädelhöhlenseitige Öffnung – Caudal opening,
42 Orbitaseitige Öffnung – Rostral opening;
43, 44 Canalis opticus:
43 Schädelhöhlenseitige Öffnung – Caudal opening,
44 Orbitaseitige Öffnung – Rostral opening;
45 Foramen ethmoidale;
46 Fossa ethmoidalis;
47 Cavum cranii, Wand – Cavum cranii, wall;
48 Os frontrale, Processus zygomaticus;
49 Sinus frontalis lateralis;
50 Sinus frontalis medialis;
51 Ectoturbinale II im Eingang zum Sinus frontalis medialis – 2nd ectoturbinate in the opening of the medial frontal sinus;
52 Ethmoturbinalien im Siebbeinlabyrinth – Ethmoturbinates in the ethmoidal labyrinth;
53 Choanenrand nach kaudal in den Processus pterygoideus mit Hamulus pterygoideus übergehend – Choanal border merging caudally into the pterygoid with the hamulus of the pterygoid bone;

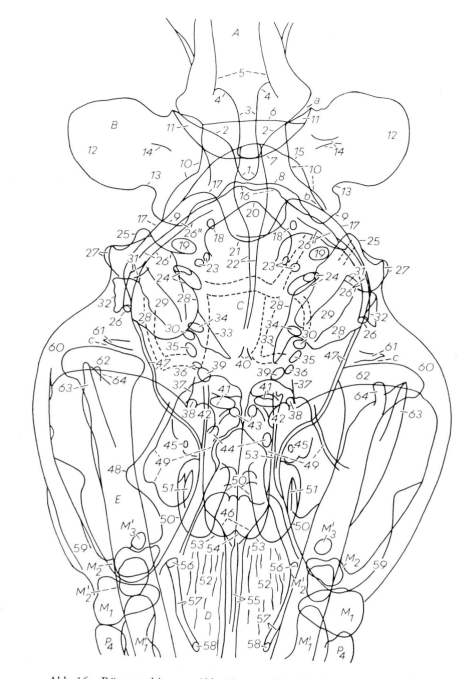

Abb. 16 Röntgenskizze zu Abb. 15   Fig. 16 X-ray sketch to fig. 15

54 Spina nasalis caudalis;
55 Vomer;
56 Foramen palatinum caudale;
57 Canalis palatinus major;
58 Foramen palatinum majus;
59 Os zygomaticum, Processus temporalis;
60 Os temporale, Processus zygomaticus;
61 Processus retroarticularis;

An der Mandibula – On the mandible:

62 Processus condylaris;
63 Processus coronoideus;
64 Processus angularis.

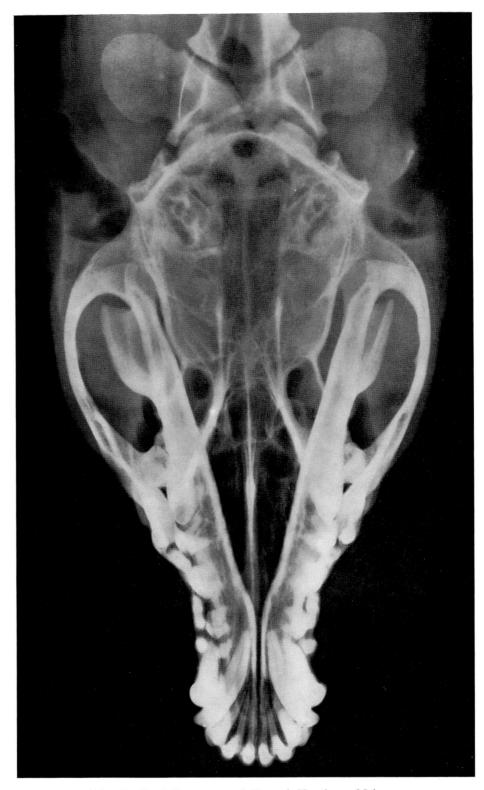

Abb. 17   Kopf. Dorso-ventral. Deutsch-Kurzhaar, 6 Jahre.
Bucky-Blende – Feinzeichnende Folie – FFA 120 cm – 65 kV – 45 mAs
Verkleinerung von 24 × 30 cm
Lagerung Abb. 11

Fig. 17   Head. Dorsoventral. German short-haired pointer, 6 years old.
Bucky diaphragm – High definition screens – FFD 120 cm – 65 kV – 45 mAs
Diminution of 24 × 30 cm
Positioning fig. 11

A   Axis;
B   Atlas;
C   Ossa cranii;
D   Ossa faciei;
E   Mandibula;
F   Maxillarer Dens molaris I – 1st maxillary molar tooth;
G   Mandibularer Dens molaris I – 1st mandibular molar tooth;
H   Maxillarer Dens caninus – Maxillary canine tooth;
J   Mandibularer Dens caninus – Mandibular canine tooth;

Anteile des Os hyoideum und der Cartilago thyreoidea sind punktiert eingezeichnet – Parts of the hyoid bone and thyroid cartilage are shown by dotted lines;

a   Articulatio atlantoaxialis;
b   Articulatio atlantooccipitalis;
c   Articulatio temporomandibularis;

Am Axis – On the axis:

1 Dens;
2 Processus articularis cranialis;
3 Processus spinosus;
4 Incisura vertebralis cranialis;
5 Foramen vertebrale, seitliche Begrenzung – Foramen vertebrale, lateral margin;
6, 7 Arcus dorsalis:
6 Kranialer Rand – Cranial border,
7 Kaudaler Rand – Caudal border;
8, 9 Arcus ventralis:
8 Kranialer Rand, abschnittsweise dargestellt – Cranial border, partly shown,
9 Kaudaler Rand – Caudal border;
10 Fovea articularis cranialis;
11 Verschattung, die sich aus deren Konkavität ergibt und in die seitliche Begrenzung des Foramen vertrebrale übergeht – Shadow formed by its concavity and merging into the lateral border of the vertebral foramen;
12 Fovea articularis caudalis;
13 Ala atlantis;
14 Incisura alaris;
15 Foramen transversarium;

Am Cranium – On the cranium;

16 Crista nuchae;
17 Squama occipitalis;
18 Tuberculum nuchale;
19 Condylus occipitalis;
20 Canalis condylaris;
21, 22 Foramen magnum:
21 Dorsaler Rand – Dorsal margin,
22 Ventraler Rand – Ventral margin;
23 Kompaktaverstärkung an der Pars basilaris des Os occipitale, nur in der linken Bildhälfte bezeichnet – Thickened compacta of the basilar part of the occipital bone, only shown on the left side of the figure;
24 Foramen mastoideum;
25 Canalis nervi hypoglossi;
26 Foramen jugulare;
27 Processus paracondylaris;
28 Meatus temporalis;
29 Foramen retroarticulare;
30 Processus mastoideus;
31 Felsenbeinpyramide, mediale Begrenzung der Pars petrosa – Petrous temporal bone, medial border of the petrous part;
32 Crista partis petrosae;
33 Canalis nervi trigemini;
34 Bulla tympanica;
35 Tuba auditiva;
36 Meatus acusticus externus;
37 Felsenbeinpyramide, äußere Begrenzung der Pars tympanica – Petrous temporal bone, lateral border of the tympanic part;
38 Processus muscularis;
39 Canalis caroticus;
40 Foramen ovale;
41 Tuberculum musculare;
42 Dorsum sellae;
43 Foramen rotundum;
44 Foramen alare caudale;
45 Canalis alaris;
46 Foramen alare rostrale;
47 Fissura orbitalis;
48 Canalis opticus;
49 Fossa ethmoidalis;
50 Cavum cranii, Wand – Cavum cranii, wall;
51 Choanenrand mit Spina nasalis caudalis – Choanal border with caudal nasal spine;
52 Processus pterygoideus;
53 Ethmoturbinalia;
54 Sinus frontalis;
55 Sutura palatina mediana et Vomer;
56 Fissura palatina;

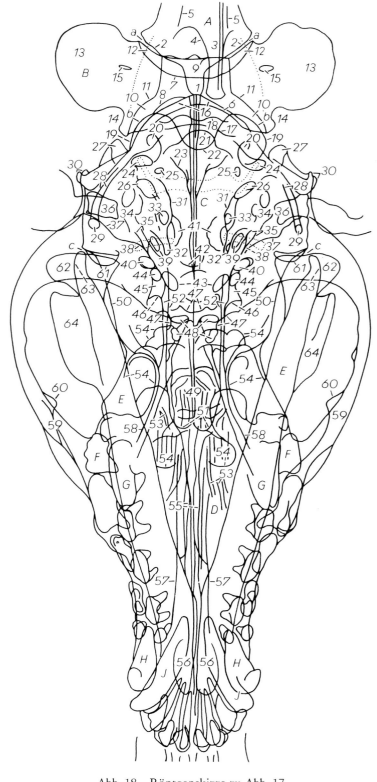

Abb. 18 Röntgenskizze zu Abb. 17
Fig. 18 X-ray sketch to fig. 17

57 Os nasale, lateraler Rand – Os nasale, lateral border;
58 Orbita, mediale Wand – Orbita, medial wall;
59 Arcus zygomaticus;
60 Os zygomaticum, Processus frontalis;
61 Processus retroarticularis;

An der Mandibula - On the mandible:

62 Processus condylaris;
63 Processus angularis;
64 Processus coronoideus.

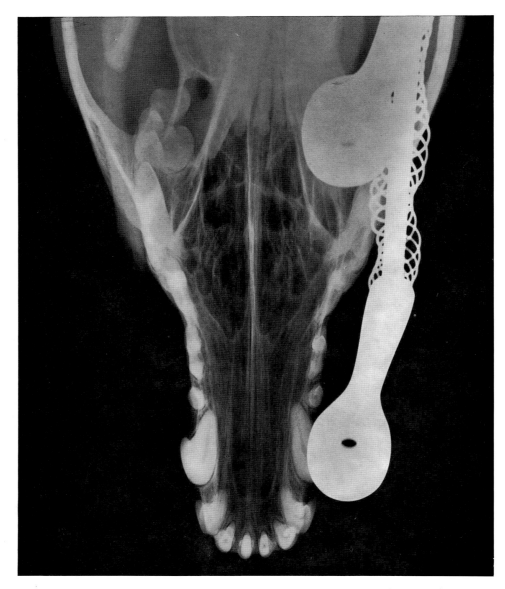

Abb. 19   Oberkiefer bei geöffneter Mundspalte. Ventro-dorsal, Schrägprojektion.
Deutscher Schäferhund, 3 Jahre.
Folienloser Film – FFA 120 cm – 70 kV – 22 mAs
Originalgröße
Lagerung Abb. 12

Fig. 19   Upper jaw with open mouth. Obliquely ventrodorsal.
German shepherd dog, 3 years old.
Non-screen film – FFD 120 cm – 70 kV – 22 mAs
Original size
Positioning fig. 12

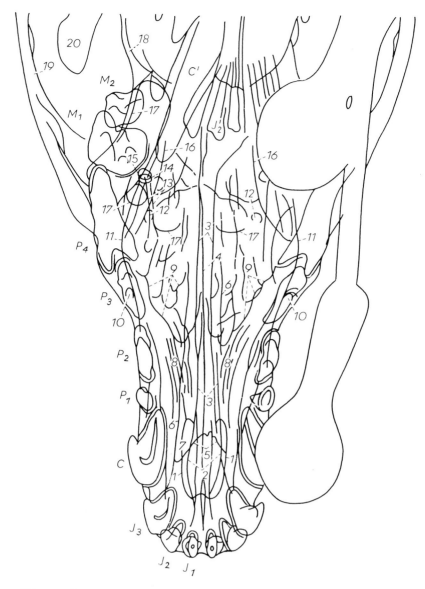

Abb. 20   Röntgenskizze zu Abb. 19   Fig. 20   X-ray sketch to fig. 19

$J_1 - J_3$	Dentes incisivi I-III;	9	Ethmoturbinalia;
C	Dens caninus;	10	Foramen infraorbitale;
$P_1 - P_4$	Dentes praemolares I-IV;	11	Canalis infraorbitalis, mediale Begrenzung – Canalis infraorbitalis, medial border;
$M_1 - M_2$	Dentes molares I-II;	12	Foramen palatinum majus;
J_2'	Dens incisivus II;	13	Canalis palatinus major;
C'	Mandibularer Dens caninus – Mandibular canine tooth;	14	Foramen palatinum caudale;
		15	Foramen sphenopalatinum;
1	Fissura palatina;	16	Recessus maxillaris, mediale Wand – Recessus maxillaris, medial wall;
2	Os incisivum, Processus palatinus;	17	Sinus frontalis rostralis, rostrale Begrenzung, abschnittsweise darge-stellt – Sinus frontalis rostralis, rostral border, partly shown;
3	Vomer;		
4	Sutura palatina mediana;		
5, 6, 7	Os nasale:	18	Cavum cranii, Wand – Cavum cranii, wall;
5	Rostraler Rand – Rostral border,	19	Arcus zygomaticus;
6	Lateraler Rand – Lateral border,	20	Mandibula, Processus coronoideus.
7	Rostrales Ende – Rostral end;		
8	Concha nasalis ventralis;		

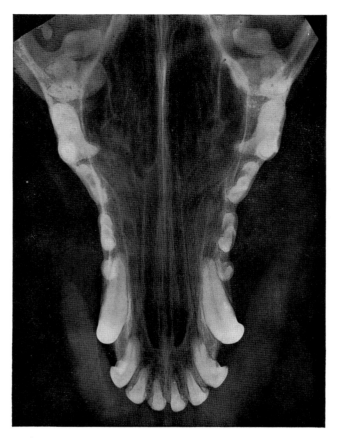

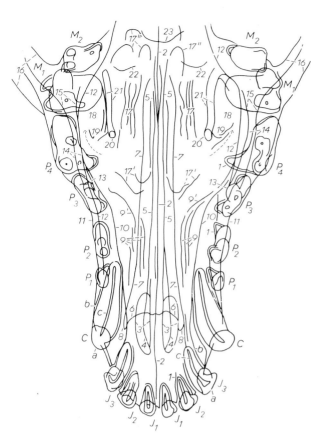

Abb. 21   Oberkiefer (Film in die Mundhöhle eingelegt).
Dorso-ventral. Deutscher Schäferhund, 3 Jahre.
Folienloser Film – FFA 100 cm – 65 kV – 30 mAs
Originalgröße
Lagerung Abb. 23

Fig. 21   Upper jaw (film placed in oral cavity). Dorso-
ventral. German shepherd dog, 3 years old.
Non-screen film – FFD 100 cm – 65 kV – 30 mAs
Original size
Positioning fig. 23

Abb. 22   Röntgenskizze zu Abb. 21
Fig. 22   X-ray sketch to fig. 21

$J_1 - J_3$ Dentes incisivi I-III;
C   Dens caninus;
$P_1 - P_4$ Dentes praemolares I-IV (P_4 Dens sectorius, Reißzahn) – (P_4 Carnassial tooth);
$M_1 - M_2$ Dentes molares I-II;

a   Corona dentis;
b   Radix dentis;
c   Cavum dentis, an J_3 und C beschriftet – Cavum dentis, labelled only on J_3 and C;

1   Alveolus dentalis, beziffert an J_2, P_2 und P_4 – Alveolus dentalis, labelled only on J_2, P_2 and P_4;
2   Sutura palatina mediana;
3   Fissura palatina;
4   Os incisivum, Processus palatinus, übergehend in 5 – Os incisivum, Processus palatinus merging into 5;
5   Vomer;
6, 7, 8   Os nasale:
6   Rostraler Rand - Rostral border,
7   Lateraler Rand - Lateral border,
8   Rostrales Ende - Rostral end;
9   Concha nasalis ventralis, in der Röntgenskizze nur zum Teil dargestellt - Concha nasalis ventralis only partly shown in the X-ray sketch;
9′   Crista conchalis;

10, 11   Processus alveolaris:
10   Linguale Begrenzung – Lingual border,
11   Bukkale Begrenzung – Buccal border;
12   Kompaktaschatten, der sich aus der Angesichtsfläche der Ossa faciei oberhalb der Alveolen ergibt – Shadow of compacta formed by the external surface of the facial bones above the alveoli;
13   Foramen infraorbitale;
14   Canalis infraorbitalis;
15   Foramen maxillare, das sich trichterförmig in die Fossa pterygopalatina öffnet – Foramen maxillare opening funnel-like into the pterygopalatine fossa;
16   Os zygomaticum, Processus temporalis, Arcus zygomaticus;
17   Ethmoturbinalia, in der Röntgenskizze nur zum Teil dargestellt – Ethmoturbinates only partly shown in the X-ray sketch:
17′   Rostrale Begrenzungen – Rostral limits,
17″   Kaudale Begrenzungen – Caudal limits;
18   Recessus maxillaris, Pfeil kennzeichnet die Apertura nasomaxillaris – Recessus maxillaris, arrow indicates the nasomaxillary aperture;
19   Processus uncinatus;
20   Foramen palatinum majus;
21   Canalis palatinus major; Foramen palatinum caudale nicht identifizierbar – Canalis palatinus major; caudal palatine foramen not visible;
22   Sinus frontalis, rostrale Begrenzung – Sinus frontalis, rostral limit;
23   Fossa ethmoidalis, rostrale Begrenzung – Fossa ethmoidalis, rostral limit.

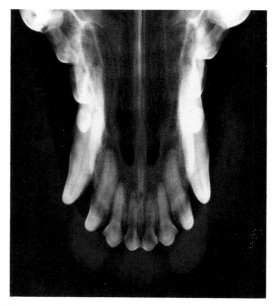

Abb. 23 Oberkiefer (Film in die Mundhöhle eingelegt).
Dorso-ventral, Schrägprojektion (60°). Pudel, 8 Jahre.
Folienloser Film – FFA 120 cm – 63 kV – 25 mAs
Originalgröße
Lagerung Abb. 25

Fig. 23 Upper jaw (film placed in oral cavity).
Obliquely dorsoventral (60°). Poodle, 8 years old.
Non-screen film – FFD 120 cm – 63 kV – 25 mAs
Original size
Positioning fig. 25

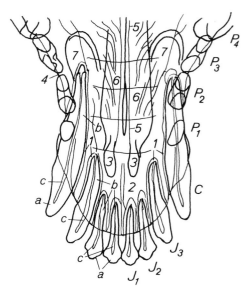

Abb. 24 Röntgenskizze zu Abb. 23
Fig. 24 X-ray sketch to fig. 23

$J_1 - J_3$ Dentes incisivi I–III;
C   Dens caninus;
$P_1 - P_4$ Dentes praemolares I–IV;

a   Corona dentis;
b   Radix dentis;
c   Cavum dentis;

1   Alveolus dentalis;
2   Sutura palatina mediana;
3   Fissura palatina;
4   Margo alveolaris;
5   Vomer;
6   Concha nasalis ventralis, in der Röntgenskizze nur zum Teil darge-
    stellt – Concha nasalis ventralis, only partly shown in the X-ray
    sketch;
7   Recessus maxillaris.

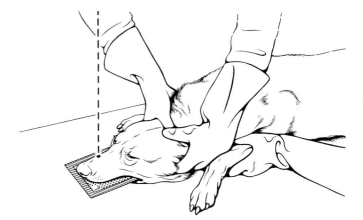

Abb. 25 Lagerung zur Aufnahme des Oberkiefers und der maxillaren
Schneidezähne (Film in die Mundhöhle eingelegt). Dorso-ventral.

Die Mundspalte ist nach dem Einlegen des folienlosen Filmes und des
Bleilatzes vorsichtig zu schließen. Bei unruhigen Tieren sollte der Fang
mit einer Bindenschlinge geschlossen werden. Um einer Verkantung vorzubeugen und zur besseren Fixierung ist der Kopf mit beiden Unterkieferknochen der Unterlage aufzulegen.

Oberkiefer: Der Zentralstrahl sollte den Nasenrücken in der Mittellinie in
Höhe des 3. Prämolaren treffen und im rechten Winkel auf den in der
Mundhöhle liegenden Film einfallen.
Schneidezähne: Der Zentralstrahl sollte den Margo alveolaris in der Me-
dianebene treffen und im Winkel von 60° auf den in der Mundhöhle lie-
genden Film einfallen.

Fig. 25 Positioning of upper jaw and maxillar incisors (film placed in
oral cavity). Dorsoventral.

After inserting the non-screen film with lead shielding, the mouth should
be closed carefully. In restless animals the fang should be closed with a
snare. In order to prevent tilting and to achieve a better fixation, the head
must be placed firmly with both halves of the mandible upon the table.
Upper jaw: The central beam should strike the dorsum nasi in the midline
on an level of the 3rd premolar and fall at the right angle on the film.
Incisors: The central beam should strike the alveolar border in the midline
and fall at an angle of 60° on the film.

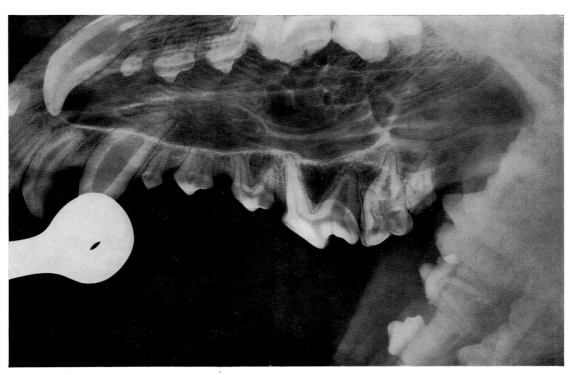

Abb. 26   Oberkiefer bei geöffneter Mundspalte. Schräglagerung. Medio-lateral. Deutscher Schäfer-
hund, 3 Jahre.
Folienloser Film – FFA 120 cm – 70 kV – 24 mAs
Originalgröße (Ausschnitt aus 13 × 18 cm)
Lagerung Abb. 27

Fig. 26   Upper jaw with open mouth. Oblique positioning. Mediolateral. German shepherd dog,
3 years old.
Non-screen film – FFD 120 cm – 70 kV – 24 mAs
Original size (section of 13 × 18 cm)
Positioning fig. 27

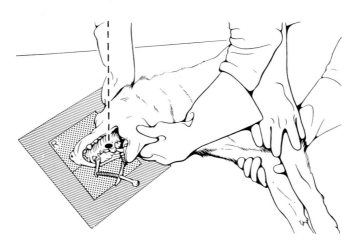

Abb. 27   Schräglagerung des Oberkiefers zur Aufnahme mit geöffneter
Mundspalte. Medio-lateral.

Fig. 27   Oblique positioning of upper jaw with open mouth.
Mediolateral.

Der Hund liegt in Seiten- oder halber Rückenlage. Der Kopf ist so zu lagern, daß sich der harte Gaumen in einem Winkel von 45° zum Film befindet und Jochbogen-Nase dem Film anliegt. Bei Verwendung eines folienlosen Filmes ist zur Vermeidung von Streustrahlung unter den Film
ein Bleilatz zu legen.
Der Zentralstrahl sollte den Alveolarrand in Höhe des 1. Molaren treffen
und im rechten Winkel auf den Film einfallen.

The dog lies in a lateral or a semidorsal position. The head should be
placed in such a manner that the hard palate forms an angle of 45° with
the film and the zygomatic arch and the nose lie adjacent to the film. If a
non-screen film is used, a lead shield should be placed under the film to
prevent scattered radiation.
The central beam should strike the alveolar border on a level of the 1st
molar and fall at the right angle on the film.

Abb. 28*   Röntgenskizze zu Abb. 26   Fig. 28*   X-ray sketch to fig. 26

J₂ – J₃   Dentes incisivi II-III;
C   Dens caninus;
P₁ – P₄   Dentes praemolares I-IV,
          P₄ des Oberkiefers - 4th maxillary premolar tooth,
          M₁ des Unterkiefers, Sectorius (Reißzahn) - 1st mandibular mo-
          lar tooth (carnassial tooth);
M₁ – M₃   Dentes molares I-III;

a   Corona dentis;
b   Radix dentis;

An dreiwurzeligen Molaren des Oberkiefers – On the molar teeth of the upper jaw (three roots):

b₁   Bukkale Wurzeln - Buccal roots;
b₂   Linguale Wurzeln - Lingual roots;
c   Cavum dentis;
d   Foramen apicis dentis;
e   Sattel zwischen bukkaler und lingualer Wurzel des M₁ – Ridge be-
          tween the buccal and lingual roots of the 1st molar tooth;

1   Margo alveolaris;
2   Kompaktaschatten, der sich aus dem Ansatz des knöchernen Gau-
          mens ergibt – Shadow of the compacta formed by the attachment of
          the hard palate;
3   Fissura palatina;
4   Sutura palatina mediana;
5   Os incisivum, rostraler Rand – Os incisivum, rostral border;
6   Cartilago nasi;
7   Os nasale, Processus nasalis;
8   Crista conchalis, benachbarte Verschattungen, die sich aus den Lamel-
          len der Concha nasalis ventralis ergeben, sind in der Röntgenskizze
          nicht eingezeichnet – Crista conchalis, adjacent shadows formed by
          the lamellae of the ventral nasal concha are not illustrated in the X-
          ray sketch;

9   Foramen infraorbitale;
10   Canalis infraorbitalis;
11   Foramen maxillare, das sich trichterförmig in die Fossa pterygopala-
          tina öffnet – Foramen maxillare opening funnel-like into the ptery-
          gopalatine fossa;
12   Sulcus palatinus major;
13   Foramen palatinum majus;
14   Canalis palatinus major;
15   Foramen palatinum caudale;
16   Foramen sphenopalatinum;
17   Canalis lacrimalis,
18   Rostrale Mündung ventral der Crista conchalis – Rostral opening
          ventral to the conchal crest;
19   Fossa sacci lacrimalis;
20   Processus uncinatus;
21   Apertura nasomaxillaris;
22   Recessus maxillaris;
23   Vomer;
24   Os ethmoidale, Basis – Os ethmoidale, base;
25   Ethmoturbinalia;
26   Choanenrand – Border of the choane;
27, 28   Os zygomaticum:
27   Processus temporalis,
28   Processus frontalis;
29   Orbita, ventraler Rand – Orbita, ventral border;
30   Processus pterygoideus, ventraler Rand – Processus pterygoideus,
          ventral border;

An der Mandibula – On the mandible:

31   Margo alveolaris;
32   Processus coronoideus.

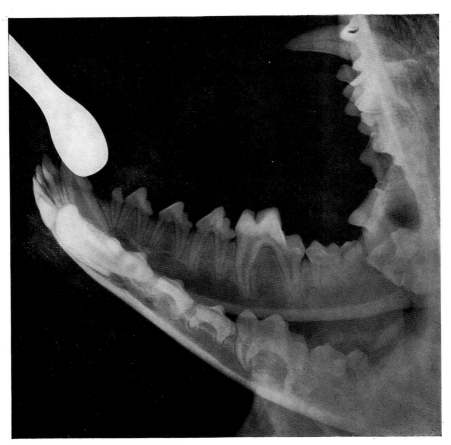

Abb. 29   Unterkiefer bei geöffneter Mundspalte. Schräglagerung.
Medio-lateral. Pudel, 1 Jahr.
Folienloser Film – FFA 100 cm – 60 kV – 18 mAs
Originalgröße (Ausschnitt aus 13 × 18 cm)
Lagerung Abb. 35

Fig. 29   Lower jaw with open mouth. Oblique positioning.
Mediolateral. Poodle, 1 year old.
Non-screen film – FFD 100 cm – 60 kV – 18 mAs
Original size (section of 13 × 18 cm)
Positioning fig. 35

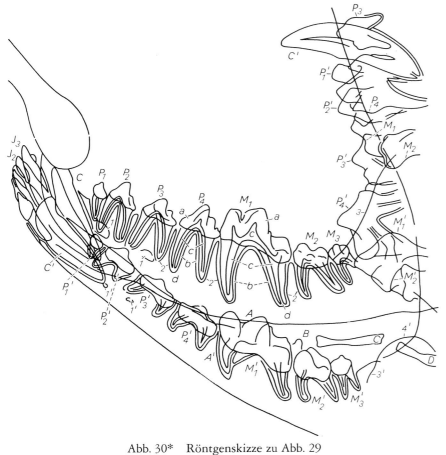

Abb. 30*   Röntgenskizze zu Abb. 29

Fig. 30*   X-ray sketch to fig. 29

A	Mandibula;
B	Ceratohyoideum;
C	Epihyoideum;
D	Stylohyoideum;
J_2 –J_3	Dentes incisivi II–III;
C	Dens caninus;
P_1 –P_4	Dentes praemolares I–IV;
M_1 – M_3	Dentes molares I–III, M_3 im Unterkiefer zweiwurzelig (meist einwurzelig) – 3rd molar tooth of the lower jaw with two roots (usually only one is present);
a	Corona dentis;
b	Radix dentis;
c	Cavum dentis;
d	Foramen apicis dentis;
1	Foramina mentalia;
2	Alveolus dentalis;
3	Ramus mandibulae, rostraler Rand – Ramus mandibulae, rostral border;
4	Processus coronoideus.

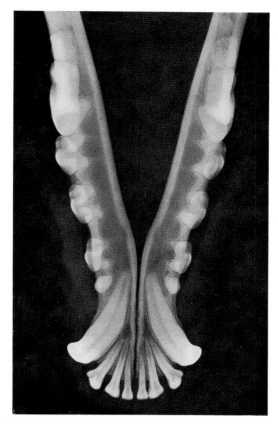

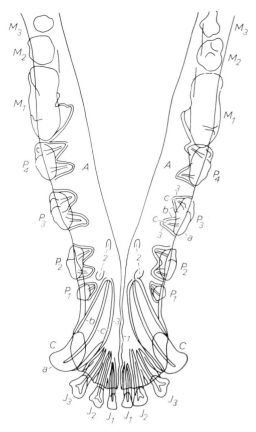

Abb. 31   Unterkiefer (Film in die Mundhöhle eingelegt). Ventro-dorsal.
Deutscher Schäferhund, 4 Jahre.
Folienloser Film – FFA 120 cm – 65 kV – 18 mAs
Originalgröße
Lagerung Abb. 36

Fig. 31   Lower jaw (film placed in oral cavity). Ventrodorsal.
German shepherd dog, 4 years old.
Non-screen-Film – FFD 120 cm – 65 kV – 18 mAs
Original size
Positioning fig. 36

Abb. 32   Röntgenskizze zu Abb. 31
Fig. 32   X-ray sketch to fig. 31

A	Corpus mandibulae;
$J_1 – J_3$	Dentes incisivi I – III;
C	Dens caninus;
$P_1 – P_4$	Dentes praemolares I – IV;
$M_1 – M_3$	Dentes molares I – III (nur in Abb. 32 – only in fig. 32);
a	Corona dentis;
b	Radix dentis;
c	Cavum dentis;
1	Synchondrosis intermandibularis;
2	Foramina mentalia;
3	Alveolus dentalis;
4	Margo alveolaris (nur bei Abb. 34 – only in fig. 34)

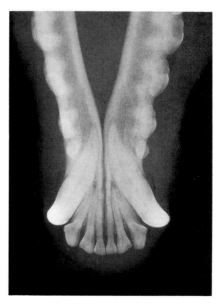

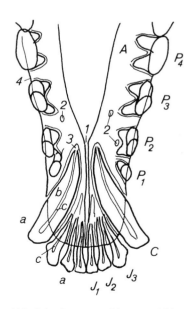

Abb. 33   Unterkiefer (Film in die Mundhöhle eingelegt).
Ventro-dorsal. Schrägprojektion (60°).
Pudel, 2 Jahre.
Folienloser Film – FFA 120 cm – 63 kV – 20 mAs
Originalgröße
Lagerung Abb. 36

Fig. 33   Lower jaw (film placed in oral cavity).
Obliquely ventrodorsal (60°).
Poodle, 2 years old.
Non-screen-film – FFD 120 cm – 63 kV – 20 mAs
Original size
Positioning fig. 36

Abb. 34   Röntgenskizze zu Abb. 33
Fig. 34   X-ray sketch to fig. 33

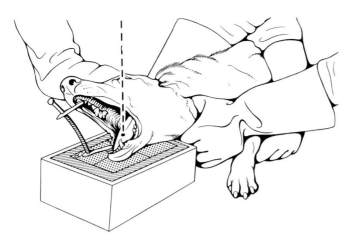

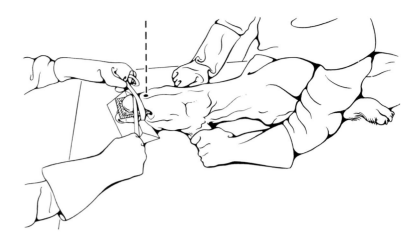

Abb. 35   Schräglagerung des Unterkiefers zur Aufnahme mit geöffneter Mundspalte. Medio-lateral.

Der Kopf des in Brust-Bauchlage befindlichen Hundes ist so zu lagern, daß der Zentralstrahl den Alveolarrand in Höhe des 1. Molaren in einem Winkel von 45° trifft und im rechten Winkel auf den Film einfällt. Bei Verwendung eines folienlosen Filmes ist zur Vermeidung von Streustrahlung unter den Film ein Bleilatz zu legen.

Fig. 35   Oblique positioning of lower jaw with open mouth. Mediolateral.

The head of the dog lying on its abdomen should be aligned in such a manner that the central beam strikes the alveolar border at an angle of 45° and falls perpendicularly on the film. If a non-screen film is used a lead shield should be placed underneath the film to avoid scattered radiation.

Abb. 36   Lagerung zur Aufnahme des Unterkiefers und der mandibularen Schneidezähne (Film in die Mundhöhle eingelegt). Ventro-dorsal.

Der Fang ist nach Einlegen des folienlosen Filmes und des Bleilatzes mit einer Bindenschlinge zu schließen. Der Kopf des auf dem Rücken liegenden Tieres ist so zu lagern, daß der harte Gaumen parallel zur Tischoberfläche liegt. Mit einem unter den Nasenrücken gelegten Schaumgummikeil läßt sich der Kopf leichter in die erforderliche Position bringen und fixieren.
Unterkiefer: Der Zentralstrahl sollte den Kehlgang in der Mittellinie in Höhe des 3. Prämolaren treffen und im rechten Winkel auf den in der Mundhöhle liegenden Film einfallen.
Schneidezähne: Der Zentralstrahl sollte den Margo alveolaris in der Mittellinie treffen und im Winkel von 60° auf den in der Mundhöhle liegenden Film einfallen.

Fig. 36   Positioning of lower jaw and mandibular incisors (film placed in oral cavity). Ventrodorsal.

After inserting the non-screen film and the lead shielding, the fang must be closed with a snare. With the animal lying on the back, the hard palate should be aligned parallel to the table. Supporting the dorsum nasi with a foam rubber pad facilitates proper alignment and stabilisation of the head.
Lower jaw: The central beam should strike the intermandibular space in the midline on a level of the 3rd premolar and fall at the right angle on the film.
Incisors: The central beam should strike the alveolar border in the midline and fall at an angle of 60° on the film.

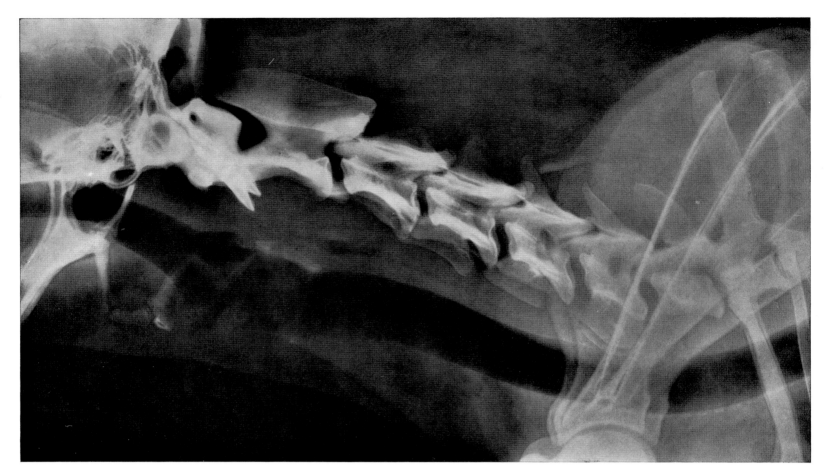

Abb. 37 Halswirbelsäule. Latero-lateral. Deutscher Schäferhund, 5 Jahre.
Bucky-Blende – Feinzeichnende Folie – FFA 120 cm – 68 kV – 58 mAs
Verkleinerung von 15 × 40 cm
Lagerung Abb. 41

Fig. 37 Cervical vertebral column. Laterolateral. German shepherd dog, 5 years old.
Bucky diaphragm – High definition screens – FFD 120 cm – 68 kV – 58 mAs
Diminution of 15 × 40 cm
Positioning fig. 41

A Ossa cranii;
B Mandibula;
C Atlas;
D Axis;
E 3. Vertebra cervicalis;
F 4. Vertebra cervicalis;
G 5. Vertebra cervicalis;
H 6. Vertebra cervicalis;
J 7. Vertebra cervicalis;
K 1. Vertebra thoracica;
L 1. Os costale;
M Scapula;
N Caput humeri;
O Stylohyoideum;
P Epihyoideum;
Q Ceratohyoideum;
R Basihyoideum;
S Thyreohyoideum;
T Epiglottis, dorsale Fläche - Epiglottis, dorsal surface;
U Cartilago thyreoidea;
V Cartilago cricoidea;
W Cartilago arytaenoidea, nur teilweise differenzierbar – Cartilago ary-taenoidea only partly visible;
X Teile einzelner Cartilagines tracheales – Parts of individual tracheal cartilages;

Am Cranium - On the cranium:

1 Basis cranii;
2 Processus pterygoideus et Hamulus pterygoideus;
3 Processus retroarticularis;

4 Bulla tympanica;
5 Meatus acusticus externus;
6 Processus paracondylaris;
7 Condylus occipitalis;
8 Tuberculum nuchale;
9 Squama occipitalis;
10 Protuberantia occipitalis externa;
11 Crista nuchae;
12 Tentorium cerebelli osseum;
13 Basis mit Canalis sinus transversi – Base with canal of the transverse sinus;
14 Os temporale, Processus zygomaticus;

An der Mandibula – On the mandible:

15 Processus condylaris;
16 Processus angularis;
17 Processus coronoideus;

Am Atlas – On the atlas:

18 Arcus dorsalis, dorsaler Abschnitt orthograph getroffen – Arcus dor-salis, dorsal part is struck orthographically;
19 Tuberculum dorsale;
20 Fovea articularis cranialis, Rand – Fovea articularis cranialis, border;
21 Arcus ventralis, ventraler Abschnitt orthograph getroffen – Arcus ventralis, ventral part is struck orthographically;
22 Tuberculum ventrale;
23 Fovea articularis caudalis, Rand – Fovea articularis caudalis, border;
24 Ala atlantis;

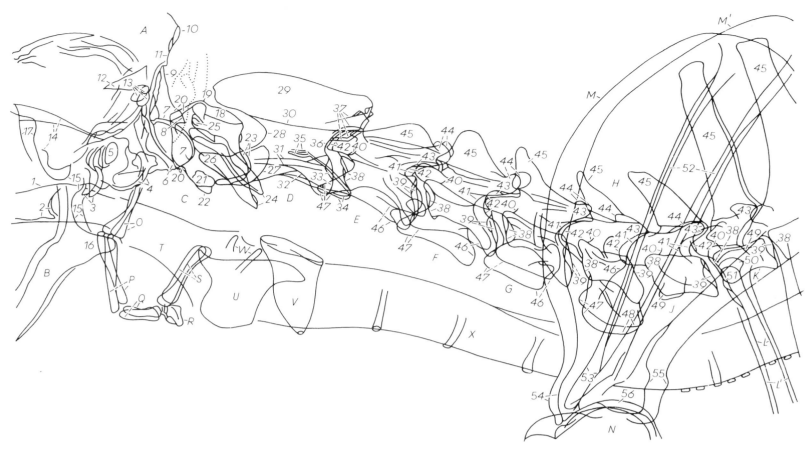

Abb. 38*   Röntgenskizze zu Abb. 37   Fig. 38*   X-ray sketch to fig. 37

25   Foramina vertebralia lateralia dextrum et sinistrum ineinander proji-
     ziert – Foramina vertebralia lateralia dextrum et sinistrum projected
     into one another;

Am Axis – On the axis:

26   Dens;
27   Processus articularis cranialis;
28   Incisura vertebralis cranialis;
29   Processus spinosus;
30, 31   Foramen vertebrale:
30   Dorsale Begrenzung, die sich aus der Konkavität des Arcus vertebrae
     ergibt – Dorsal margin formed by the concavity of the vertebral arch,
31   Orthograph getroffener Boden – Bottom is struck orthographically;
32   Crista ventralis;
33   Extremitas caudalis;
34   Processus transversi dexter et sinister ineinander projiziert – Proces-
     sus transversi dexter et sinister projected into one another;
35   Foramen transversarium, dorsale Begrenzung - Foramen transversa-
     rium, dorsal margin;
36   Incisura vertebralis caudalis;
37   Processus articulares caudales dexter et sinister ineinander projiziert
     – Processus articulares caudales dexter et sinister projected into
     another;

**An der 3. Vertebra cervicalis bis zur 1. Vertebra thoracica – On the 3rd
cervical to the 1st thoracic vertebrae:**

38   Extremitas cranialis;

39   Extremitas caudalis mit abschnittsweise intensiverem Kompakta-
     schatten, der sich aus ihrer Wölbung ergibt – Caudal extremity with
     regionally more intensive shadowing of the compacta formed by its
     curvature;
40   Incisura vertebralis cranialis;
41   Incisura vertebralis caudalis;
42   Processus articularis cranialis;
43   Processus articularis caudalis;
44   Muskelhöcker – Tubercle;
45   Processus spinosus;
46 – 49   Processus transversus:
46   Tuberculum dorsale,
47   Tuberculum ventrale,
48   Plattenförmige Verbreiterung, Lamina ventralis, an der 6. Vertebra
     cervicalis – Platelike expansion (Lamina ventralis) of the 6th cervical
     vertebra;

Am 1. Os costale – On the 1st rib:

50   Tuberculum costae;
51   Caput costae;

An der Scapula – On the scapula:

52   Spina scapulae;
53   Acromion;
54   Tuberculum supraglenoidale;
55   Tuberculum infraglenoidale;
56   Cavitas glenoidalis.

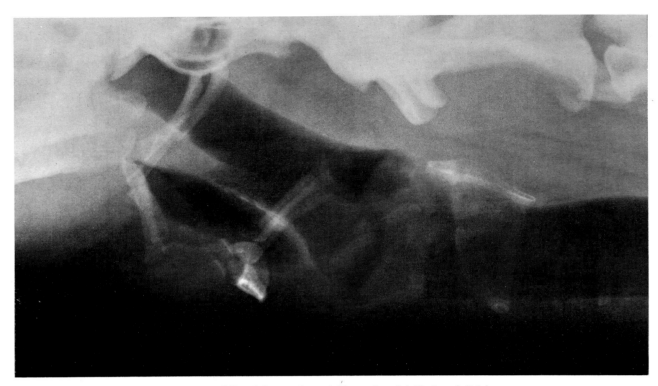

Abb. 39 Kehlkopf. Latero-lateral. Deutscher Schäferhund, 8 Jahre.
Bucky-Blende – Feinzeichnende Folie – FFA 120 cm – 60 kV – 20 mAs
Originalgröße (Ausschnitt aus 24 × 30 cm)
Lagerung Abb. 41

Fig. 39 Larynx. Laterolateral. German shepherd dog, 8 years old.
Bucky diaphragm – High definition screens – FFD 120 cm – 60 kV – 20 mAs
Original size (section of 24 × 30 cm)
Positioning fig. 41

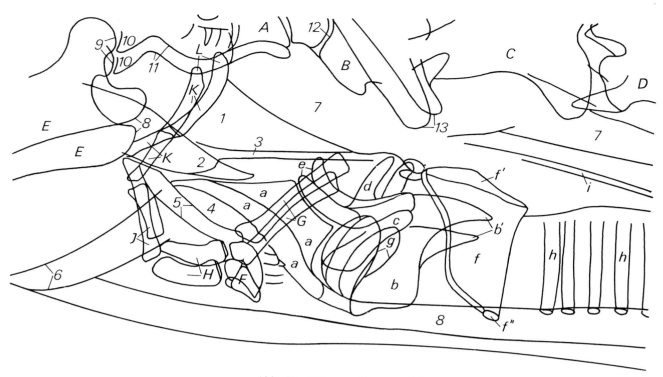

Abb. 40   Röntgenskizze zu Abb. 39

Fig. 40   X-ray sketch to fig. 39

A   Os occipitale;
B   Atlas;
C   Axis;
D   3. Vertebra cervicalis;
E   Mandibula;

Am Zungenbein – On the hyoid:

F   Basihyoideum;
G   Thyreohyoideum;
H   Ceratohyoideum;
J   Epihyoideum;
K   Stylohyoideum;
L   Tympanohyoideum;

a   Articulatio epiglottica;
b   Articulatio thyreohyoidea;
b′ Cornua caudalia;
c   Articulationes arytaenoideae;
d   Processus corniculatus;
e   Processus cuneiformis;
f   Cartilago cricoidea,
f′ Lamina cartilaginis cricoidea, orthograph getroffen – Lamina cartilaginis cricoidea, struck orthographically,

f″ Arcus cartilaginis cricoidea, orthograph getroffen – Arcus cartilaginis cricoidea, struck orthographically;
g   Ventriculus laryngis lateralis;
h   Cartilago trachealis;
i   Oesophagus;

1   Pars nasalis pharyngis;
2   Velum palatinum;
3   Ostium intrapharyngeum, seitliche Begrenzung – Ostium intrapharyngeum, lateral border;
4   Pars oralis pharyngis;
5   Zungengrund – Retrolingual region;
6   Begrenzung des Kehlgangs – Border of the laryngeal duct;
7   Verschattung, die sich aus der ventralen Kopf- und Halsmuskulatur ergibt – Shadow formed by the ventral musculature of the head and the neck;
8   Angulus mandibulae;
9   Processus condylaris;
10 Processus retroarticularis;
11 Bulla tympanica;
12 Articulatio atlantooccipitalis;
13 Alae atlantis.

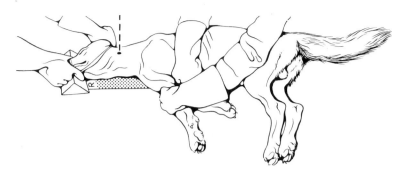

Abb. 41   Lagerung zur Aufnahme der Halswirbelsäule und des Kehlkopfs. Latero-lateral.

Person 1 lagert und fixiert den Kopf des Hundes in Seitenlage mit einer Hand am obenliegenden Ohr, mit der anderen Hand am Fang. Die am Rücken des Hundes stehende 2. Person zieht die Schultergliedmaßen gering nach kaudal und drückt mit den Unterarmen den Hund auf den Tisch. Wichtig ist, daß die Medianebene des Kopfes und die des Rumpfes parallel zur Unterlage ausgerichtet sind. Deshalb ist unter den rostralen Kopfbereich und bei fettleibigen Tieren auch unter die aufliegende Schultergliedmaße in Höhe der haltenden Hand ein Schaumgummikeil zu legen. Erforderliche Korrekturen besorgen die haltenden Hände am Ohr und an den obenliegenden Schultergliedmaßen.
Zur Reduzierung der Streustrahlung ist eine Bucky-Blende oder eine Kassette mit stehendem Raster erforderlich.
Halswirbelsäule: Der Zentralstrahl sollte den Hals ein- bis zweifingerbreit dorsal der Trachea in Höhe des 3. Halswirbels treffen und im rechten Winkel auf die Kassette einfallen.
Kehlkopf: Der Zentralstrahl sollte den Kehlkopf in Höhe des 1. Halswirbels treffen und im rechten Winkel auf die Kassette einfallen.

Fig. 41   Positioning of cervical vertebral column and larynx. Laterolateral.

One assistant places and secures the head of the dog in the lateral position by holding it with one hand at the earbase and with the other one at the fang. The second assistant, standing at the back of the animal, pulls the thoracic limbs slightly caudally, pressing the dog with his forearms against the table. It is important to align the median plane of the head and the trunk parallel to the table. For this reason, foam rubber pads should be placed under the muzzle and in obese animals also under the front limbs. Hands holding the animal at the earbase and the upper front leg should make the necessary corrections.
To reduce scattered radiation, a Bucky diaphragm or a cassette with a stationary grid is necessary.
Cervical column: The central beam should strike the neck one to two fingers width dorsally to the trachea on a level of the 3rd vertebra, and fall at the right angle on the cassette.
Larynx: The central beam should strike the larynx on a level of the 1st cervical vertebra, and fall at the right angle on the cassette.

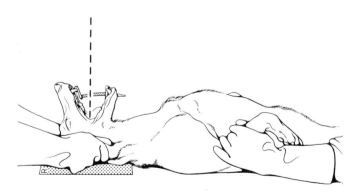

Abb. 42   Lagerung zur Aufnahme des 1. und 2. Halswirbels durch die geöffnete Mundhöhle. Ventro-dorsal.

Person 1 fixiert den Hund mit nach kaudal gestreckten Schultergliedmaßen in Rückenlage. Person 2 faßt den Kopf mit beiden Händen an den Ohren und richtet ihn so auf, daß der Nasenrücken bei geöffneter Mundhöhle annähernd einen rechten Winkel mit der Kassette bildet. Wichtig ist, daß die Medianebene des Kopfes und die des Rumpfes senkrecht zur Kassette stehen. Beim narkotisierten Hund ist die Zunge zur Erleichterung der Atmung vorzulagern.

Zur Reduzierung der Streustrahlung ist eine Bucky-Blende oder eine Kassette mit stehendem Raster zweckmäßig.
Der Zentralstrahl sollte die geöffnete Mundhöhle in der Mittellinie in Höhe des Zungenwulstes treffen und im rechten Winkel auf die Kassette einfallen.

Fig. 42   Positioning of 1st and 2nd cervical vertebrae through the open mouth. Ventrodorsal.

One assistant secures the dog on its back with the thoracic limbs stretched out caudally. The second assistant holds the head with both hands at the ears in such a manner that a right angle is formed approximately between the cassette and the dorsum nasi with the mouth opened. It is important to align the median plane of the head and trunk perpendicular to the cassette. In the anesthetized dog the tongue should be pulled out to facilitate respiration.
A Bucky diaphragm or a cassette with a stationary grid is used to reduce scattered radiation.
The central beam should strike the open oral cavity in the midline on a level of the root of the tongue and fall at the right angle on the cassette.

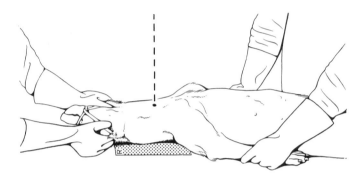

Abb. 43   Lagerung zur Aufnahme der Halswirbelsäule. Ventro-dorsal.

Person 1 hält den Hund mit nach kaudal gestreckten Schultergliedmaßen in Rückenlage. Person 2 streckt zunächst den Kopf so weit, daß die Kehlränder des Unterkiefers parallel zur Kassette liegen, und fixiert ihn in dieser Lage mit beiden Händen. Es ist darauf zu achten, daß die Medianebene des Kopfes und die des Rumpfes senkrecht zur Kassette stehen.
Zur Reduzierung der Streustrahlung ist eine Bucky-Blende oder eine Kassette mit stehendem Raster zweckmäßig.
Der Zentralstrahl sollte den Hals in der Mittellinie in Höhe des 3. Halswirbels treffen und im rechten Winkel auf die Kassette einfallen.

Fig. 43   Positioning of cervical vertebral column. Ventrodorsal.

One assistant secures the dog on its back with the thoracic limbs stretched out caudally. The second assistant extends the head to bring both borders of the mandible parallel to the cassette. The head is secured with both hands in this position. The median plane of the head and the trunk should be perpendicular to the cassette.
A Bucky diaphragm or a cassette with a stationary grid is used to reduce scattered radiation.
The central beam should strike the neck in the midline on an level of the 3rd cervical vertebra and fall at the right angle on the cassette.

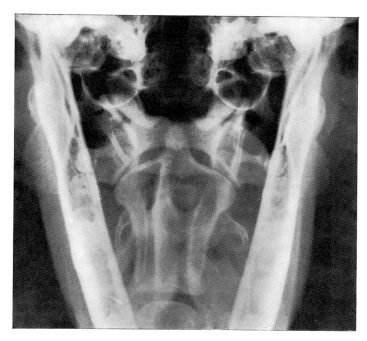

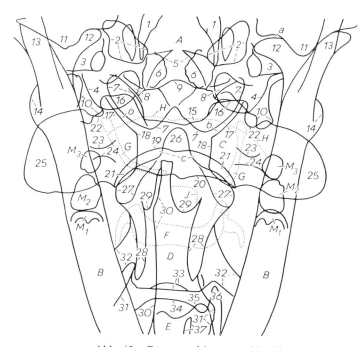

Abb. 44   1. und 2. Halswirbel durch die geöffnete Mundhöhle.
Ventro-dorsal. Deutscher Schäferhund, 4 Jahre.
Bucky-Blende – Mittelverstärkende Folie – FFA 120 cm –
65 kV – 36 mAs
Originalgröße (Ausschnitt aus 13 × 18 cm)
Lagerung Abb. 42

Fig. 44   1st and 2nd cervical vertebrae through the open mouth.
Ventrodorsal. German shepherd dog, 4 years old.
Bucky diaphragm – Standard screens – FFD 120 cm –
65 kV – 36 mAs
Original size (section of 13 × 18 cm)
Positioning fig. 42

Abb. 45   Röntgenskizze zu Abb. 44
Fig. 45   X-ray sketch to fig. 44

A   Ossa cranii;
B   Mandibula;
C   Atlas;
D   Axis;
E   3. Vertebra cervicalis;
F   Basihyoideum;
G   Stylohyoideum;
H   Dorsum linguae;
J   Epiglottis;
M_1 Mandibularer Dens molaris I – 1st mandibular molar tooth;
M_2 Mandibularer Dens molaris II – 2nd mandibular molar tooth;
M_3 Mandibularer Dens molaris III – 3rd mandibular molar tooth;

a   Articulatio temporomandibularis;
b   Articulatio atlantooccipitalis;
c   Articulatio atlantoaxialis;

Am Cranium – On the cranium:

1   Processus pterygoideus et Hamulus pterygoideus;
2   Os temporale, Pars petrosa;
3   Meatus acusticus externus;
4   Os temporale, Pars tympanica;
5   Crista partis petrosae;
6   Verschattung, die sich aus dem Tuberculum nuchale und dem Tuberculum musculare ergibt – Shadow formed by the nuchal and muscular tubercles;
7   Condylus occipitalis;
8   Knochenbrücke des Os occipitale zur Felsenbeinpyramide, die zugleich das Foramen jugulare kaudal begrenzt – Osseous connection of the occipital and petrous temporal bones demarcating the jugular foramen caudally;
9   Foramen magnum, dorsaler Rand – Foramen magnum, dorsal margin;
10   Processus paracondylaris;
11   Processus retroarticularis;

An der Mandibula – On the mandible:

12   Processus condylaris;
13   Processus coronoideus;
14   Processus angularis;

Am Atlas – On the atlas:

15   Vordere Begrenzung des Atlas; kranialer Rand des Arcus dorsalis und des Arcus ventralis sind ineinander projiziert – Cranial border of the atlas; cranial borders of the dorsal and ventral arches are projected into one another;
16   Fovea articularis cranialis, Rand - Fovea articularis cranialis, border;
17, 18   Verschattungen, die sich aus der Konkavität der Fovea articularis cranialis (17) und der des Foramen vertebrale (18), zugleich Begrenzung des Canalis vertebrae, ergeben – Shadows formed by the concavity of the cranial articular fovea (17) and the vertebral foramen (18), bordering the vertebral canal;
19   Arcus ventralis, kaudaler Rand – Arcus ventralis, caudal border;
20   Arcus dorsalis, kaudaler Rand – Arcus dorsalis, caudal border;
21   Fovea articularis caudalis;
22   Incisura alaris;
23, 24   Foramen transversarium:
23   Ventrale Begrenzung – Ventral margin,
24   Dorsale Begrenzung – Dorsal margin;
25   Ala atlantis;

Am Axis – On the axis:

26   Dens;
27   Processus articularis cranialis;
28   Pediculus arcus vertebrae, nach medial zugleich seitliche Begrenzung des Foramen vertebrale – Pediculus arcus vertebrae medially forming the lateral margin of the vertebral foramen;
29   Incisura vertebralis cranialis;
30   Processus spinosus;
31   Processus articularis caudalis;
32   Processus transversus;
33   Extremitas caudalis;

An der 3. Vertebra cervicalis – On the 3rd cervical vertebra:

34   Extremitas cranialis;
35   Processus articularis cranialis;
36   Processus transversus, Tuberculum ventrale;

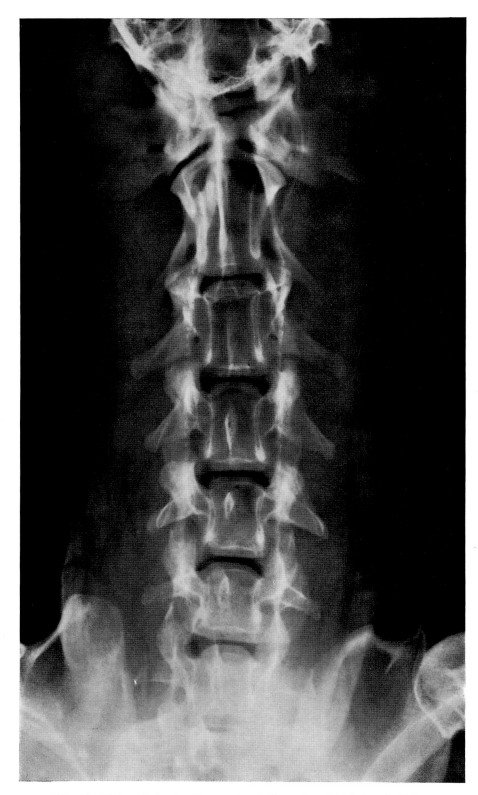

Abb. 46   Halswirbelsäule. Ventro-dorsal. Deutscher Schäferhund, 4 Jahre.
Bucky-Blende – Feinzeichnende Folie – FFA 120 cm – 68 kV – 50 mAs
Verkleinerung von 15 × 40 cm
Lagerung Abb. 43

Fig. 46   Cervical vertebral column. Ventrodorsal. German shepherd dog, 4 years old.
Bucky diaphragm – High definition screens – FFD 120 cm – 68 kV – 50 mAs
Diminution of 15 × 40 cm
Positioning fig. 43

A	Ossa cranii;	K	1. Os costale;
B	Atlas;	L	Scapula;
C	Axis;		
D	3. Vertebra cervicalis;	a	Articulatio atlantooccipitalis;
E	4. Vertebra cervicalis;	b	Articulatio atlantoaxialis;
F	5. Vertebra cervicalis;		
G	6. Vertebra cervicalis;		**Am Cranium – On the cranium:**
H	7. Vertebra cervicalis;	1	Crista sagittalis externa;
J	1. Vertebra thoracica;	2	Crista nuchae;

3 Condylus occipitalis;
4 Processus paracondylaris;
5 Processus mastoideus;
6 Meatus acusticus externus;
7 Foramen magnum, ventraler Rand – Foramen magnum, ventral margin;

Am Atlas – On the atlas:

8, 9 Arcus dorsalis:
8 Kranialer Rand – Cranial border,
9 Kaudaler Rand – Caudal border;
10, 11 Arcus ventralis:
10 Kranialer Rand – Cranial border,
11 Kaudaler Rand – Caudal border;
12 Tuberculum ventrale;
13 Fovea articularis cranialis;
14 Verschattung, die sich aus deren Konkavität ergibt und in die seitliche Begrenzung des Foramen vertebrale (15) übergeht – Shadow formed by the concavity of the cranial articular fovea and blending with the lateral margin of the vertebral foramen (15);
16 Ala atlantis;
17 Incisura alaris;
18 Foramen vertebrale laterale;
19, 20, 21 Foramen transversarium:
20 Kraniale Begrenzung auf der Dorsalfläche – Cranial limit on the dorsal surface,
21 Kaudale Begrenzung auf der Ventralfläche des Atlasflügels, Fossa atlantis – Caudal limit on the ventral surface of the wing of atlas, atlantal fossa;
22 Fovea articularis caudalis;

Am Axis – On the axis:

23 Dens mit randständiger Verschattung, projektionsbedingt – Shadowed margin of the dens caused by projection;
24, 25, 26 Processus articularis cranialis:
24 Weiteste Vorwölbung – Most cranial curvature,
25 Dorsale Begrenzung – Dorsal border,
26 Ventrale Begrenzung – Ventral border;
27 Processus spinosus;
28 Incisura vertebralis cranialis;
29 Processus articularis caudalis;
30 Verbreiterter Processus spinosus, kaudaler Rand – Caudal border of the expanded spinous process;
31 Extremitas caudalis;
32 Kompaktaschatten, der sich aus deren Konkavität ergibt – Shadow of compacta formed by its concavity;
33 Crista ventralis;
34 Processus transversus;
35 Foramen transversarium;
36 Pediculus arcus vertebrae, nach medial zugleich seitliche Begrenzung des Foramen vertebrale – Pediculus arcus vertebrae medially forming the lateral margin of the vertebral foramen;

An der 3. Vertebra cervicalis bis zur 1. Vertebra thoracica – On the 3rd cervical to the 1st thoracic vertebrae:

37 Extremitas cranialis,
38 Kompaktaschatten, der sich aus deren Wölbung ergibt – Shadow of compacta formed by its curvature;
39 Extremitas caudalis, ventrale Begrenzung – Extremitas caudalis, ventral border;
40 Kompaktaschatten, die sich aus der Konkavität und der dorsalen Begrenzung der Extremitas caudalis ergeben – Shadows of compacta formed by the concavity and the dorsal border of the caudal extremity;
41, 42 Arcus vertebrae:
41 Kranialer Rand – Cranial border,
42 Kaudaler Rand – Caudal border;
43 Processus articularis cranialis;
44 Processus articularis caudalis;
45 Muskelhöcker am Processus articularis caudalis, nur an der 6. und 7. Vertebra cervicalis verifizierbar – Tubercle of the caudal articular process, only visible on the 6th and 7th cervical vertebrae;
46 Processus spinosus;
47, 48, 49 Processus transversus:
47 Tuberculum dorsale,
48 Tuberculum ventrale,

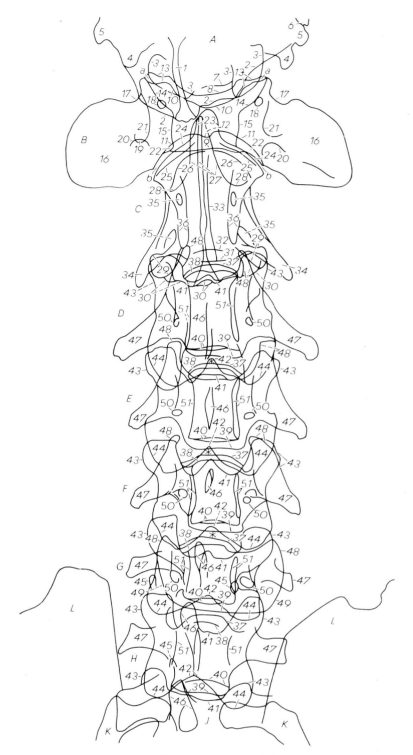

Abb. 47 Röntgenskizze zu Abb. 46
Os hyoideum, Cartilagines laryngis et trachea sind in der Röntgenskizze nicht berücksichtigt.

Fig. 47 X-ray sketch to fig. 46
Hyoid bone, cartilages of the larynx and trachea are disregarded in the X-ray sketch.

49 Plattenförmige Verbreiterung, Lamina ventralis, an der 6. Vertebra cervicalis – Plate-like expansion (ventral lamina) of the 6th cervical vertebra;
50 Foramen transversarium, außer 7. Vertebra cervicalis – Foramen transversarium, absent in the 7th cervical vertebra;
51 Pediculus arcus vertebrae, nach medial zugleich seitliche Begrenzung des Foramen vertebrale, nach lateral des Foramen transversarium – Pediculus arcus vertebrae medially forming the margin of the vertebral foramen and laterally the margin of the transverse foramen.
* Spatium interarcuale ohne Überlagerung dargestellt – Interarcual space not superimposed.

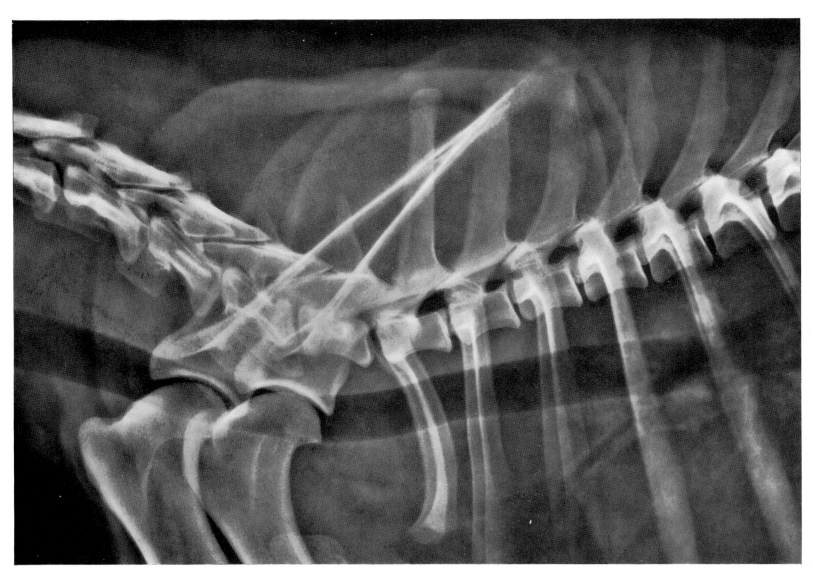

Abb. 48 Hals-Brustwirbelsäule. Latero-lateral. Pudel, 11 Jahre.
Bucky-Blende – Feinzeichnende Folie – FFA 120 cm – 62 kV – 30 mAs
Verkleinerung von 18 × 24 cm
Lagerung Abb. 54

Fig. 48 Cervical and thoracic vertebral column. Laterolateral. Poodle, 11 years old.
Bucky diaphragm – High definition screens – FFD 120 cm – 62 kV – 30 mAs
Diminution of 18 × 24 cm
Positioning fig. 54

A	3. Vertebra cervicalis;		R	5. Os costale;
B	4. Vertebra cervicalis;		S	6. Os costale;
C	5. Vertebra cervicalis;		T	Manubrium sterni mit Gelenkgruben für den Knorpel des 1. Os costale – Manubrium sterni with articular facets for the cartilage of the 1st rib;
D	6. Vertebra cervicalis;			
E	7. Vertebra cervicalis;			
F	1. Vertebra thoracica;			
G	2. Vertebra thoracica;		U	Scapula;
H	3. Vertebra thoracica;		V	Humerus;
J	4. Vertebra thoracica;		W	Trachea;
K	5. Vertebra thoracica;			
L	6. Vertebra thoracica;		a	Articulatio humeri;
M	7. Vertebra thoracica;			
N	1. Os costale;			
O	2. Os costale;		1	Extremitas cranialis;
P	3. Os costale;		2	Extremitas caudalis;
Q	4. Os costale;		3	Foramen vertebrale, ventrale bzw. dorsale Begrenzung – Foramen vertebrale, ventral and dorsal margins respectively;

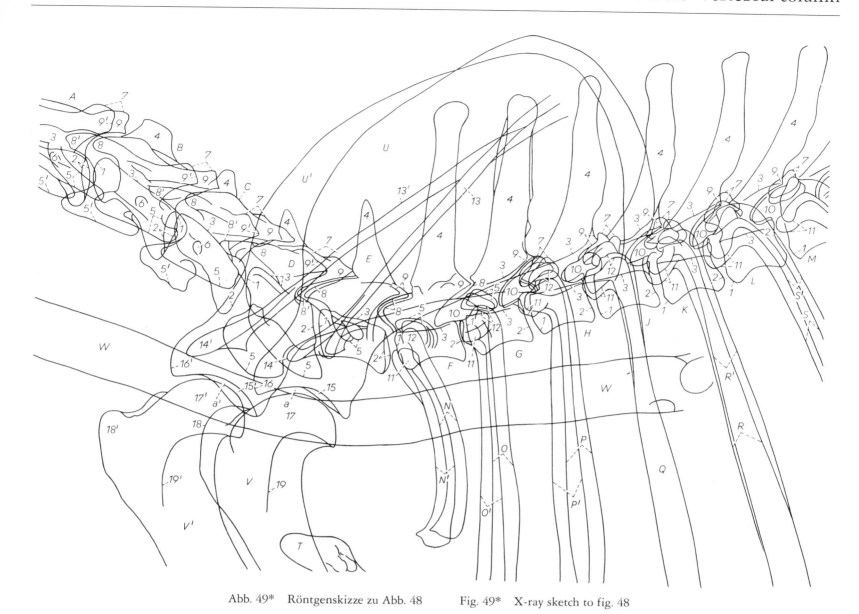

Abb. 49* Röntgenskizze zu Abb. 48   Fig. 49* X-ray sketch to fig. 48

4   Processus spinosus;
5   Processus transversus, bis zur 2. Vertebra thoracica bezeichnet; von der 7. Vertebra cervicalis bis zur 7. Vertebra thoracica stellt sich nur die orthograph getroffene Kompakta an der Basis des Processus transversus dar, wobei plattennah und plattenfern nicht sicher differenzierbar sind – Processus transversus, labelled to the 2nd thoracic vertebra; the orthographically struck compacta from the 7th cervical to the 7th thoracic vertebrae appears only at the base of the transverse process; parts next to the film and next to the tube cannot be identified with certainty;
6   Foramen transversarium der 3., 4. und 5. Vertebra cervicalis – Foramen transversarium of the 3rd, 4th, and 5th cervical vertebrae;
7   Muskelhöcker an der 3. bis 7. Vertebra thoracica auf der Basis des Processus transversus, siehe Anmerkung zu 5 – Tubercle on the base of the transverse process from the 3rd to the 7th thoracic vertebrae, see note at 5;

8   Processus articularis cranialis, von der 3. bis 7. Vertebra thoracica wegen Überlagerung nicht sichtbar — Processus articularis cranialis from the 3rd to the 7th thoracic vertebrae not visible because of superimposition;
9   Processus articularis caudalis;
10 Foramina intervertebralia, ineinander projiziert — Foramina intervertebralia projected into one another;
11 Caput costae;
12 Tuberculum costae;
13 Spina scapulae;
14 Acromion;
15 Cavitas glenoidalis;
16 Tuberculum supraglenoidale;
17 Caput humeri;
18 Tuberculum majus humeri;
19 Crista humeri.

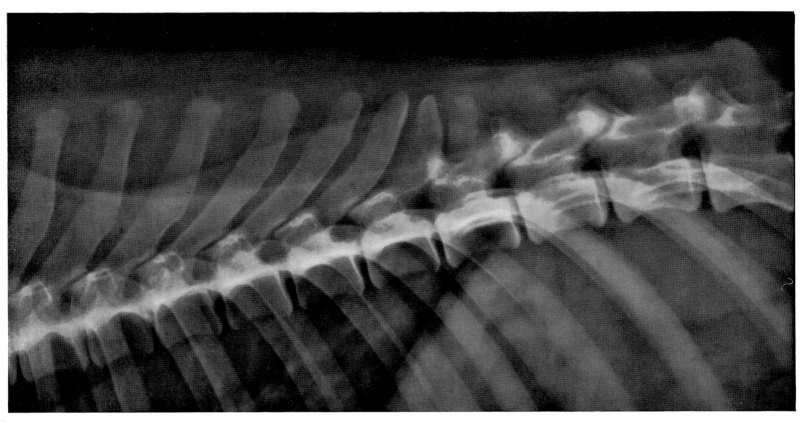

Abb. 50 Brustwirbelsäule. Latero-lateral.
Deutscher Schäferhund, 12 Jahre.
Bucky-Blende – Hochverstärkende Folie – FFA 120 cm – 58 kV – 30 mAs
Originalgröße (Ausschnitt aus 24 × 30 cm)
Lagerung Abb. 54

Fig. 50 Thoracic vertebral column. Laterolateral.
German shepherd dog, 12 years old.
Bucky diaphragm – High speed screens – FFD 120 cm – 58 kV – 30 mAs
Original size (selection of 24 × 30 cm)
Positioning fig. 54

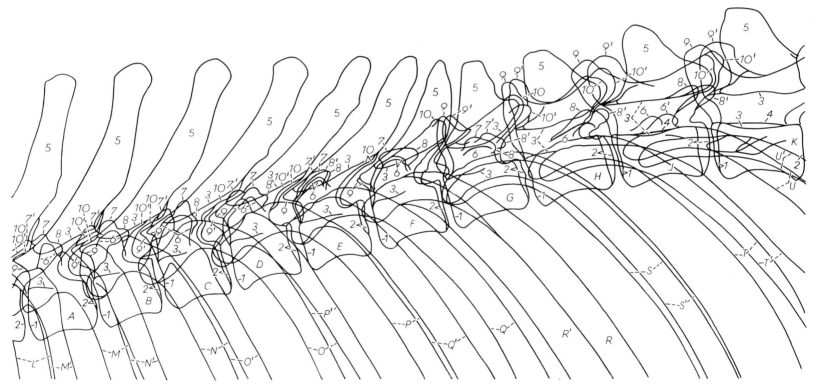

Abb. 51* Röntgenskizze zu Abb. 50

Fig. 51* X-ray sketch to fig. 50

A	5. Vertebra thoracica;	L	4. Os costale;
B	6. Vertebra thoracica;	M	5. Os costale;
C	7. Vertebra thoracica;	N	6. Os costale;
D	8. Vertebra thoracica;	O	7. Os costale;
E	9. Vertebra thoracica;	P	8. Os costale;
F	10. Vertebra thoracica;	Q	9. Os costale;
G	11. Vertebra thoracica;	R	10. Os costale;
H	12. Vertebra thoracica;	S	11. Os costale;
J	13. Vertebra thoracica;	T	12. Os costale;
K	1. Vertebra lumbalis;	U	13. Os costale;

1 Extremitas cranialis;
2 Extremitas caudalis;
3 Foramen vertebrale, ventrale bzw. dorsale Begrenzung, wegen Überlagerungen nur abschnittsweise dargestellt – Foramen vertebrale, ventral and dorsal margins respectively, because of superimposition only parts can be seen;
4 Crista dorsalis;
5 Processus spinosus;
6 Verschattung, die sich aus der Basis des Processus transversus ergibt – Shadow formed by the base of the transverse process;
7 Processus articularis caudalis, Muskelhöcker – Processus articularis caudalis, tubercle;
8 Processus accessorius;
9 Processus articularis cranialis, der in den Schatten der Incisura vertebralis cranialis übergeht – Processus articularis cranialis merging into the shadow of the cranial vertebral notch;
10 Processus articularis caudalis, der in den Schatten der Incisura vertebralis caudalis übergeht – Processus articularis caudalis merging into the shadow of the caudal vertebral notch.

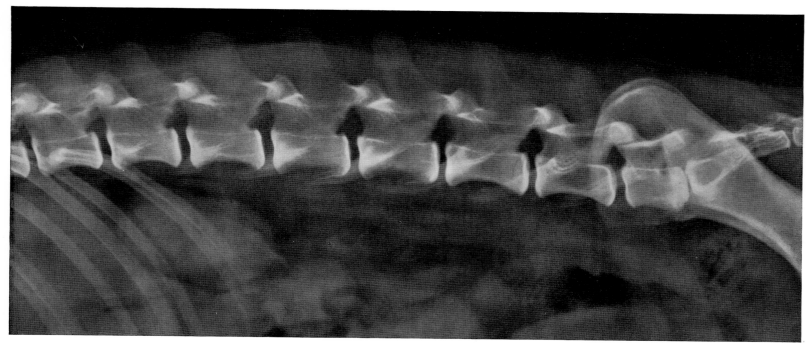

Abb. 52   Lendenwirbelsäule. Latero-lateral.
Irish Terrier, 9 Jahre.
Bucky-Blende – Mittelverstärkende Folie – FFA 120 cm – 55 kV – 35 mAs
Verkleinerung von 24 × 30 cm
Lagerung Abb. 54

Fig. 52   Lumbar vertebral column. Laterolateral.
Irish terrier, 9 years old.
Bucky diaphragm – Standard screens – FFD 120 cm – 55 kV – 35 mAs
Diminution of 24 × 30 cm
Positioning fig. 54

Abb. 53*   Röntgenskizze zu Abb. 52

Fig. 53*   X-ray sketch to fig. 52

A	13. Vertebra thoracica;	
B	1. Vertebra lumbalis;	
C	2. Vertebra lumbalis;	
D	3. Vertebra lumbalis;	
E	4. Vertebra lumbalis;	
F	5. Vertebra lumbalis;	
G	6. Vertebra lumbalis;	
H	7. Vertebra lumbalis;	
J	Os sacrum;	
K	1. Vertebra caudalis;	
L	Os ilium;	
M	10. Os costale;	
N	11. Os costale;	
O	12. Os costale;	
P	13. Os costale;	

1   Extremitas cranialis;
2   Extremitas caudalis;
3   Foramen vertebrale, ventrale bzw. dorsale Begrenzung – Foramen vertebrale, ventral and dorsal margins respectively;
4   Crista dorsalis;
5   Processus spinosus;
6   Verschattung, die sich aus der Basis des Processus transversus ergibt – Shadow formed by the base of the transverse process;
7   Processus transversus;
8   Ala ossis sacri;
9   Processus accessorius;
10   Processus articularis cranialis;
11   Processus articularis caudalis;
12   Processus articularis, rudimentär – Processus articularis, rudimentary;
13   Foramina intervertebralia.

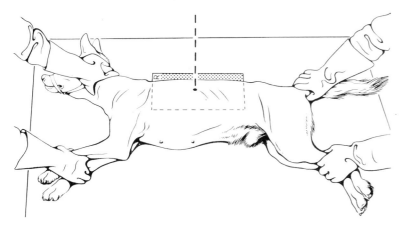

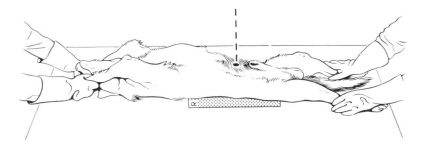

Abb. 54   Lagerung zur Aufnahme der Wirbelsäule. Latero-lateral.

Person 1 fixiert mit einer Hand den Hals und mit der anderen Hand die leicht nach kranial gezogenen Schultergliedmaßen. Person 2 hält mit der einen Hand die leicht nach kaudal gestreckten Beckengliedmaßen und zieht mit der anderen Hand die Rute auf den Tisch. Wichtig ist, daß die Medianebene des Rumpfes parallel zur Kassette liegt. Um dieses zu erreichen, sind bei fettleibigen, ggf. auch bei kurzbeinigen Hunden Schaumgummiteile unter die aufliegenden Gliedmaßen in Höhe der haltenden Hände zu legen.
Zur Reduzierung der Streustrahlung ist eine Bucky-Blende oder eine Kassette mit stehendem Raster erforderlich.
Der Zentralstrahl sollte die Wirbelsäule zur Darstellung des Hals-Brustbereichs in Höhe des 1. Brustwirbels, zur Darstellung des Brustbereichs in Höhe des 10. Brustwirbels und zur Darstellung des Lendenbereichs in Höhe des 3. Lendenwirbels treffen und im rechten Winkel auf die Kassette einfallen.

Fig. 54   Positioning of vertebral column. Laterolateral.

One assistant secures the neck with one hand and the slightly cranially extended thoracic limbs with the other one. The second assistant holds the slightly caudally extended pelvic extremities with one hand, pulling the tail on to the table with the other hand. It is essential to align the median plane of the trunk parallel to the cassette. In obese and short legged dogs this can be facilitated by placing foam rubber pads underneath the legs at the points where they are grasped by the hands.
To reduce scattered radiation, a Bucky diaphragm or a cassette with a stationary grid should be employed.
The central beam should strike the spine on a level of the 1st thoracic vertebra for radiography of the neck-chest region, on a level of the 10th thoracic vertebra for radiography of the thorax, and on a level of the 3rd lumbar vertebra for radiography of the lumbar region. It should fall at the right angle on the cassette.

Abb. 55   Lagerung zur Aufnahme der Wirbelsäule. Ventro-dorsal.

Bei der Lagerung in Rückenlage ist darauf zu achten, daß die Medianebene senkrecht zur Kassette steht. Es ist zweckmäßig, die Beckengliedmaßen in leichter Abduktion so weit nach hinten zu strecken, daß sich die Hände, die die Gliedmaßen im Bereich der Sprunggelenke halten, auf die Tischplatte stützen können.
Zur Verminderung der Streustrahlung ist eine Bucky-Blende oder eine Kassette mit stehendem Raster notwendig. Nur bei kleinen, nicht fettleibigen Hunden kann auf den Raster verzichtet werden.
Der Zentralstrahl sollte im rechten Winkel auf die Kassette auftreffen.
Einstellungen des Zentralstrahls:
Brustwirbelsäule: 7. Brustwirbel;
Lendenwirbelsäule: 4. Lendenwirbel.

Fig. 55   Positioning of vertebral column. Ventrodorsal.

When positioning the animal on its back, care must be exercised to align the median plane perpendicular to the cassette. The hind legs should be extended and slightly abducted with the hands holding them at the hocks on the table.
To reduce scattered radiation, a Bucky diaphragm or a cassette with a stationary grid should be used. A grid is not necessary for small lean dogs.
The central beam should fall at the right angle on the cassette.
Directing the central beam:
Thoracic vertebral column: 7th thoracic vertebra;
Lumbar vertebral column: 4th lumbar vertebra.

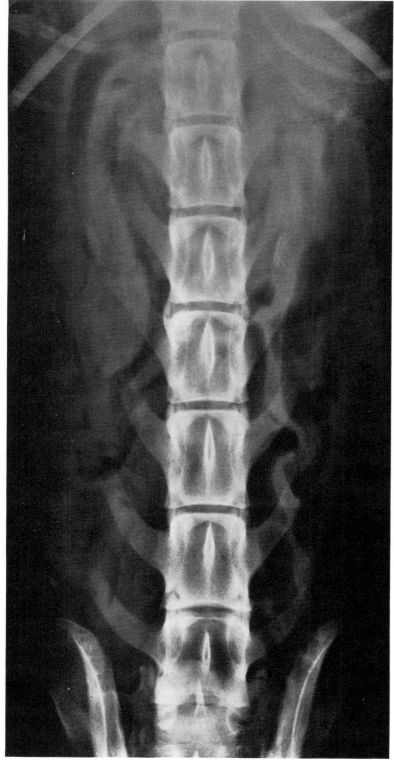

Abb. 56 Lendenwirbelsäule. Ventro-dorsal. Dalmatiner, ♂, 2 Jahre.
Bucky-Blende – Feinzeichnende Folie – FFA 120 cm – 65 kV – 48 mAs
Verkleinerung von 24 × 30 cm
Lagerung Abb. 55

Fig. 56 Lumbar vertebral column. Ventrodorsal. Dalmatian, ♂,
2 years old.
Bucky diaphragm – High definition screens – FFD 120 cm – 65 kV –
48 mAs
Diminution of 24 × 30 cm
Positioning fig. 55

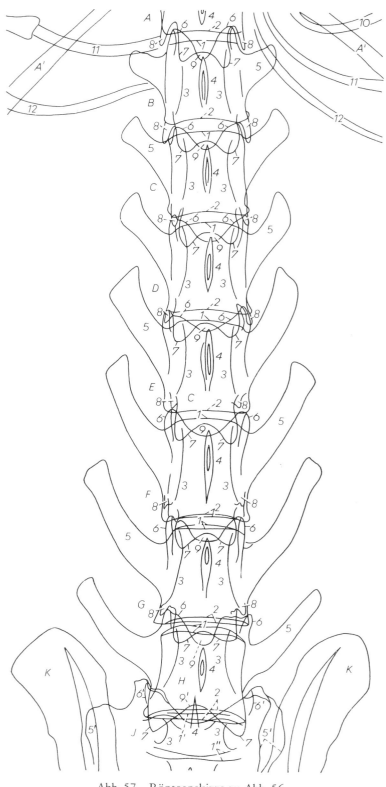

Abb. 57 Röntgenskizze zu Abb. 56
Fig. 57 X-ray sketch to fig. 56

A	13. Vertebra thoracica;	F	5. Vertebra lumbalis;
A'	13. Os costale;	G	6. Vertebra lumbalis;
B	1. Vertebra lumbalis;	H	7. Vertebra lumbalis;
C	2. Vertebra lumbalis;	J	Os sacrum;
D	3. Vertebra lumbalis;	K	Ala ossis ilii;
E	4. Vertebra lumbalis;		

1 Extremitas cranialis;
1' 1'' Basis ossis sacri:
1' Dorsale Kontur – Dorsal border,
1'' Ventrale Kontur, Promontorium – Ventral border, promontory;

2 Extremitas caudalis; an der 5., 6. und 7. Vertebra lumbalis ist die deutlicher sichtbare doppelte Begrenzung, die sich aus der Konkavität der Extremitas caudalis ergibt, eingezeichnet – Extremitas caudalis; the more clearly visible double border formed by the concavity of the caudal extremity, is illustrated on the 5th, 6th, and 7th lumbar vertebrae;

3 Pediculus arcus vertebrae, zugleich seitliche Begrenzung des Foramen vertebrale – Pediculus arcus vertebrae forming the lateral margin of the vertebral foramen;

4 Processus spinosus;
5 Processus transversus;
5' Ala ossis sacri;
6 Processus articularis cranialis et Processus mamillaris;
6' Os sacrum, Processus articularis cranialis;
7 Processus articularis caudalis;
8 Processus accessorius;
9 Spatium interarcuale;
9' Spatium interarcuale lumbosacrale;
10, 11, 12 Cartilagines des 10., 11. bzw. 12. Os costale – Cartilagines of the 10th, 11th, and 12th ribs respectively.

49

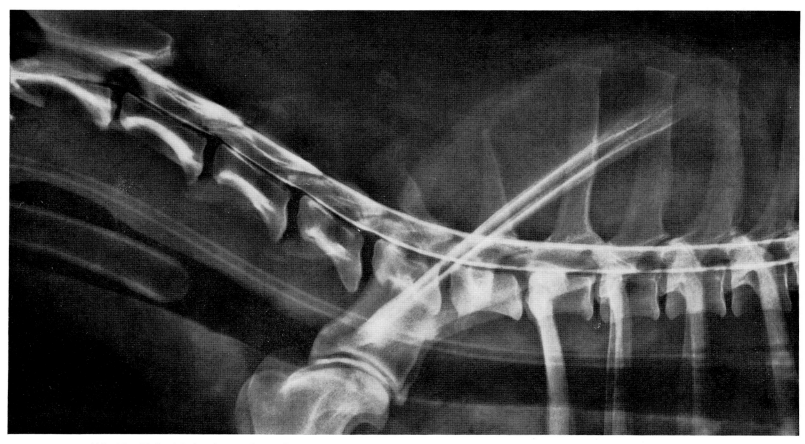

Abb. 58   Halswirbelsäule, Myelographie
(Amipaque, Schering). Latero-lateral.
Bayerischer Gebirgsschweißhund, 8 Jahre.
Bucky-Blende – Feinzeichnende Folie – FFA 115 cm – 70 kV – 60 mAs
Verkleinerung von 24 × 30 cm
Lagerung Abb. 54

Fig. 58   Cervical vertebral column, myelography
(Amipaque, Schering). Laterolateral.
Bavarian mountain bloodhound, 8 years old.
Bucky diaphragm – High definition screens – FFD 115 cm – 70 kV – 60 mAs
Diminution of 24 × 30 cm
Positioning fig. 54

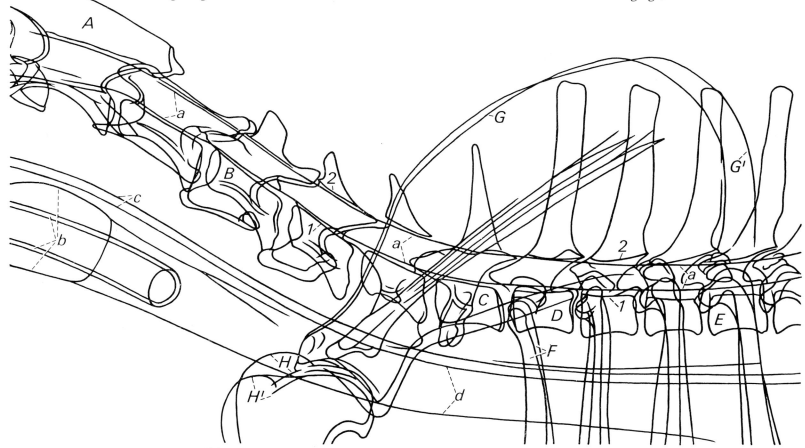

Abb. 59*   Röntgenskizze zu Abb. 58   Fig. 59*   X-ray sketch to fig. 58

A	Axis;	E	4. Vertebra thoracica;
B	4. Vertebra cervicalis;	F	1. Os costale;
C	7. Vertebra cervicalis;	G	Scapula;
D	1. Vertebra thoracica;	H	Humerus;

a   Kontrastmittel — Contrast medium;

b   Intratrachealtubus mit aufgeblasener Manschette — Intratracheal
tube with distended sleeve;
c   Phonendoskop im Oesophagus — Phonendoscope in esophagus;
d   Trachea;
1,   2   Foramen vertebrale:
1   Ventrale Begrenzung – Ventral margin,
2   Dorsale Begrenzung – Dorsal margin.

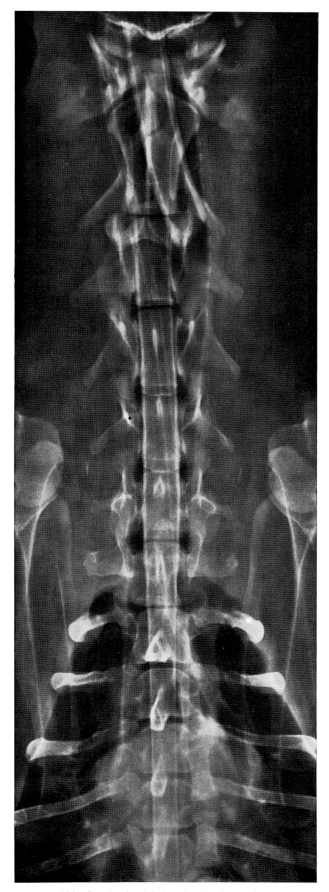

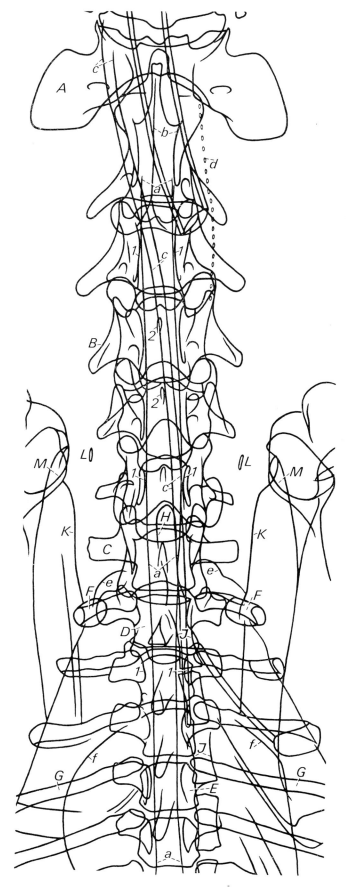

Abb. 60 Halswirbelsäule, Myelographie
(Amipaque, Schering). Ventro–dorsal.
Bayerischer Gebirgsschweißhund, 8 Jahre.
Bucky-Blende – Feinzeichnende Folie – FFA 115 cm – 70 kV – 60 mAs
Verkleinerung von 15 × 40 cm
Lagerung Abb. 55

Fig. 60 Cervical vertebral column, myelography
(Amipaque, Schering). Ventrodorsal.
Bavarian mountain bloodhound, 8 years old.
Bucky diaphragm – High definition screens –
FFD 115 cm – 70 kV – 60 mAs
Diminution of 15 × 40 cm
Positioning fig. 55

Abb. 61 Röntgenskizze zu Abb. 60
Fig. 61 X-ray sketch to fig. 60

A	Atlas;	G	4. Os costale;
B	4. Vertebra cervicalis;	H	Manubrium sterni;
C	7. Vertebra cervicalis;	J	Sternebrae;
D	1. Vertebra thoracica;	K	Scapula;
E	4. Vertebra thoracica;	L	Clavicula;
F	1. Os costale;	M	Caput humeri;

a   Kontrastmittel – Contrast medium;
b   Intratrachealtubus – Intratracheal tube;
c   Phonendoskop im Oesophagus – Phonendoscope in esophagus;
d   Trachea;   e Cupula pleurae;   f Cor;

1   Basis des Pediculus arcus vertebrae, zugleich seitliche Begrenzung des
    Foramen vertebrale – Base of the pedicle of vertebral arch forming
    the lateral margin of the vertebral foramen.

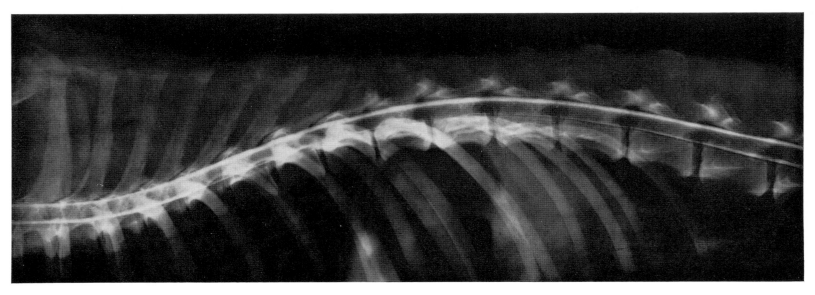

Abb. 62   Brust-Lendenwirbelsäule, Myelographie (Amipaque, Schering). Latero-lateral.
Bayerischer Gebirgsschweißhund, 8 Jahre.
Bucky-Blende – Feinzeichnende Folie – FFA 115 cm – 70 kV – 60 mAs
Verkleinerung von 15 × 40 cm
Lagerung Abb. 54

Fig. 62   Thoracic and lumbar vertebral column, myelography (Amipaque, Schering). Laterolateral.
Bavarian mountain bloodhound, 8 years old.
Bucky diaphragm – High definition screens – FFD 115 cm – 70 kV – 60 mAs
Diminution of 15 × 40 cm
Positioning fig. 54

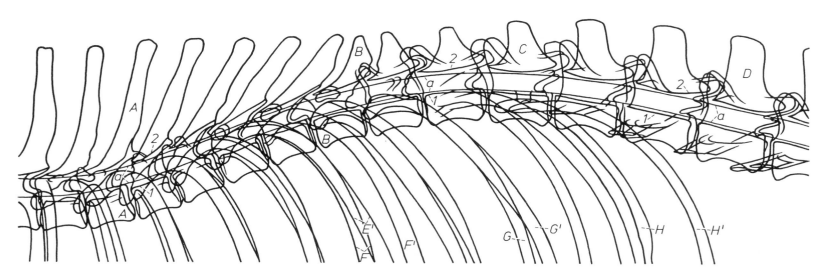

Abb. 63*   Röntgenskizze zu Abb. 62   Fig. 63*   X-ray sketch to fig. 62

A	5. Vertebra thoracica;	H	13. Os costale;
B	10. Vertebra thoracica;		
C	13. Vertebra thoracica;	a	Kontrastmittel – Contrast medium;
D	3. Vertebra lumbalis;		
E	8. Os costale;	1, 2	Foramen vertebrale:
F	9. Os costale;	1	Ventrale Begrenzung – Ventral margin,
G	11. Os costale;	2	Dorsale Begrenzung – Dorsal margin.

Myelographie

Ein raumeinengender Prozeß im Bereich des Wirbelkanals, extra- und/oder subdural gelegen, der nicht schattengebend ist (Tumor, Prolapsus nuclei pulposi, Hämatom) kann durch ein in den Subarachnoidealraum appliziertes Kontrastmittel nachgewiesen werden.

Vorbereitung: Der Patient sollte 12 Stunden vor der Untersuchung kein Futter und keine Flüssigkeit mehr aufnehmen. Zur Untersuchung ist der Patient zu narkotisieren. Am besten geeignet ist eine Inhalationsnarkose (Lachgas–Halothan). Bei einer Injektionsnarkose sollte intubiert werden, um bei einem Atemstillstand sofort beatmen zu können. Von einer Neuroleptanalgesie ist abzusehen, weil die Schwelle für epileptogene Anfälle gesenkt wird, also Konvulsionen entstehen können.
Immer sollten beide Injektionsstellen, die zervikale und die lumbale, für die Punktion vorbereitet werden.

Kontrastmittel: Das derzeit bestverträgliche Kontrastmittel für die Myelographie ist Metrizamid (Amipaque, Schering). Eine Lösung von 170 mg J/ml ist liquorisotonisch. Es ist wasserlöslich, dissoziiert jedoch nicht und hat daher einen wesentlich geringeren osmotischen Druck als ionische Kontrastmittel. Die Neurotoxizität und die epileptogene Wirkung sind gering.

Dosis: Kleine Hunde 0,3 ml/kg KGW, mittelgroße Hunde 0,2 ml/kg KGW, große Hunde 0,1 ml/kg KGW. Für den Fall, daß Konvulsionen einsetzen sollten, sollte Diazepam (0,6 mg/kg KGW – ggf. etwas nachdosieren; Valium, Hoffmann–La Roche) bereitgestellt werden.

Applikation: Das Kontrastmittel kann nach Okzipitalpunktion in die Cisterna cerebellomedullaris und/oder nach Punktion des Subarachnoidealraums durch das Spatium interarcuale zwischen 4. und 5. Lendenwirbel ohne Liquorentnahme zügig injiziert werden. Da das spezifische Gewicht des Kontrastmittels größer als das des Liquors ist, ist die Applikationsstelle, d. h. der kraniale oder der kaudale Bereich der Wirbelsäule, nach der Injektion für 5 bis 10 Minuten höher zu lagern. Gegebenenfalls ist der Fluß des Kontrastmittels unter der Durchleuchtung zu verfolgen. In der Regel ist die Okzipitalpunktion in rechter Seitenlage einfacher als die Lumbalpunktion in Brust-Bauchlage des Patienten.
Fließt das Kontrastband nicht oder nicht ausreichend an dem Hindernis vorbei, d.h. daß die kraniale bzw. kaudale Begrenzung und damit die Ausdehnung des Prozesses nicht gedeutet werden kann, sollte die Applikation ggf. in der gleichen Dosis, und zwar zervikal bei vorheriger lumbaler bzw. lumbal bei vorheriger zervikaler Applikation, wiederholt werden.

Untersuchungstechnik: Aufnahmen in zwei Ebenen sollten 5 bis 10 Minuten nach Applikation und Hochlagerung des kranialen bzw. kaudalen Bereichs der Wirbelsäule, ggf. gezielt aus der Durchleuchtung, angefertigt werden. Weitere Aufnahmen ergeben sich aus der Fragestellung. Ausreichender Kontrast ist für etwa 60 Minuten vorhanden.

Myelography

A space occupying lesion situated extradurally or subdurally in the vertebral canal and producing no shadows (tumour, prolapse of the nucleus pulposus, haematoma) can be demonstrated by introducing a contrast medium into the subarachnoid space.

Preparation of the patient: The patient should be starved and deprived of water 12 hours prior to the examination. Inhalation anesthesia using nitrous oxide – halothane is the most effective. In injection anesthesia is used the animal should be intubated to facilitate breathing in the event of respiratory arrest. Neuroleptanalgesia is not recommended since the threshold for epileptogenic attacks is lowered, resulting in convulsions. Both cervical and lumbar puncture sites should always be prepared for injection.

Contrast medium: The best tolerated contrast medium currently used for myelography is Metrizamide (Amipaque, Schering). A solution of 170 mg J/ml is isotonic with the cerebrospinal fluid. It is water soluble but does not dissociate and has therefore a much lower osmotic pressure than ionic contrast media. The neurotoxicity and epileptogenic action are low.

Dosage: Small dogs 0,3 ml/kg body weight, medium sized dogs 0,2 ml/kg and large dogs 0,1 ml/kg. In the event of possible convulsions Diazepam (0,6 mg/kg body weight – giving more if necessary; Valium, Hoffmann – La Roche) should be kept available.

Application: The contrast medium can be injected into the cerebello-medullary cistern after occipital puncture and/or into the subarachnoid space after puncturing the interarcual space between the 4th and 5th lumbar vertebrae. Cerebrospinal fluid should not be withdrawn and the injection should be given uninterruptedly. Since the specific gravity of the contrast medium is higher than that of the cerebrospinal fluid, the injection site on either the cranial or caudal part of the vertebral column should be raised for 5 – 10 minutes after the injection. If necessary the flow of the contrast medium may be followed by fluoroscopy. Occipital puncture in right lateral recumbency is usually easier than lumbar puncture in abdominal recumbency.
If the contrast medium does not flow satisfactorily past the obstruction i.e. if the cranial or caudal limit and therefore the extent of the lesion cannot be determined, the injection should be repeated at the alternative injection site using the same dosage.

Technique: Radiographs should be taken in two different planes 5 – 10 minutes after the injection and raising of the vertebral column. Specific radiographs can be taken with the aid of fluoroscopy. Depending on the case in question additional radiographs can be taken. The contrast medium is sufficient to last approximately 60 minutes.

Abb. 64 Brust–Lendenwirbelsäule, Myelographie (Amipaque, Schering). Latero-lateral.
Bayerischer Gebirgsschweißhund, 8 Jahre.
Bucky-Blende – Feinzeichnende Folie – FFA 115 cm – 70 kV – 60 mAs
Verkleinerung von 15 × 40 cm
Lagerung Abb. 54

Fig. 64 Thoracic and lumbar vertebral column, myelography (Amipaque, Schering). Laterolateral.
Bavarian mountain bloodhound, 8 years old.
Bucky diaphragm – High definition screens – FFD 115 cm – 70 kV – 60 mAs
Diminution of 15 × 40 cm
Positioning fig. 54

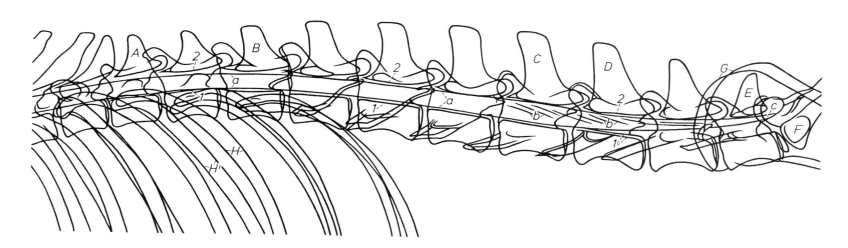

Abb. 65* Röntgenskizze zu Abb. 64   Fig. 65* X-ray sketch to fig. 64

A	11. Vertebra thoracia;	a	Kontrastmittel – Contrast medium;
B	13. Vertebra thoracica;	b	Radices nervi;
C	4. Vertebra lumbalis;	c	Cavum subdurale, Ende – Cavum subdurale, end;
D	5. Vertebra lumbalis;		
E	7. Vertebra lumbalis;		
F	Os sacrum;	1, 2	Foramen vertebrale:
G	Os ilium;	1	Ventrale Begrenzung – Ventral margin,
H	11. Os costale;	2	Dorsale Begrenzung – Dorsal margin.

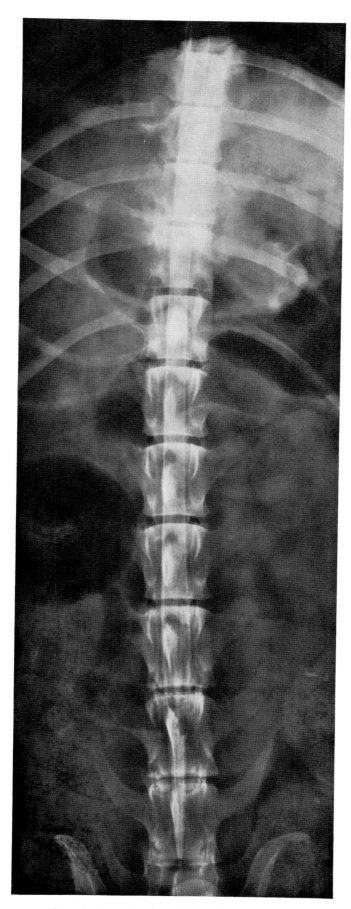

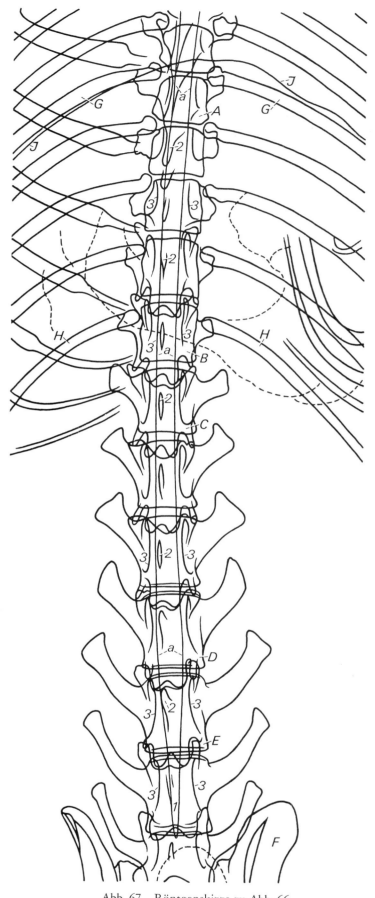

Abb. 66 Brust-Lendenwirbelsäule, Myelographie
(Amipaque, Schering). Ventro-dorsal.
Bayerischer Gebirgsschweißhund, ♀, 8 Jahre.
Bucky-Blende – Feinzeichnende Folie – FFA 115 – 70 kV – 60 mAs
Verkleinerung von 15 × 40 cm
Lagerung Abb. 55

Fig. 66 Thoracic and lumbar vertebral column, myelography
(Amipaque, Schering). Ventrodorsal.
Bavarian mountain bloodhound, ♀, 8 years old.
Bucky diaphragm – High definition screens –
FFD 115 cm – 70 kV – 60 mAs
Diminution of 15 × 40 cm
Positioning fig. 55

Abb. 67 Röntgenskizze zu Abb. 66
Fig. 67 X-ray sketch to fig. 66

A	9. Vertebra thoracica;	F	Os ilium;
B	13. Vertebra thoracica;	G	9. Os costale;
C	1. Vertebra lumbalis;	H	13. Os costale;
D	4. Vertebra lumbalis;	J	Diaphragma;
E	5. Vertebra lumbalis;		

a Kontrastmittel – Contrast medium;

1 Cavum subdurale, Ende – Cavum subdurale, end;
2 Processus spinosus;
3 Basis des Pediculus arcus vertebrae, zugleich seitliche Begrenzung des
   Foramen vertebrale – Base of the pedicle of vertebral arch forming
   the lateral margin of the vertebral foramen.

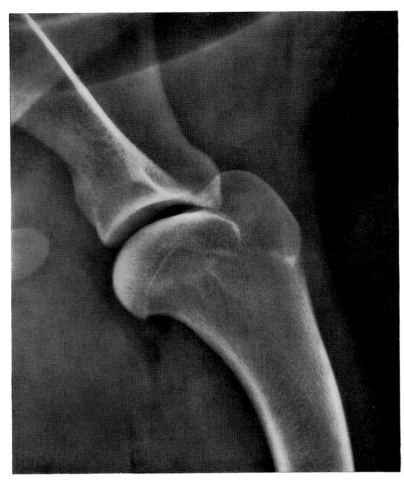

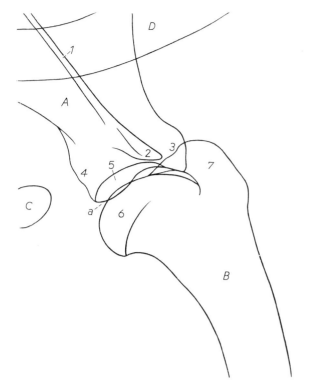

Abb. 69   Röntgenskizze zu Abb. 68
Fig. 69   X-ray sketch to fig. 68

Abb. 68   Linkes Schultergelenk. Medio-lateral.
Dalmatiner, 2 Jahre.
Bucky-Blende – Feinzeichnende Folie – FFA 120 cm – 65 kV – 40 mAs
Originalgröße (Ausschnitt aus 24 × 30 cm)
Lagerung Abb. 76

Fig. 68   Left shoulder joint. Mediolateral.
Dalmation, 2 years old.
Bucky diaphragm – High definition screens – FFD 120 cm – 65 kV –
40 mAs
Original size (section of 24 × 30 cm)
Positioning fig. 76

A   Scapula;
B   Humerus;
C   Manubrium sterni;
D   Trachea;

a   Articulatio humeri;

1   Spina scapulae;
2   Acromion;
3   Tuberculum supraglenoidale;
4   Tuberculum infraglenoidale;
5   Cavitas glenoidalis;
6   Caput humeri;
7   Tuberculum majus.

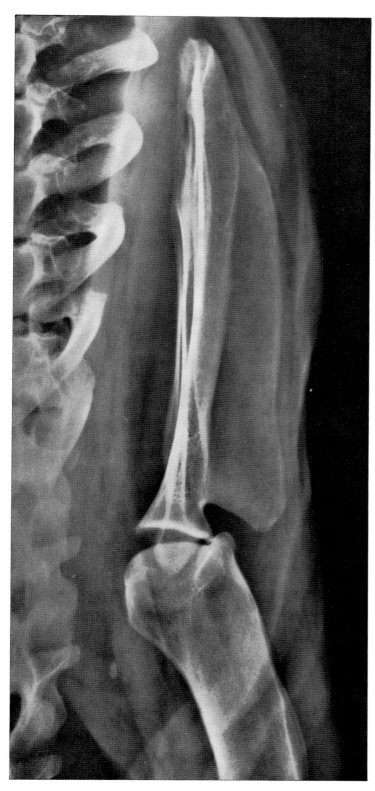

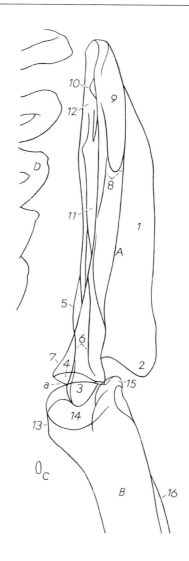

Abb. 70   Linkes Schultergelenk. Kaudo–kranial.
Deutscher Schäferhund, 7 Jahre.
Bucky-Blende – Hochverstärkende Folie – FFA 120 cm – 60 kV –
30 mAs
Verkleinerung von 24 × 30 cm
Lagerung Abb. 77

Fig. 70   Left shoulder joint. Caudocranial.
German shepherd dog, 7 years old.
Bucky diaphragm – High speed screens – FFD 120 cm – 60 kV –
30 mAs
Diminution of 24 × 30 cm
Positioning fig. 77

A   Scapula;
B   Humerus;
C   Clavicula;
D   3. Os costale;

a   Articulatio humeri;

1   Spina scapulae;
2   Acromion;
3   Tuberculum supraglenoidale et Processus coracoideus;
4   Cavitas glenoidalis;
5   Margo cranialis;
6   Fortsetzung von 5, der im Bereich des Collum scapulae in die laterale
     und mediale Begrenzung des Tuberculum supraglenoidale übergeht
     bzw. einläuft – Continuation of 5 merging at the neck of the scapula
     into the lateral and medial borders of the supraglenoid tuberosity;
7   Collum scapulae, mediale Begrenzung – Collum scapulae, medial bor-
     der;
8   Margo caudalis;
9   Facies muscularis des M. teres major am Margo caudalis – The area
     of attachment of M. teres major on the caudal border of the scapula;
10   Angulus caudalis;
11   Fossa subscapularis mit Muskelleisten – Fossa subscapularis with
     muscular lines;
12   Facies serrata;
13   Tuberculum minus;
14   Caput humeri;
15   Tuberculum majus;
16   Tuberositas deltoidea.

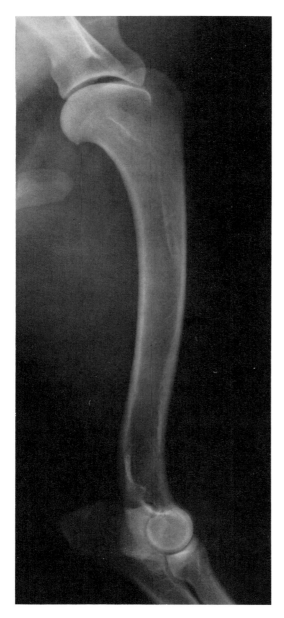

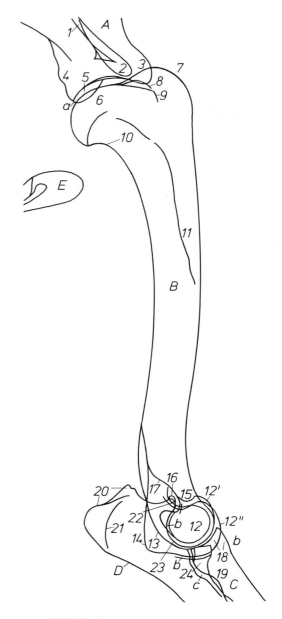

Abb. 72   Linker Oberarm. Medio-lateral.
Deutscher Schäferhund, 2 Jahre.
Bucky-Blende – Feinzeichnende Folie – FFA 120 cm – 66 kV – 32 mAs
Verkleinerung von 24 × 33 cm
Lagerung Abb. 76

Fig. 72   Left arm. Mediolateral.
German shepherd dog, 2 years old.
Bucky diaphragm – High definition screens –
FFD 120 cm – 66 kV – 32 mAs
Diminution of 24 × 30 cm
Positioning fig. 76

Abb. 73   Röntgenskizze zu Abb. 72
Fig. 73   X-ray sketch to fig. 72

A   Scapula;
B   Humerus;
C   Radius;
D   Ulna;
E   Manubrium sterni;

a   Articulatio humeri;
b   Articulatio cubiti;
c   Articulatio radioulnaris proximalis;

An der Scapula – On the scapula:

1   Spina scapulae;
2   Acromion;
3   Tuberculum supraglenoidale;
4   Tuberculum infraglenoidale;
5   Cavitas glenoidalis;

Am Humerus – On the humerus:

6   Caput humeri;
7   Tuberculum majus;
8   Tuberculum minus;

9   Sulcus intertubercularis;
10   Linea m. tricipitis;
11   Tuberositas deltoidea;
12   Condylus humeri,
12'   Lateraler Rand – lateral border,
12"   Medialer Rand – Medial border;
13   Epicondylus lateralis;
14   Epicondylus medialis;
15   Fossa radialis;
16   Foramen supratrochleare;
17   Fossa olecrani;

Am Radius – On the radius:

18   Caput radii mit Fovea capitis radii – Caput radii with the fovea of the radial caput;
19   Collum radii;

An der Ulna – On the ulna:

20   Tuber olecrani;
21   Mediale Knochenleiste – Medial crest;
22   Processus anconaeus;
23   Incisura trochlearis;
24   Processus coronoideus medialis.

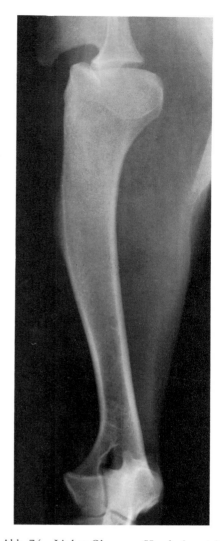

Abb. 74   Linker Oberarm. Kaudo-kranial.
Deutscher Schäferhund, 2 Jahre.
Feinzeichnende Folie – FFA 120 cm – 60 kV – 8 mAs
Verkleinerung von 24 × 30 cm
Lagerung Abb. 77, 78

Fig. 74   Left arm. Caudocranial. German shepherd dog, 2 years old.
High definition screens – FFD 120 cm – 60 kV – 8 mAs
Diminution of 24 × 30 cm
Positioning fig. 77, 78

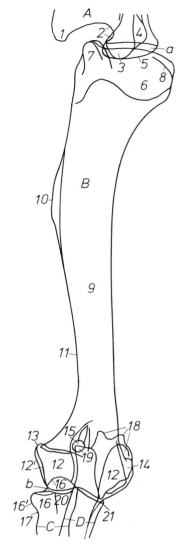

Abb. 75   Röntgenskizze zu Abb. 74
Fig. 75   X-ray sketch to fig. 74

A   Scapula;
B   Humerus;
C   Radius;
D   Ulna;

a   Articulatio humeri;
b   Articulatio cubiti;

An der Scapula – On the scapula:

1   Acromion;
2   Collum scapulae;
3   Tuberculum supraglenoidale;
4   Processus coracoideus;
5   Cavitas glenoidalis, kaudaler Rand – Cavitas glenoidalis,
    caudal border;

Am Humerus – On the humerus:

6   Caput humeri;
7   Tuberculum majus;
8   Tuberculum minus;
9   Corpus humeri;
10   Tuberositas deltoidea;
11   Sulcus m. brachialis;
12   Condylus humeri;

12′   Laterale Bandgrube – Lateral depression for ligamentous attachment;
13   Epicondylus lateralis;
14   Epicondylus medialis;
15   Foramen supratrochleare;

Am Radius – On the radius:

16   Caput radii mit Fovea capitis radii – Caput radii with the fovea of the
    radial caput;
16′   Bandhöcker – Eminence for ligamentous attachment;
17   Collum radii;

An der Ulna – On the ulna:

18   Tuber olecrani;
19   Processus anconaeus;
20   Processus coronoideus lateralis;
21   Processus coronoideus medialis.

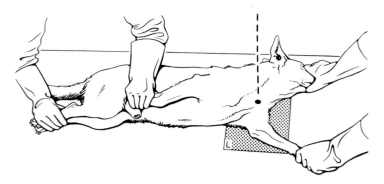

Abb. 76   Lagerung zur Aufnahme des Schultergelenks
und des Oberarms. Medio-lateral.

Die zu untersuchende (untenliegende) Schultergliedmaße ist so weit nach vorn und die obenliegende Schultergliedmaße so weit nach hinten zu ziehen, daß der jeweilige Bereich ohne Überlagerung dargestellt wird. Um eine Überlagerung mit dem Hals zu vermeiden, sind Kopf und Hals zum Rücken hin zu fixieren.
Zur Verringerung der Streustrahlung ist eine Bucky-Blende oder eine Kassette mit stehendem Raster notwendig.
Schultergelenk: Der Zentralstrahl sollte das Schultergelenk treffen und im rechten Winkel auf die Kassette einfallen.
Oberarm: Der Zentralstrahl sollte den Humerus in Schaftmitte treffen und im rechten Winkel auf die Kassette einfallen.

Abb. 76   Positioning of shoulder joint and arm. Mediolateral.

The thoracic limb to be examined (the lower one) must be pulled forward and the upper one backward as far as possible in order to ensure radiography of specific areas without any overlaying.
To avoid an overlaying with the neck, the head and neck must be extended and secured.
A Bucky diaphragm or a cassette with a stationary grid is necessary to reduce scattered radiation.
Shoulder joint: The central beam should strike the shoulder joint and fall at the right angle on the cassette.
Arm: The central beam should strike the mid-shaft of the humerus and fall at the right angle on the cassette.

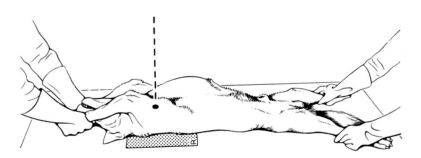

Abb. 77   Lagerung zur Aufnahme des Schultergelenks. Kaudo-kranial

Die gestreckten Schultergliedmaßen des auf dem Rücken liegenden Hundes sind so weit nach vorn zu ziehen, daß sich die die Gliedmaßen haltenden Hände auf der Tischplatte stützen können. Dabei werden die Gliedmaßen gegen den Kopf gedrückt, um diesen mit zu fixieren. Es ist zu beachten, daß die Ellbogenhöcker nach oben zeigen und die Medianebene des Rumpfes senkrecht zur Kassette steht.
Zur Reduzierung der Streustrahlung ist eine Bucky-Blende oder eine Kassette mit stehendem Raster notwendig.
Der Zentralstrahl sollte die Mitte der Gliedmaßenhinterfläche in Höhe des Tuberculum majus humeri treffen und im rechten Winkel auf die Kassette einfallen.

Fig. 77   Positioning of shoulder joint. Caudocranial.

The dog is placed on its back. The thoracic limbs should be pulled forward far enough to allow the hands to rest upon the table. At the same time the limbs should be pressed against the head to secure its position. The tips of the elbows must point upward and the median plane of the trunk should be perpendicular to the cassette.
A Bucky diaphragm or a cassette with a stationary grid is necessary to reduce scattered radiation.
The central beam should strike the middle of the caudal aspect of the limb on a level of the tuberculum majus humeri and fall at the right angle on the cassette.

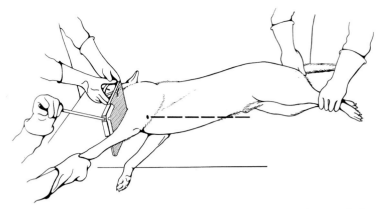

Abb. 78   Lagerung zur Aufnahme des Oberarms bei Frakturverdacht.
Kaudo-kranial.

Die obenliegende Schultergliedmaße des auf der gesunden Seite gelagerten Hundes wird horizontal vorgezogen und so gehalten, Kopf und Hals sind zum Rücken hin gestreckt zu fixieren und die Kassette kranial am Oberarm anzulegen.
Der Zentralstrahl sollte den Humerus in Schaftmitte treffen und im rechten Winkel auf die Kassette einfallen.

Fig. 78   Positioning of the arm in case of suspected fracture of the humerus. Caudocranial.

The dog ist pulled in lateral position, with healthy limb on the table. The limb to be examined should be stretched horizontally and held so. The cassette is placed cranially, close to the arm.
The central beam should strike the mid-shaft of the humerus and fall at the right angle on the cassette.

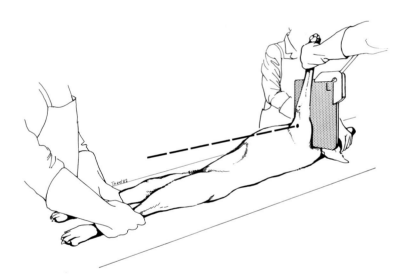

Abb. 79   Lagerung zur Aufnahme des Oberarms bei Frakturverdacht.
Rückenlage. Kaudo-kranial.

Die zu untersuchende Schultergliedmaße des auf dem Rücken liegenden Hundes wird vertikal hochgezogen und so gehalten. Die Kassette wird kranial am Oberarm angelegt.
Der Zentralstrahl sollte den Humerus in Schaftmitte treffen und im rechten Winkel auf die Kassette einfallen.

Fig. 79   Positioning of arm in case of suspected fracture of the humerus.
Dorsal recumbency. Caudocranial.

The dog is placed in ventrodorsal recumbency, the limb to be examined pulled straight up and held so. The cassette is placed cranially, close to the arm.
The central beam should strike the mid-shaft of the humerus and fall at the right angle on the cassette.

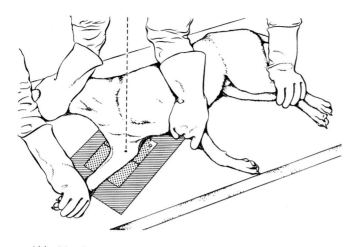

Abb. 80   Lagerung zur Aufnahme des Ellbogengelenks.
Medio-lateral.

Die am Kopf des Hundes stehende Person fixiert mit einer Hand den Kopf und mit der anderen die etwas nach vorn gezogene untenliegende Schultergliedmaße. Die am Rücken des Hundes stehende Person fixiert mit einer Hand beide Beckengliedmaßen und mit der anderen die etwas nach hinten gezogene obenliegende Schultergliedmaße und drückt dabei mit dem Unterarm den Hund auf den Tisch.
Die Aufnahme kann ohne Kassette mit einem folienlosen Film (Bleiunterlage nicht vergessen!) angefertigt werden.
Der Zentralstrahl sollte das Ellbogengelenk eine halbe bis eine Fingerbreite distal des Epicondylus medialis humeri treffen und im rechten Winkel auf den Film einfallen.

Fig. 80   Positioning of elbow joint. Mediolateral.

The assistant at the head of the dog secures the head of the animal with one hand and with the other hand the lower limb, pulling it slightly forward. The assistant at the back of the dog holds both of the pelvic limbs with one hand and with the other hand the upper thoracic limb, pulling it slightly backward. The forearm presses the animal against the table.
The radiograph should be taken with a non-screen film (lead blocker not to be forgotten!).
The central beam should strike the elbow joint one half to one finger's width distally to the medial epicondyle of the humerus and fall at the right angle on the film.

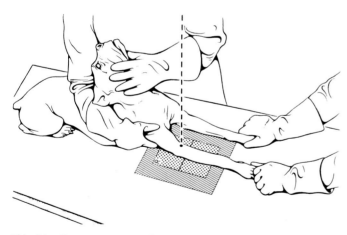

Abb. 81   Lagerung zur Aufnahme des Ellbogengelenks und des
Unterarms. Kranio-kaudal.

Person 1 fixiert mit einer Hand den Kopf, mit möglichst weit zum Rücken hin abgebogenem Hals. Mit der anderen Hand ist die zu untersuchende Schultergliedmaße in Höhe des Schultergelenks zu fixieren und dabei der Hund mit dem Unterarm auf den Tisch zu drücken. Die 2. Person fixiert die Schultergliedmaßen an den Pfoten. Wichtig ist, daß das zu untersuchende Ellbogengelenk nur mit dem Ellbogenhöcker dem Film aufliegt. Bei einer Verkantung, also bei Auflage des Olekranons und des Epicondylus lateralis bzw. medialis humeri, wird ein Epikondylus von der Ulna überlagert dargestellt. Dadurch können wichtige Details übersehen werden.
Die Aufnahme kann ohne Kassette mit einem folienlosen Film (Bleiunterlage nicht vergessen!) angefertigt werden.
Ellbogengelenk: Der Zentralstrahl sollte die Mitte der Schultergliedmaße etwa eine halbe bis eine Fingerbreite distal der Epikondylen treffen und im rechten Winkel auf den Film einfallen.
Unterarm: Der Zentralstrahl sollte den Radius in Schaftmitte treffen und im rechten Winkel auf den Film einfallen.

Fig. 81   Positioning of elbow joint and forearm. Craniocaudal.

One assistant holds the head with one hand, pulling it backward as far as possible. The other hand secures the limb to be examined at the level of the shoulder joint, at the same time pressing the dog with the forearm against the table. The second assistant secures both of the thoracic limbs at the paws. It is important that only the olecranon of the elbow joint to be examined should come in contact with the film. In case of tilting, i.e. overlaying of the olecranon and the lateral or medial epicondyle of the humerus, one of the epicondyles will be seen with the ulna superimposed. In such cases important details may be overseen.
The radiograph should be taken with a non-screen film (lead blocker not to be forgotten!).
Elbow joint: The central beam should strike the middle of the front leg approximately one half to one finger's width distally to the epicondyles and fall at the right angle on the film.
Forearm: The central beam should strike the mid-shaft of the radius and fall at the right angle on the film.

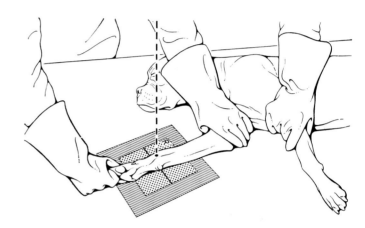

Abb. 82   Lagerung zur Aufnahme des Karpalgelenks. Medio-lateral.

Person 1 hält mit einer Hand die zu untersuchende Schultergliedmaße proximal des Ellbogengelenks und drückt dabei mit dem Unterarm Kopf und Hals des Hundes auf den Tisch. Die andere Hand fixiert die Pfote so, daß die Zehen senkrecht übereinander liegen. Die am Rücken des Hundes stehende 2. Person fixiert mit einer Hand die etwas nach hinten gezogene obenliegende Schultergliedmaße und mit der anderen Hand die Beckengliedmaßen. Dabei wird der Hund mit den Unterarmen auf den Tisch gedrückt.
Die Aufnahme ist mit einem folienlosen Film (Bleiunterlage nicht vergessen!) anzufertigen.
Der Zentralstrahl sollte die Mitte der Schultergliedmaße in Höhe des proximalen Gelenkspalts des Karpalgelenks treffen und im rechten Winkel auf den Film einfallen.

Fig. 82   Positioning of carpal joint. Mediolateral.

One assistant grasps the leg to be examined proximal to the elbow joint and presses the head and neck of the dog with the forearm against the table. With the other hand he secures the paw in such a manner that the phalanges lie perpendicularly upon each other. The assistant at the back of the dog secures the upper front limb with one hand, pulling it slightly backward. The other hand holds the pelvic limbs. At the same time he presses the dog against the table with the forearms.
The radiograph should be taken with a non-screen film (lead blocker not to be forgotten!).
The central beam should strike the middle of the front leg at the level of the proximal articular space of the carpal joint and fall at the right angle on the film.

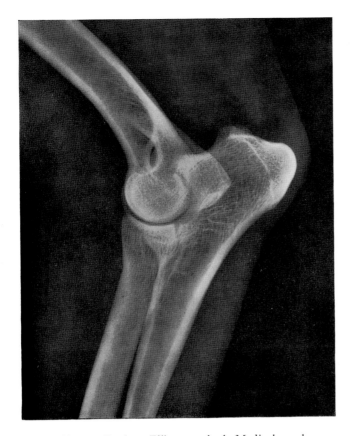

Abb. 83   Rechtes Ellbogengelenk. Medio-lateral.
Deutscher Schäferhund, 3 Jahre.
Folienloser Film – FFA 120 cm – 63 kV – 18 mAs
Originalgröße (Ausschnitt aus 13 × 18 cm)
Lagerung Abb. 80

Fig. 83   Right elbow joint. Mediolateral.
German shepherd dog, 3 years old.
Non-screen film – FFD 120 cm – 63 kV – 18 mAs
Original size (section of 13 × 18 cm)
Positioning fig. 80

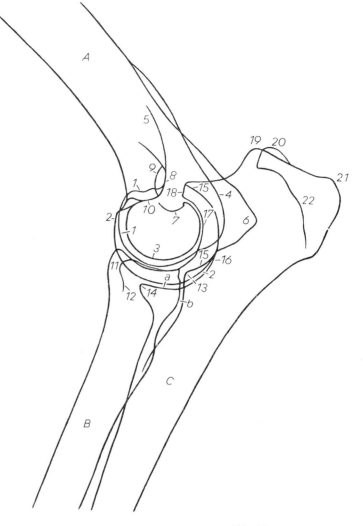

Abb. 84   Röntgenskizze zu Abb. 83
Fig. 84   X-ray sketch to fig. 83

A   Humerus;
B   Radius;
C   Ulna;

a   Articulatio cubiti;
b   Articulatio radioulnaris proximalis;

1, 2   Condylus humeri:
1   Lateraler Abschnitt, Capitulum – Lateral part, capitulum,
2   Medialer Abschnitt, Trochlea – Medial part, trochlea;
3   Kompaktaschatten, der sich aus der Konkavität der Führungsrinne zwischen 1 und 2 ergibt – Shadow of compacta formed by the concavity of the groove between 1 and 2;
4   Epicondylus lateralis;
5   Crista epicondyli lateralis;
6   Epicondylus medialis;
7   Fossa olecrani, Grund – Fossa olecrani, base;
8, 9   Foramen supratrochleare:
8   Lateraler Rand – Lateral margin,
9   Medialer Rand – Medial margin;

10   Fossa radialis, Grund – Fossa radialis, base;
11   Tuberositas radii;
12   Radius, lateraler Bandhöcker – Radius, eminence for attachment of lateral collateral ligament;
13   Processus coronoideus lateralis;
14   Processus coronoideus medialis;
15   Laterale Begrenzung der Gelenkfläche im Bereich der Incisura trochlearis – Lateral border of the articular surface in the area of the trochlear incisure;
16   Mediale Begrenzung dieser Gelenkfläche, die wegen Überlagerung nur im distalen Bereich dargestellt ist – Medial border of this articular surface, seen only in the distal part because of superimposition;
17   Incisura trochlearis;
18   Processus anconaeus;
19 – 22   Tuber olecrani:
19   Kraniolateraler Höcker – Craniolateral tubercle,
20   Kraniomedialer Höcker – Craniomedial tubercle,
21   Kaudaler Höcker – Caudal tubercle,
22   Mediale Knochenleiste – Medial ridge.

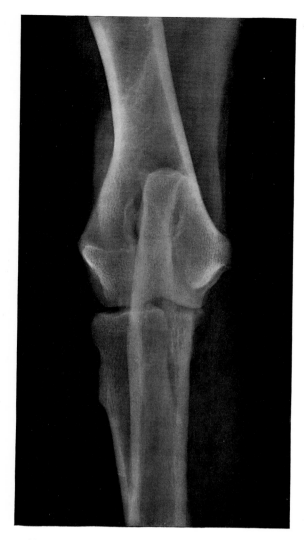

Abb. 85   Rechtes Ellbogengelenk. Kranio-kaudal.
Deutscher Schäferhund, 7 Jahre.
Folienloser Film – FFA 120 cm – 75 kV – 30 mAs
Originalgröße (Ausschnitt aus 13 × 18 cm)
Lagerung Abb. 81

Fig. 85   Right elbow joint. Craniocaudal.
German shepherd dog, 7 years old.
Non-screen film – FFD 120 cm – 75 kV – 30 mAs
Original size (section of 13 × 18 cm)
Positioning fig. 81

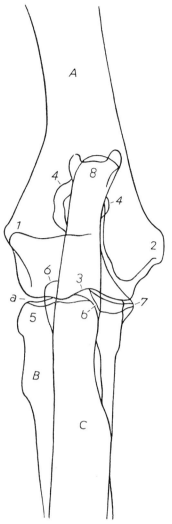

Abb. 86   Röntgenskizze zu Abb. 85
Fig. 86   X-ray sketch to fig. 85

A   Humerus;
B   Radius;
C   Ulna;

a   Articulatio cubiti;
b   Articulatio radioulnaris proximalis;

1   Epicondylus lateralis;
2   Epicondylus medialis;
3   Condylus humeri, Rand - Condylus humeri, border;
4   Fossa olecrani;
5   Caput radii;
6   Processus coronoideus lateralis;
7   Processus coronoideus medialis;
8   Dreigeteiltes Tuber olecrani - Trifid olecranon tuberosity.

63

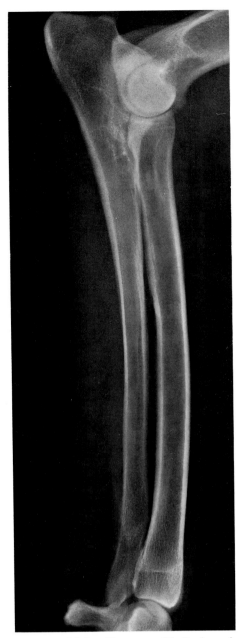

Abb. 87   Linker Unterarm. Medio-lateral.
Deutscher Schäferhund, 2 Jahre.
Bucky-Blende – Feinzeichnende Folie –
FFA 120 cm – 66 kV – 25 mAs
Verkleinerung von 24 × 30 cm
Lagerung Abb. 80

Fig. 87   Left forearm. Mediolateral.
German shepherd dog, 2 years old.
Bucky-diaphragm – High definition screens
– FFD 120 cm – 66 kV - 25 mAs
Diminution of 24 × 30 cm
Positioning fig. 80

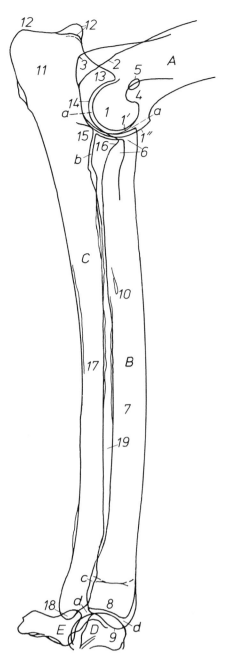

Abb. 88   Röntgenskizze zu Abb. 87
Fig. 88   X-ray-sketch to fig. 87

A   Humerus;
B   Radius;
C   Ulna;
D   Os carpi intermedioradiale und Os carpi ulnare, ineinander projiziert – Os carpi intermedioradiale and Os carpi ulnare, projected one into another;
E   Os carpi accessorium;

a   Articulatio cubiti;
b   Articulatio radioulnaris proximalis;
c   Articulatio radioulnaris distalis;
d   Articulatio antebrachiocarpea;

Am Humerus – On the humerus:

1   Condylus humeri:
1′ Lateraler Rand – Lateral border,
1″ Medialer Rand – Medial border;
2   Epicondylus lateralis,
3   Epicondylus medialis,
4   Fossa radialis;
5   Foramen supratrochleare;

Am Radius – On the radius:

6   Caput radii mit Fovea capitis radii – Caput radii with the fovea of the radial caput;
7   Corpus radii;
8   Trochlea radii;
9   Processus styloideus radii;
10 Foramen nutricium;

An der Ulna – On the ulna:

11 Olecranon;
12 Tuber olecrani;
13 Processus anconaeus;
14 Incisura trochlearis;
15 Processus coronoideus lateralis;
16 Processus coronoideus medialis;
17 Corpus ulnae;
18 Processus styloideus ulnae;
19 Spatium interosseum antebrachii.

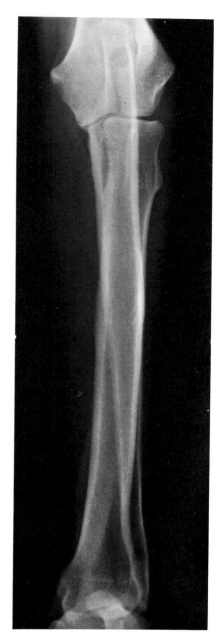

Abb. 89   Linker Unterarm. Kranio-kaudal.
Deutscher Schäferhund, 2 Jahre.
Bucky-Blende – Feinzeichnende Folie – FFA 120 cm –
66 kV – 25 mAs
Verkleinerung von 24 × 30 cm
Lagerung Abb. 81

Fig. 89   Left forearm. Craniocaudal.
German shepherd dog, 2 years old.
Bucky-diaphragm – High definition screens –
FFD 120 cm – 66 kV - 25 mAs
Diminution of 24 × 30 cm
Positioning fig. 81

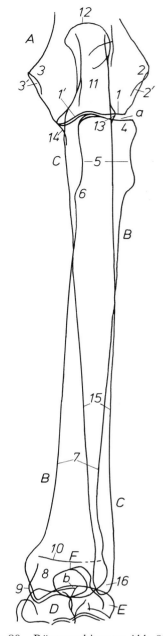

Abb. 90   Röntgenskizze zu Abb. 89
Fig. 90   X-ray-sketch to fig. 89

A   Humerus;
B   Radius;
C   Ulna;
D   Os carpi intermedioradiale;
E   Os carpi ulnare;
F   Os carpi accessorium;

a   Articulatio cubiti;
b   Articulatio antebrachiocarpea;

Am Humerus – On the humerus:

1, 1′   Condylus humeri:
1   Capitulum humeri,
1′   Trochlea humeri;
2   Epicondylus lateralis,
2′   Bandgrube – Depression for ligamentous attachment;
3   Epicondylus medialis,
3′   Bandgrube – Depression for ligamentous attachment;

Am Radius – On the radius:

4   Caput radii mit Fovea capitis radii – Caput radii with the fovea of the radial caput;
5   Collum radii;
6   Tuberositas radii;
7   Corpus radii;
8   Trochlea radii;
9   Processus styloideus;
10   Crista transversa;

An der Ulna – On the ulna:

11   Olecranon;
12   Tuber olecrani;
13   Processus coronoideus lateralis;
14   Processus coronoideus medialis;
15   Corpus ulnae;
16   Processus styloideus.

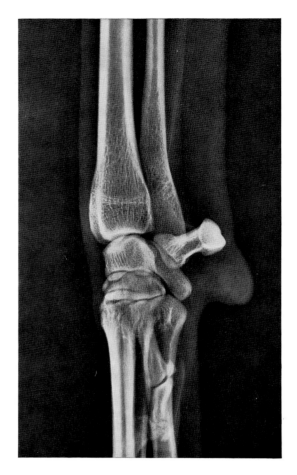

Abb. 91 Rechtes Karpalgelenk. Medio-lateral.
Deutscher Schäferhund, 3 Jahre.
Folienloser Film – FFA 120 cm – 65 kV – 24 mAs
Originalgröße (Ausschnitt aus 13 × 18 cm)
Lagerung Abb. 82

Fig. 91 Right carpal joint. Mediolateral.
German shepherd dog, 3 years old.
Non-screen film – FFD 120 cm – 65 kV – 24 mAs
Original size (section of 13 × 18 cm)
Positioning fig. 82

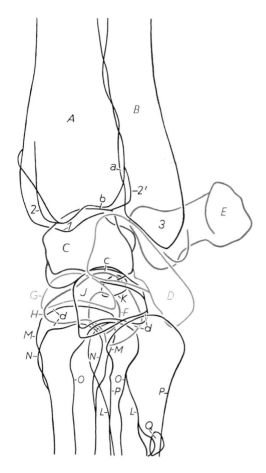

Abb. 92** Röntgenskizze zu Abb. 91
Fig. 92** X-ray sketch to fig. 91

A Radius;
B Ulna;
C Os carpi intermedioradiale;
D Os carpi ulnare;
E Os carpi accessorium;
F Os carpale I;
G Os carpale II ;
H Os carpale III;
J Os carpale IV;
K Os sesamoideum m. abductoris pollicis longi;
L Os metacarpale I;
M Os metacarpale II;
N Os metacarpale III;
O Os matacarpale IV;

P Os metacarpale V;
Q Ossa sesamoidea proximalia am Zehengrundgelenk der 1. Zehe – Ossa sesamoidea proximalia at the metacarpophalangeal joint of the 1st digit;

a Articulatio radioulnaris distalis;
b Articulatio antebrachiocarpea;
c Articulatio mediocarpea;
d Articulationes carpometacarpeae;

1 Processus styloideus radii;
2, 2′ Trochlea radii:
2 Lateralrand (plattennah) - Lateral border (next to film),
2′ Medialrand (plattenfern) - Medial border (next to tube);
3 Processus styloideus ulnae.

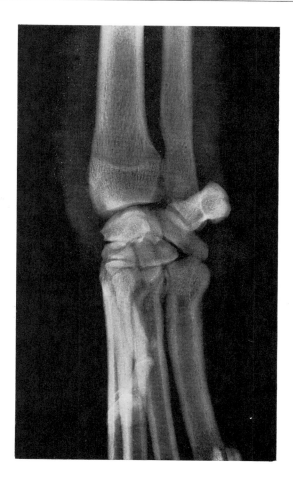

Abb. 93   Rechtes Karpalgelenk. Schräglagerung.
Mediopalmar-dorsolateral.
Boxer, 4 Jahre.
Folienloser Film – FFA 120 cm – 65 kV – 24 mAs
Originalgröße (Ausschnitt aus 13 × 18 cm)
Lagerung Abb. 99

Fig 93   Right carpal joint. Oblique positioning.
Mediopalmar-dorsolateral.
Boxer, 4 years old.
Non-screen film – FFD 120 cm – 65 kV – 24 mAs
Original size (section of 13 × 18 cm)
Positioning fig. 99

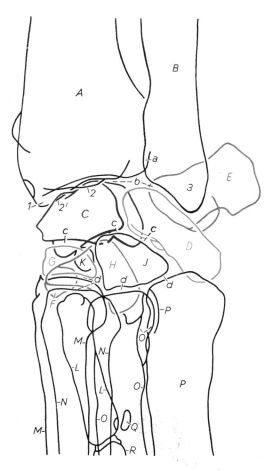

Abb. 94**   Röntgenskizze zu Abb. 93
Fig. 94**   X-ray sketch to fig. 93

A   Radius;
B   Ulna;
C   Os carpi intermedioradiale;
D   Os carpi ulnare;
E   Os accessorium;
F   Os carpale I;
G   Os carpale II;
H   Os carpale III;
J   Os carpale IV;
K   Os sesamoideum m. abductoris pollicis longi;
L   Os metacarpale I;
M   Os metacarpale II;
N   Os metacarpale III;
O   Os metacarpale IV;
P   Os metacarpale V;

Q   Ossa sesamoidea proximalia am Zehengrundgelenk der 1. Zehe – Ossa sesamoidea proximalia at the metacarpophalangeal joint of the 1st digit;
R   Phalanx proximalis;

a   Articulatio radioulnaris distalis;
b   Articulatio antebrachiocarpea;
c   Articulationes carpometacarpeae;

1   Processus styloideus radii;
2, 2′   Trochlea radii:
2   Lateralrand (plattennah) – Lateral border (next to film),
2′   Medialrand (plattenfern) – Medial border (next to tube);
3   Processus styloideus ulnae.

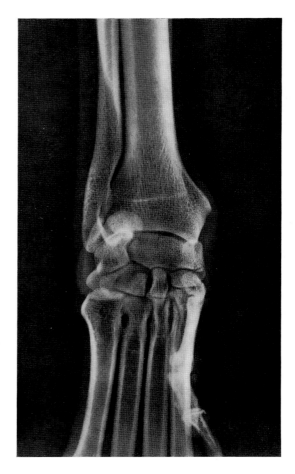

Abb. 95   Rechtes Karpalgelenk. Dorso-palmar.
Deutscher Schäferhund, 3 Jahre.
Folienloser Film – FFA 120 cm – 65 kV – 18 mAs
Originalgröße (Ausschnitt aus 13 × 18 cm)
Lagerung Abb. 100

Fig. 95   Right carpal joint. Dorsopalmar.
German shepherd dog, 3 years old.
Non-screen film – FFD 120 cm – 65 kV – 18 mAs
Original size (section of 13 × 18 cm)
Positioning fig. 100

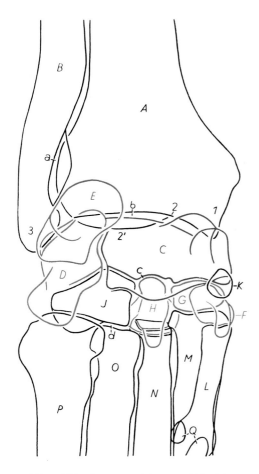

Abb. 96**   Röntgenskizze zu Abb. 95
Fig. 96**   X-ray sketch to fig. 95

A   Radius;
B   Ulna;
C   Os carpi intermedioradiale;
D   Os carpi ulnare;
E   Os carpi accessorium;
F   Os carpale I;
G   Os carpale II;
H   Os carpale III;
J   Os carpale IV;
K   Os sesamoideum m. abductoris pollicis longi;
L   Os metacarpale I;
M   Os metacarpale II;
N   Os metacarpale III;
O   Os metacarpale IV;

P   Os metacarpale V;
Q   Ossa sesamoidea proximalia am Zehengrundgelenk der 1. Zehe – Ossa sesamoidea proximalia at the metacarpophalangeal joint of the 1st digit;

a   Articulatio radioulnaris distalis;
b   Articulatio antebrachiocarpea;
c   Articulatio mediocarpea;
d   Articulationes carpometacarpeae;

1   Processus styloideus radii;
2,   2′   Trochlea radii:
2   Palmarrand (plattennah) – Palmar border (next to film),
2′   Dorsalrand (plattenfern) – Dorsal border (next to tube);
3   Processus styloideus ulnae.

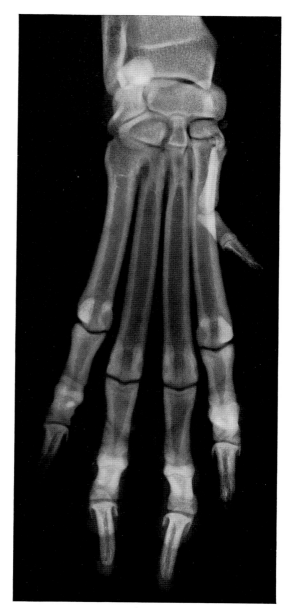

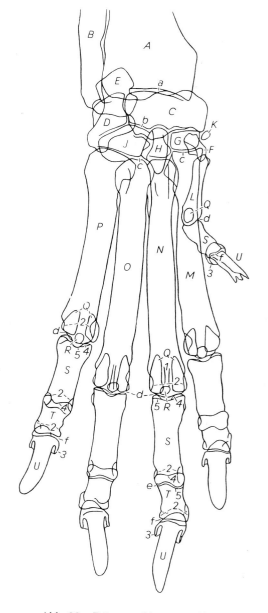

Abb. 97   Rechter Vorderfuß. Dorso-palmar.
Schnauzer, 10 Jahre.
Folienloser Film – FFA 120 cm – 60 kV – 15 mAs
Originalgröße (Ausschnitt aus 13 × 18 cm)
Lagerung Abb. 101

Fig. 97   Right forepaw. Dorsopalmar.
Schnauzer, 10 years old.
Non-screen Film – FFD 120 cm – 60 kV – 15 mAs
Original size (section of 13 × 18 cm)
Positioning fig. 101

Abb. 98   Röntgenskizze zu Abb. 97
Fig. 98   X-ray sketch to fig. 97

A   Radius;
B   Ulna;
C   Os carpi intermedioradiale;
D   Os carpi ulnare;
E   Os carpi accessorium;
F   Os carpale I
G   Os carpale II;
H   Os carpale III;
J   Os carpale IV;
K   Os semamoideum m. abductoris pollicis longi;
L   Os metacarpale I;
M   Os metacarpale II;
N   Os metacarpale III;
O   Os metacarpale IV;
P   Os metacarpale V;
Q   Ossa sesamoidea proximalia am Zehengrundgelenk – Ossa sesamoidea proximalia at the metacarpophalangeal joint;
R   Ossa sesamoidea dorsalia;
S   Phalanx proximalis der 1. bis 5. Zehe – Phalanx proximalis of the 1st to 5th digits;

T   Phalanx media der 2. bis 5. Zehe – Phalanx media of the 2nd to 5th digits;
U   Phalanx distalis der 1. bis 5. Zehe – Phalanx distalis of the 1st to 5th digits;

a   Articulatio antebrachiocarpea;
b   Articulatio mediocarpea;
c   Articulationes carpometacarpeae;
d   Articulatio metacarpophalangea;
e   Articulatio interphalangea proximalis manus;
f   Articulatio interphalangea distalis manus;

1   Sagittalkamm – Sagittal crest;
2   Bandgruben an den Phalangen – Depressions for ligamentous attachment;
3   Crista unguicularis;
4, 5   Fovea articularis des proximalen bzw. mittleren Zehenglieds – Fovea articularis of the proximal and middle phalanges respectively:
4   Palmare Begrenzung – Palmar border,
5   Dorsale Begrenzung – Dorsal border.

69

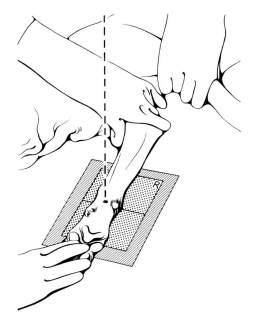

Abb. 99   Schräglagerung zur Aufnahme des Karpalgelenks.
Mediopalmar-dorsolateral.

Person 1 hält mit einer Hand die zu untersuchende Schultergliedmaße proximal des Ellbogengelenks und drückt dabei Kopf und Hals des Hundes mit dem Unterarm auf den Tisch. Mit der anderen Hand wird die Pfote fixiert, indem die dorsale Fläche auf die Unterlage gedrückt wird. Durch die Drehung im Karpalgelenk werden dessen laterale Anteile übersichtlich dargestellt. Die Aufnahme ist mit einem folienlosen Film (Bleiunterlage nicht vergessen!) anzufertigen. Der Zentralstrahl sollte die Mitte des Vorderfußes in Höhe des Karpalballens treffen und im rechten Winkel auf den Film einfallen.

Fig. 99   Oblique positioning of carpal joint. Mediopalmar-dorsolateral.

One assistant grasps the leg to be examined proximal to the elbow joint and presses the head and neck of the dog with the forearm against the table. With the other hand he secures the paw by pressing its dorsal part against the table. The lateral aspect of the carpal joint is shown better by this rotation.
The radiograph should be taken with a non-screen film (lead blocker not to be forgotten!).
The central beam should strike the middle of the forepaw at the level of the carpal pad and fall at the right angle on the film.

Die Aufnahme ist mit einem folienlosen Film (Bleiunterlage nicht vergessen!) anzufertigen.
Der Zentralstrahl sollte die Mitte der Schultergliedmaße in Höhe des proximalen Gelenkspalts des Karpalgelenks treffen und im rechten Winkel auf den Film einfallen.

Fig. 100   Positioning of carpal joint. Dorsopalmar.

One assistant holds the head with one hand, pulling it backward as far as possible. The other hand holds the limb to be examined at the elbow joint. The dog is pressed with the forearm against the assistant and the table. The second assistant secures both of the thoracic limbs at the paws.
The radiograph should be taken with a non-screen film (lead blocker not to be forgotten!).
The central beam should strike the middle of the front leg at the level of the proximal articular space of the carpal joint and fall at the right angle on the film.

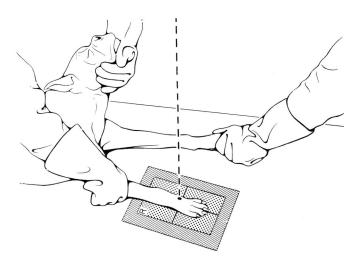

Abb. 101   Lagerung zur Aufnahme des Vorderfußes. Dorso-palmar.

Person 1 fixiert mit einer Hand den rückenwärts abgebogenen Kopf und Hals. Mit der anderen Hand wird die zu untersuchende Schultergliedmaße am Unterarm fixiert und dabei der Hund mit dem Unterarm auf den Tisch gedrückt. Person 2 fixiert die andere Pfote.
Die Aufnahme ist mit einem folienlosen Film (Bleiunterlage nicht vergessen!) anzufertigen.
Der Zentralstrahl sollte die Mitte des Vorderfußes in halber Länge der Ossa metacarpalia treffen und im rechten Winkel auf den Film einfallen.

Fig. 101   Positioning of forepaw. Dorsopalmar.

One assistant secures the head and neck with one hand by pulling it backward. The other hand holds the limb to be examined at the forearm, pressing the dog against the table. The second assistant secures the other leg.
The radiograph should be taken with a non-screen film (lead blocker not to be forgotten!).
The central beam should strike the center of the forepaw at the middle of the length of the metacarpal bones and fall at rhe right angle on the film.

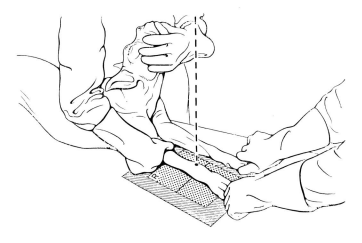

Abb. 100   Lagerung zur Aufnahme des Karpalgelenks. Dorso-palmar.

Person 1 fixiert mit einer Hand den Kopf, wobei der Hals zum Rücken hin abgebogen wird. Mit der anderen Hand wird die zu untersuchende Schultergliedmaße in Höhe des Ellbogengelenks gehalten und dabei der Hund mit dem Arm an die Person und auf den Tisch gedrückt. Person 2 fixiert beide Schultergliedmaßen an den Pfoten.

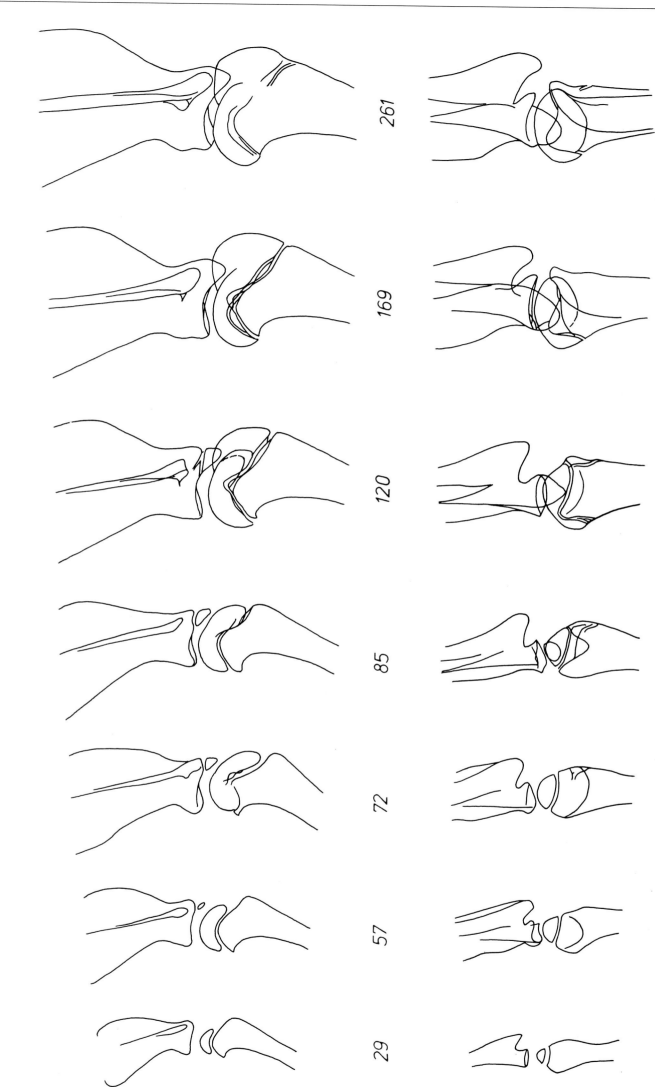

Fig. 102   Left shoulder joint. Mediolateral. German shepherd dog.
Postnatal development. Age in days.

Abb. 102   Linkes Schultergelenk. Medio-lateral. Deutscher Schäferhund.
Postnatale Entwicklung. Lebensalter in Tagen.

Fig. 103   Left shoulder joint. Caudocranial. German shepherd dog.
Postnatal development. Age in days.

Abb. 103   Linkes Schultergelenk. Kaudo-kranial. Deutscher Schäferhund.
Postnatale Entwicklung. Lebensalter in Tagen

261   169   120   85   72   57   29

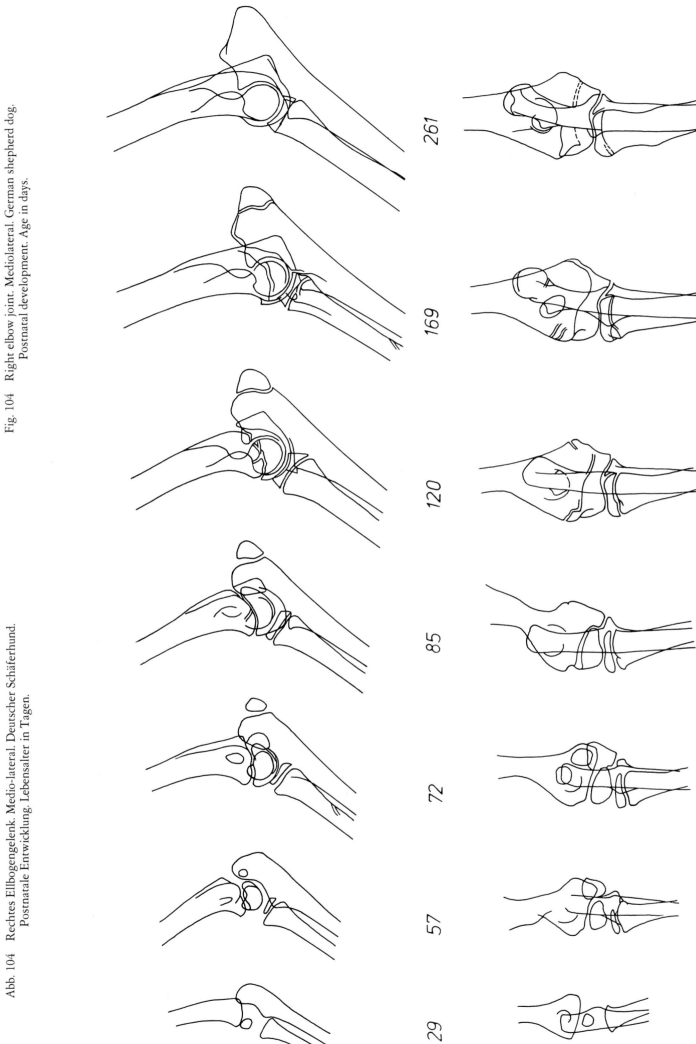

Fig. 104   Right elbow joint. Mediolateral. German shepherd dog.
Postnatal development. Age in days.

Abb. 104   Rechtes Ellbogengelenk. Medio-lateral. Deutscher Schäferhund.
Postnatale Entwicklung. Lebensalter in Tagen.

Fig. 105   Right elbow joint. Craniocaudal. German shepherd dog.
Postnatal development. Age in days.

Abb. 105   Rechtes Ellbogengelenk. Kranio-kaudal. Deutscher Schäferhund.
Postnatale Entwicklung. Lebensalter in Tagen.

29

57

72

85

120

169

261

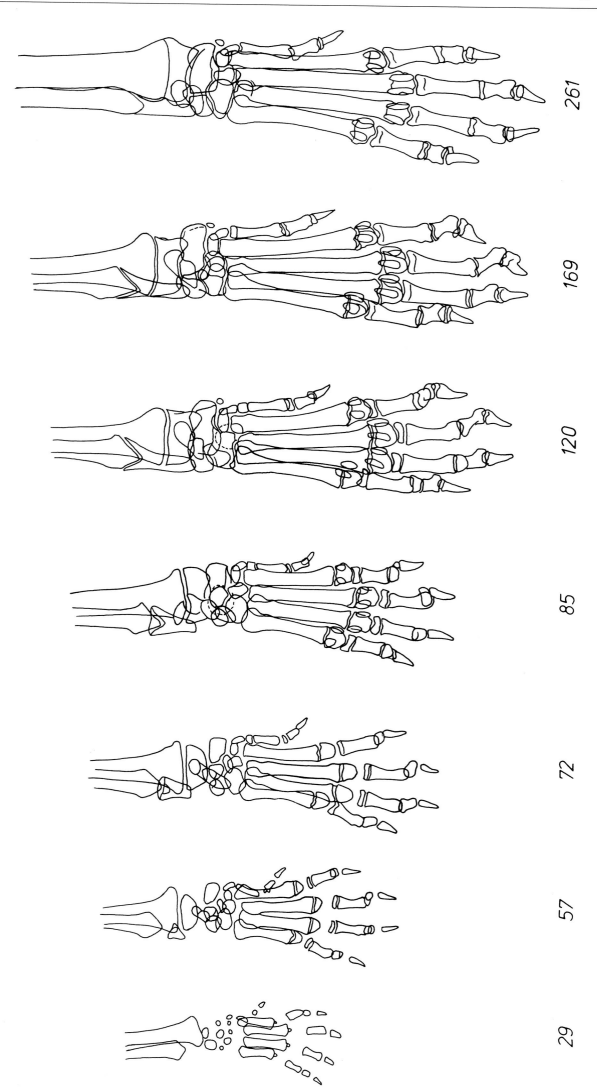

261

169

120

85

72

57

29

Fig. 106   Right forepaw. Dorsopalmar. German shepherd dog. Postnatal development. Age in days.

Abb. 106   Rechter Vorderfuß. Dorso-palmar. Deutscher Schäferhund. Postnatale Entwicklung. Lebensalter in Tagen.

Tabelle 1:   Zeitliches Auftreten der Ossifikationspunkte sowie des Apo- und Epiphysenfugenschlusses am Skelett der Schultergliedmaße vom Deutschen Schäferhund (nach SCHROEDER 1978)

Table 1:   Time-table of the appearance of ossification centers and closures of apo- and epiphyseal lines of the thoracic limb in German shepherd dog (after SCHROEDER 1978)

Ossifikationspunkte Apophysen und Epiphysen / Ossification centers Apophyses and Epiphyses	Auftreten der Ossifikationspunkte (Angabe in Tagen) / Appearance of the ossification centers (in days)	Verschmelzen der Ossifikationskerne (Angabe in Monaten) / Fusion of the ossification centers (in months)	Apo- und Epiphysenfugenschluß (Angabe in Monaten) / Fusion of apo- and epiphyseal lines (in months)
SCAPULA			
Tuberculum supraglenoidale	49 – 65		5 – 6
HUMERUS			
Epiphysis proximalis humeri	14 – 16		10,5 – 12
Epicondylus medialis	49 – 65		
Condylus lateralis	14 – 22		5,5 – 6,5
Condylus medialis	21 – 43		
RADIUS			
Epiphysis proximalis radii	28 – 43		9 – 11
Epiphysis distalis radii	14 – 29		9 – 11
ULNA			
Apophysis proximalis ulnae	49 – 72		6,5 – 9,5
Epiphysis distalis ulnae	49 – 65		9 – 11
OSSA CARPI			
Os carpi radiale	28 – 29	Os carpi intermedioradiale 3 – 4	
Os carpi intermedium	16 – 22		
Os carpi centrale	28 – 36		
Os carpi ulnare	28 – 36		
Os carpi accessorium	14 – 16		
Apophysis ossis carpi accessorii	49 – 72		4 – 5
Os carpale I	21 – 29		
Os carpale II	28 – 36		
Os carpale III	28 – 36		
Os carpale IV	21 – 29		
OSSA METACARPALIA			
Epiphysis proximalis ossis metacarpalis I	49 – 57		5,5 – 6,5
Epiphyses distales ossium metacarpalium II – V	28 – 36		6,5 – 7,5
OSSA DIGITORUM MANUS			
Epiphysis proximalis phalangis proximalis I	28 – 65		5,5 – 5,5
Epiphyses proximales phalangium proximalium II – V	28 – 43		5,5 – 6,5
phalangium mediarum II – V	28 – 65		5,5 – 6,5
OSSA SESAMOIDEA			
Ossa sesamoidea proximalia	63 – 92		
Ossa sesamoidea dorsalia	91 – 141		
Os sesamoideum m. abductoris pollicis longi	120		

Tabelle 2: Zeitliches Auftreten der Ossifikationspunkte sowie des Apo- und Epiphysenfugenschlusses am Skelett der Beckengliedmaße vom Deutschen Schäferhund (nach WIDMER 1978)

Table 2: Time-table of the appearance of ossification centers and closures of apo- and epiphyseal lines of the pelvic limb in German shepherd dog (after WIDMER 1978)

Ossifikationspunkte Apophysen und Epiphysen	Auftreten der Ossifikationspunkte (Angabe in Tagen)	Apo- und Epiphysen- fugenschluß (Angabe in Monaten)
Ossification centers Apophyses and Epiphyses	Appearance of the ossification centers (in days)	Fusion of apo- and epiphyseal lines (in months)
OS COXAE		
Os ilium	⎫ Bei Beginn der	
Os ischii	⎬ Unters.-Reihe	
Os pubis	⎭ bereits vorh.	
Os acetabuli	49 – 85	5 – 6
Tuberculum ischiadicum	50 – 85	5 – 6
Crista iliaca	120 – 141	10 – 11
„Os interischiadicum"	147 – 197	
Arcus ischiadicus	141 – 173	10 – 12
OS FEMORIS		
Caput ossis femoris	14 – 29	11 – 12
Trochanter major	35 – 50	11
Trochanter minor	35 – 78	11 – 12
Epiphysis distalis ossis femoris	14 – 22	11
TIBIA		
Epiphysis proximalis tibiae	14 – 22	11 – 12
Tuberositas tibiae	49 – 78	11 – 12
Malleolus medialis	77 – 92	4 – 5
Epiphysis distalis tibiae	14 – 29	8,5 – 11
FIBULA		
Epiphysis proximalis fibulae	49 – 72	10 – 12
Epiphysis distalis fibulae	35 – 43	10 – 11
OSSA TARSI		
Tuber calcanei	49 – 65	6,5 – 7,5
Os tarsi centrale	14 – 22	
Os tarsale I	36 – 49	
Os tarsale II	29 – 36	
Os tarsale III	21 – 35	
OSSA METATARSALIA		
Os metatarsale I	49 – 78	
Epiphyses distales ossium metatarsalium II – V	29 – 36	7 – 8
OSSA DIGITORUM PEDIS		
Epiphyses proximales		
phalangium proximalium II – V	35 – 43	6,5 – 7,5
phalangium mediarum II – V	35 – 57	6,5 – 7,5
OSSA SESAMOIDEA		
Patella	49 – 85	
Ossa sesamoidea m. gastrocnemii	91 – 100	
Os sesamoideum m. poplitei	126 – 169	
Ossa sesamoidea proximalia (plantaria)	63 – 92	
Ossa sesamoidea dorsalia	126 – 169	

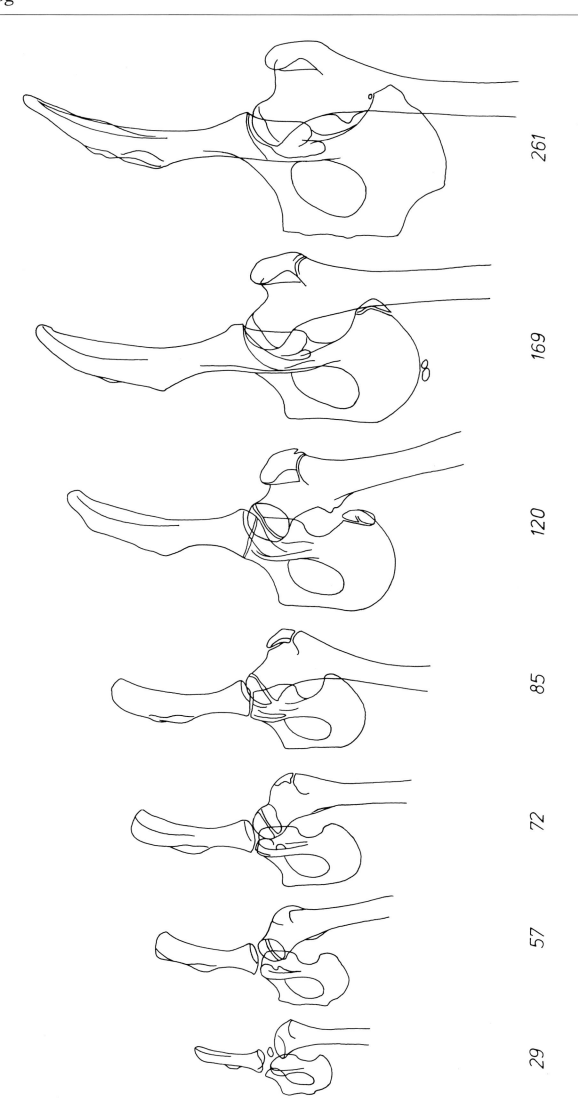

261

169

120

85

72

57

29

Fig. 107 Left half of the pelvis and left hip joint.
Ventrodorsal. German shepherd dog.
Postnatal development. Age in days.

Abb. 107 Linke Beckenhälfte und linkes Hüftgelenk. Ventro-dorsal.
Deutscher Schäferhund.
Postnatale Entwicklung. Lebensalter in Tagen.

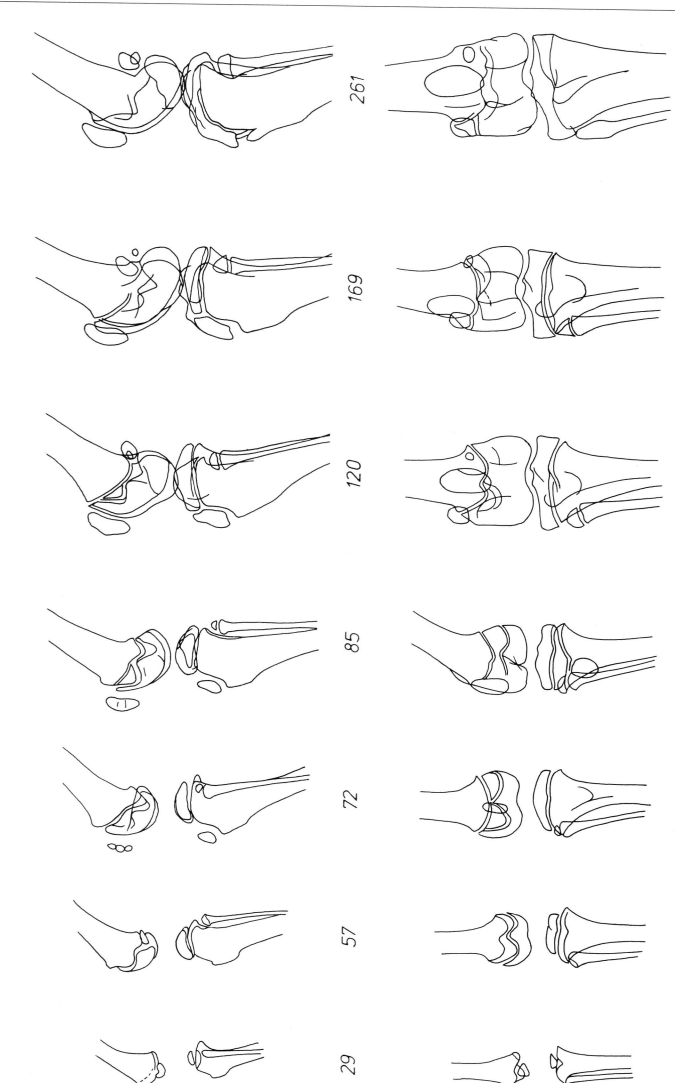

261

169

120

85

72

57

29

Fig. 108 Right stifle joint. Mediolateral. German shepherd dog. Postnatal development. Age in days.

Abb. 108 Rechtes Kniegelenk. Medio-lateral. Deutscher Schäferhund. Postnatale Entwicklung. Lebensalter in Tagen.

Fig. 109 Right stifle joint. Craniocaudal. German shepherd dog. Postnatal development. Age in days.

Abb. 109 Rechtes Kniegelenk. Kranio-kaudal. Deutscher Schäferhund. Postnatale Entwicklung. Lebensalter in Tagen.

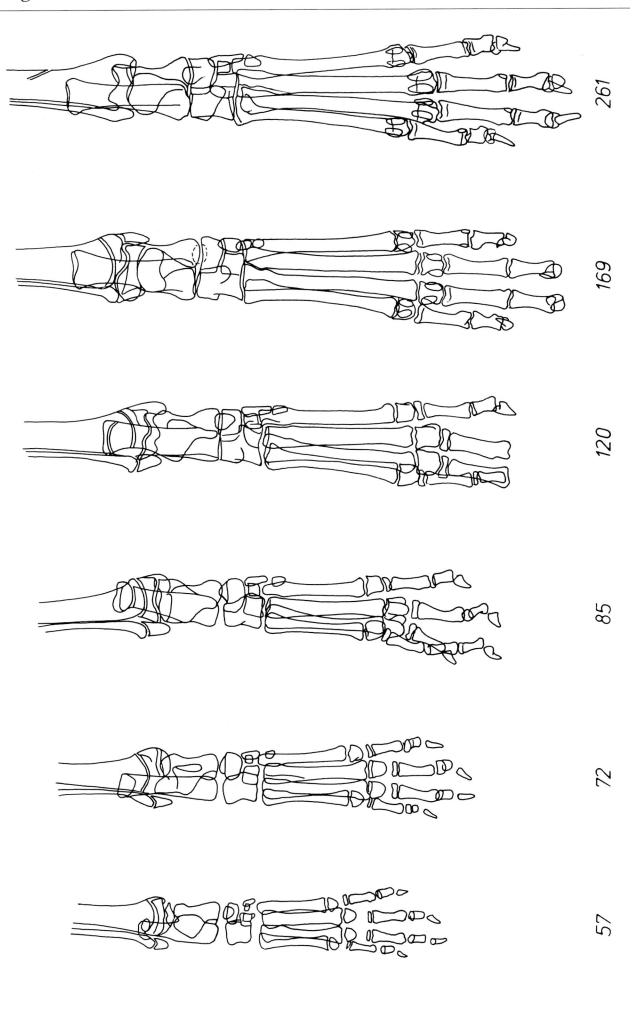

261

169

120

85

72

57

29

Abb. 110 Rechter Hinterfuß. Dorso-plantar. Deutscher Schäferhund. Postnatale Entwicklung. Lebensalter in Tagen.

Fig. 110 Right hindpaw. Dorsoplantar. German shepherd dog. Postnatal development. Age in days.

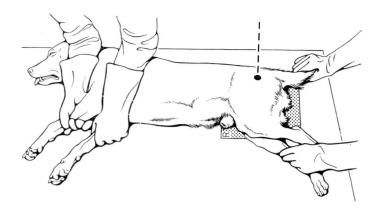

Abb. 111 Lagerung zur Aufnahme des Beckens. Latero-lateral.

Um eine gleichmäßige Abbildung des Beckens zu erreichen, muß die Medianebene des Rumpfes parallel zur Kassette liegen. Bei fettleibigen Tieren kann die korrekte Lagerung durch Schaumgummikeile unter den aufliegenden Gliedmaßen in Höhe der haltenden Hände ggf. auch unter dem Brustbein erreicht werden. Die Lagerung läßt sich durch Zug an der Rute erleichtern. Die Beckengliedmaßen sind nur mäßig nach hinten gestreckt zu fixieren.
Abgesehen von kleinen, nicht fettleibigen Hunden ist eine Bucky-Blende oder eine Kassette mit stehendem Raster zur Reduzierung der Streustrahlung notwendig.
Der Zentralstrahl sollte den Trochanter major der obenliegenden Gliedmaße treffen und im rechten Winkel auf die Kassette einfallen.

Fig. 111   Positioning of pelvis. Laterolateral.

In order to obtain a uniform radiograph of the pelvis in a lateral position, the median plane of the trunk must be parallel to the cassette. In obese animals, proper positioning can be achieved with the help of foam rubber pads placed under the limbs resting on the table and if necessary also under the sternum. Pull on the tail will also facilitate the positioning. The pelvic limbs should be secured by slight stretching in a caudal direction.
A Bucky diaphragm or a cassette with a stationary grid is necessary to reduce scattered radiation, except in small, lean dogs.
The central beam should strike the trochanter major of the upper limb and fall at the right angle on the cassette.

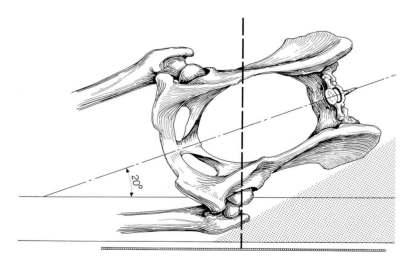

Abb. 112   Lagerung zur Aufnahme des Beckens. Latero-lateral, Schrägprojektion.

Zur Beurteilung der operativen Rekonstruktion nach Fraktur des Darmbeins, des Azetabulums und/oder des Sitzbeins ist eine Schrägprojektion erforderlich. Die zu beurteilende Beckenhälfte ist plattennah zu lagern. Durch Unterlegen eines Schaumstoffkeils unter den Rumpf im Bereich des Beckens wird die Medianebene um etwa 20° zur Unterlage hin gekantet. Die Lagerung wird durch Zug an der Rute erleichtert. Die Beckengliedmaßen sind nur mäßig nach hinten gestreckt zu fixieren.
Zur Reduzierung der Streustrahlung ist eine Bucky-Blende oder eine Kassette mit stehendem Raster notwendig.
Der Zentralstrahl sollte fingerbreit dorsal des Trochanter major die obenliegende Gliedmaße treffen und im rechten Winkel auf die Kassette einfallen.

Fig. 112   Positioning of pelvis. Obliquely laterolateral.

An oblique projection is necessary in order to evaluate a surgical reconstruction of a fracture of ilium, of acetabulum and/or of os ischii. The median plane of the pelvis must be tilted approximately 20° toward the table with the fracture side close to the film. Pelvic limbs are moderately stretched backward. A slight pull on the tail will facilitate the positioning.
A Bucky diaphragm or a cassette with stationary grid is necessary in order to reduce scattered radiation.
The central beam should strike the upper limb about one finger's width dorsally to the trochanter major and fall at the right angle on the cassette.

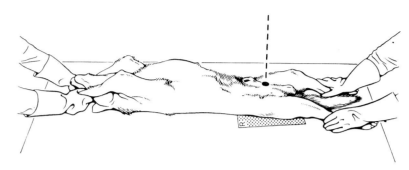

Abb. 113   Lagerung zur Aufnahme des Beckens. Ventro-dorsal.

Um eine symmetrische Abbildung beider Beckenhälften zu erhalten, muß die Medianebene des Rumpfes senkrecht zur Kassette stehen. Die Beckengliedmaßen sind so weit nach hinten zu strecken, daß die Sprunggelenke der Unterlage aufliegen. Die Beckengliedmaßen sollten in leichter Abduktion und geringer Außenrotation von der Hilfsperson fixiert werden. Diese Gliedmaßenlagerung ist mit Ausnahme der Untersuchung der Hüftgelenke zur Feststellung einer Dysplasie zweckmäßig.
Für die Untersuchung der Hüftgelenke zur Feststellung einer Dysplasie sind die Beckengliedmaßen parallel und in leichter Innenrotation zu lagern, so daß sich die Pfoten berühren und die Kniescheiben exakt über den Gelenkflächen liegen.
Zur Reduzierung der Streustrahlung ist eine Bucky-Blende oder eine Kassette mit stehendem Raster erforderlich.
Der Zentralstrahl sollte die Mittellinie in Höhe der Trochanteren treffen und im rechten Winkel auf die Kassette einfallen.

Fig. 113   Positioning of pelvis. Ventrodorsal.

In order to obtain a symmetrical radiograph of both halves of the pelvis, the median plane of the trunk must be perpendicular to the cassette. The pelvic limbs should be pulled backward to bring the hocks in contact with the table. Both limbs should be secured by an assistant in moderate abduction and slight outward rotation. Except for the evaluation of hip dysplasia this positioning is suitable for radiography of the hip joints.
In radiography for evaluation of hip dysplasia, the limbs should be positioned parallel and in slight inward rotation. The paws should touch each other und the patellae must be situated exactly over the articular surface.
The Bucky diaphragm or a cassette with a stationary grid should be used to reduce scattered radiation.
The central beam should strike the midline on a level of the trochanters and fall at the right angle on the cassette.

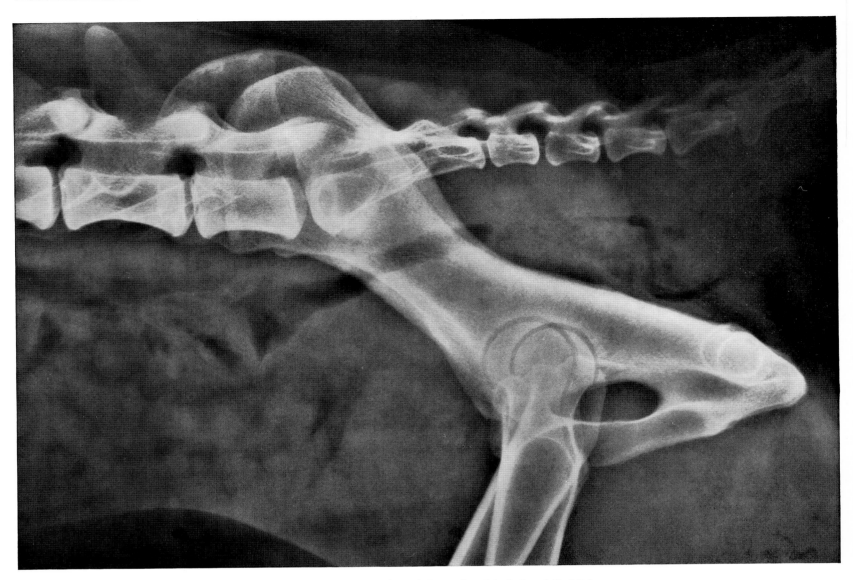

Abb. 114   Becken. Latero-lateral. Deutscher Schäferhund, ♀, 3 Jahre
Bucky-Blende – Feinzeichnende Folie – FFA 120 cm – 67 kV – 50 mAs
Verkleinerung von 24 × 30 cm
Lagerung Abb. 111

Fig. 114   Pelvis. Laterolateral. German shepherd dog, ♀, 3 years old.
Bucky diaphragm – High definition screens – FFD 120 cm – 67 kV – 50 mAs
Diminution of 24 × 30 cm
Positioning fig. 111

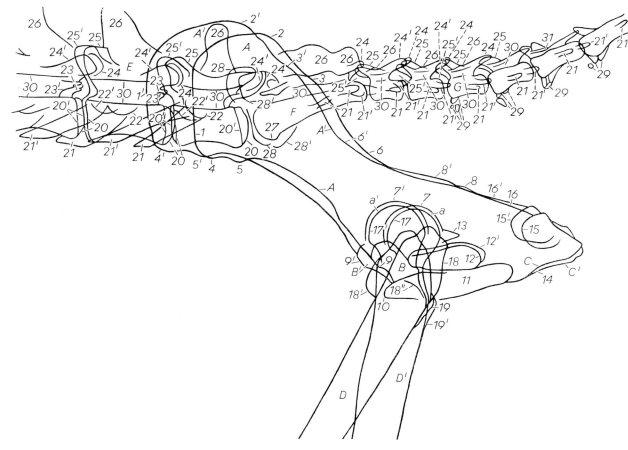

Abb. 115*   Röntgenskizze zu Abb. 114   Fig. 115*   X-ray sketch to fig. 114

A   Os ilium;
B   Os pubis;
C   Os ischii;
D   Os femoris;
E   6. Vertebra lumbalis;
F   Os sacrum;
G   3. Vertebra caudalis;

a   Articulatio coxae;

Am Becken – On the pelvis:

1   Ala ossis ilii;
2, 3   Tuber sacrale:
2   Spina iliaca dorsalis cranialis,
3   Spina iliaca dorsalis caudalis;
4   Tuber coxae, Spina iliaca ventralis cranialis;
5   Spina alaris;
6   Incisura ischiadica major corporis ossis ilii;
7   Acetabulum;
8   Spina ischiadica;
9   Eminentia iliopubica;
10   Pecten ossis pubis;
11   Symphysis pelvina;
12   Foramen obturatum;
13   Incisura acetabuli;

14   Arcus ischiadicus;
15   Tuber ischiadicum;
16   Incisura ischiadica minor;

Am Os femoris – On the femur:

17   Caput ossis femoris;
18   Trochanter major;
18″   Fossa trochanterica, Grund, plattenfern – Fossa trochanterica, base next to tube;
19   Trochanter minor;

An der Wirbelsäule – On the vertebral column:

20   Corpus vertebrae;
21   Processus transversus;
22   Basis processus transversi;
23   Processus accessorius;
24   Processus articularis caudalis;
25   Processus articularis cranialis;
26   Processus spinosus;
27   Promontorium;
28   Ala ossis sacri;
29   Arcus haemalis;
30   Canalis vertebralis;
31   Öffnung und zugleich Ende des Canalis vertebralis – Combined entrance and termination of the vertebral canal.

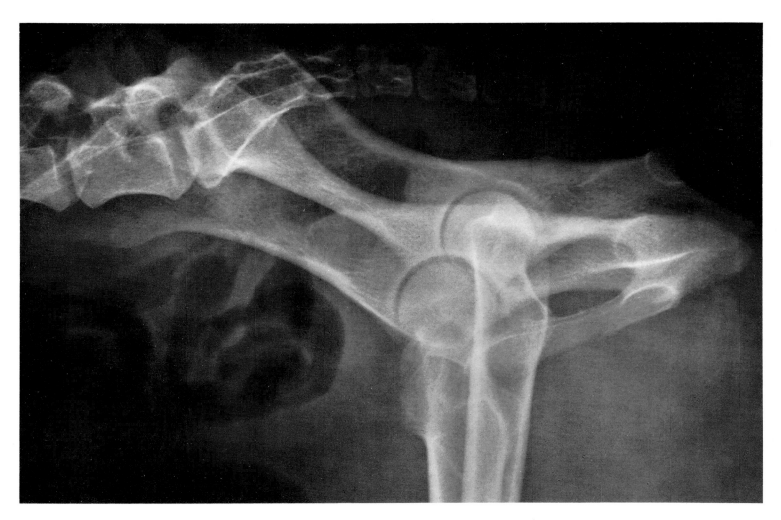

Abb. 116   Becken. Latero-lateral, Schrägprojektion (20°).
Deutsch-Kurzhaar, ♀, 1 1/2 Jahre.
Feinzeichnende Folie – FFA 120 cm – 70 kV – 64 mAs
Originalgröße (Ausschnitt aus 24 × 30 cm)
Lagerung Abb. 112

Fig. 116   Pelvis. Obliquely laterolateral (20°).
German short-haired pointer, ♀, 1 1/2 years old.
High definition screens – FFD 120 cm – 70 kV – 64 mAs
Original size (section of 24 × 30 cm)
Positioning fig. 112

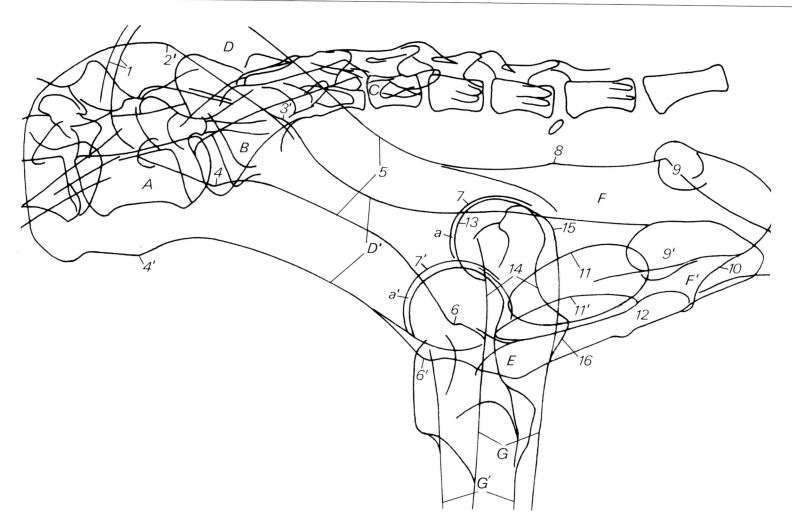

Abb. 117* Röntgenskizze zu Abb. 116
Fig. 117* X-ray sketch to fig. 116

A   7. Vertebra lumbalis;
B   Os sacrum;
C   1. Vertebra caudalis;
D   Os ilium;
E   Os pubis;
F   Os ischii;

a   Articulatio coxae;

Am Becken – On the pelvis:

1   Crista iliaca;
2, 3 Tuber sacrale:
2   Spina iliaca dorsalis cranialis,
3   Spina iliaca dorsalis caudalis;

4   Tuber coxae, Spina ventralis caudalis;
5   Corpus ossis ilii;
6   Eminentia iliopubica;
7   Acetabulum;
8   Spina ischiadica;
9   Tuber ischiadicum;
10   Arcus ischiadicus;
11   Foramen obturatum;
12   Symphysis pelvina;

Am Os femoris – On the femur:

13   Caput ossis femoris;
14   Collum ossis femoris;
15   Trochanter major;
16   Trochanter minor.

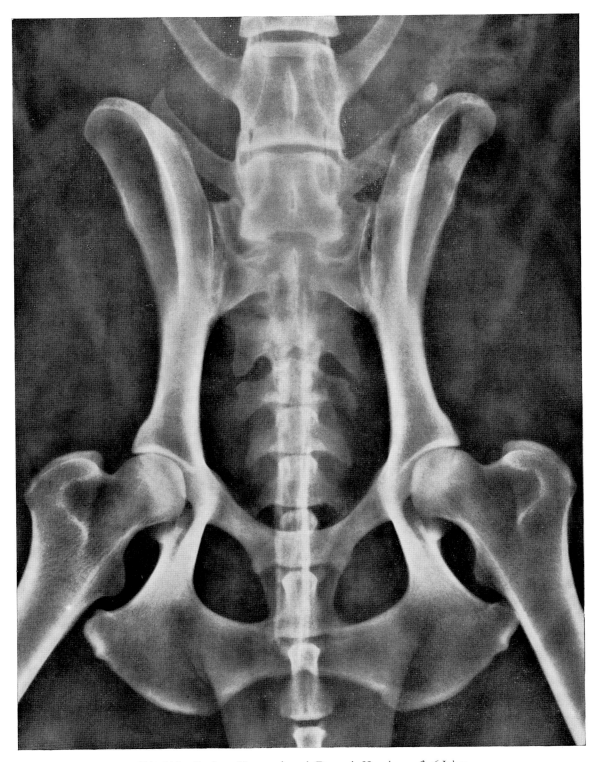

Abb. 118   Becken. Ventro-dorsal. Deutsch-Kurzhaar, ♂, 6 Jahre.
Bucky-Blende – Feinzeichnende Folie – FFA 120 cm – 65 kV – 45 mAs
Verkleinerung von 24 × 30 cm
Lagerung Abb. 113

Fig. 118   Pelvis. Ventrodorsal. German short-haired pointer, ♂, 6 years old.
Bucky diaphragm – High definition screens – FFD 120 cm – 65 kV – 45 mAs
Diminution of 24 × 30 cm
Positioning fig. 113

A   Os ilium;
B   Os pubis;
C   Os ischii;
D   Os femoris;
E   6. Vertebra lumbalis;
F   Os sacrum;
G   3. Vertebra caudalis;
H   Os penis;

a   Articulatio coxae;
b   Articulatio sacroiliaca;

Am Becken – On the pelvis:

1   Ala ossis ilii;
2   Tuber sacrale, Spina iliaca dorsalis caudalis;
3   Linea glutaea;
4   Kompaktaschatten, der sich aus der Konkavität der Facies glutaea ergibt – Compacta shadow formed by concavity of the gluteal surface;
5   Crista iliaca;
6   Tuber coxae, Spina iliaca ventralis cranialis;
7   Spina alaris;

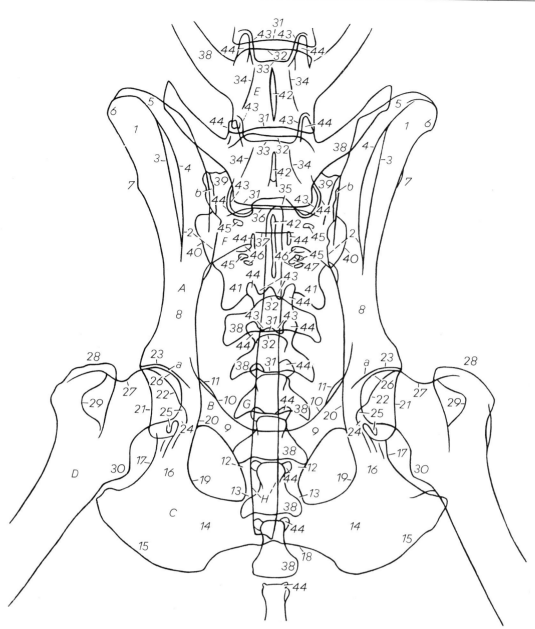

Abb. 119   Röntgenskizze zu Abb. 118

Fig. 119   X-ray sketch to fig. 118

8   Corpus ossis ilii;
9   Ramus cranialis ossis pubis;
10 Pecten ossis pubis;
11 Eminentia iliopubica;
12 Ramus caudalis ossis pubis;
13 Ramus ossis ischii;
14 Tabula ossis ischii;
15 Tuber ischiadicum;
16 Corpus ossis ischii;
17 Incisura ischiadica minor;
18 Arcus ischiadicus;
19 Foramen obturatum;
20 Spina ischiadica;
21, 22 Acetabulum:
21 Dorsaler Rand – Dorsal border,
22 Ventraler Rand – Ventral border;
23 Verschattung, die sich aus der Konkavität der Facies lunata ergibt –
Shadow formed by the concavity of the lunate surface;
24 Incisura acetabuli;
25 Fossa acetabuli;

Am Os femoris – On the femur:

26 Caput ossis femoris;
27 Collum ossis femoris;
28 Trochanter major;
29 Fossa trochanterica;
30 Trochanter minor;

An der Wirbelsäule – On the vertebral column:

31 Extremitas caudalis;
32 Extremitas cranialis;
33 Dorsaler Rand von 32; Anmerkung: Der scheinbare Übergang von
33 in 43 ergibt sich aus der Lagerung – Dorsal border of 32; note:
The apparent transition from 33 to 43 is caused by the positioning;
34 Canalis vertebralis, seitliche Begrenzung – Canalis vertebralis, lateral
border;
35 Aufhellung: Wirbelbogen des 7. Lendenwirbels hier ohne Überlagerungen, nicht zu verwechseln mit Spatium interarcuale lumbosacrale
– Bright area: vertebral arch of the 7th lumbar vertebra without superimposition here, is not to be confused with the lumbosacral interarcual space;
36, 37 Basis ossis sacri:
36 Dorsaler Rand – Dorsal border,
37 Ventraler Rand, Promontorium – Ventral border, promontory;
38 Processus transversus;
39, 40 Ala ossis sacri:
39 Dorsaler Anteil – Dorsal part,
40 Ventraler Anteil – Ventral part;
41 Os sacrum, Pars lateralis;
42 Processus spinosus;
43 Processus articularis caudalis;
44 Processus articularis cranialis;
45 Foramina sacralia dorsalia;
46 Foramina sacralia pelvina;
47 Foramen intervertebrale.

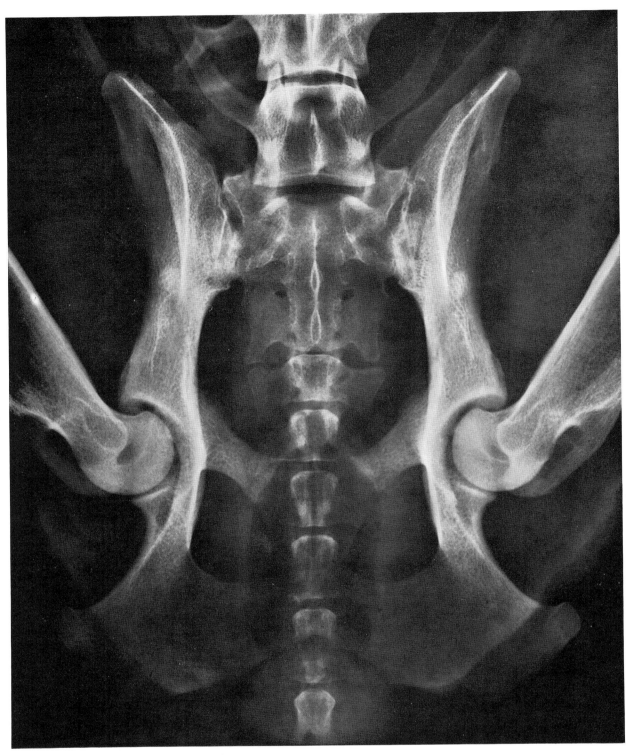

Abb. 120   Becken. Ventro-dorsal. Deutscher Schäferhund, ♀, 4 Jahre.
Bucky-Blende – Feinzeichnende Folie – FFA 115 cm – 70 kV – 80 mAs
Ausschnitt und Verkleinerung von 30 × 40 cm
Lagerung Abb. 124

Fig. 120   Pelvis. Ventrodorsal. German shepherd dog, ♀, 4 years old.
Bucky diaphragm – High definition screens – FFD 115 cm – 70 kV – 80 mAs
Section and diminution of 30 × 40 cm
Positioning fig. 124

A   Os ilium;
B   Os pubis;
C   Os ischii;
D   Os femoris;

a   Articulatio coxae;
b   Articulatio sacroiliaca;

Am Becken – On the pelvis:

1   Ala ossis ilii;
2   Kompaktaschatten, der sich aus der Konkavität der Facies glutaea ergibt – Compacta shadow formed by concavity of the gluteal surface;

E   7. Vertebra lumbalis;
F   Os sacrum;
G   3. Vertebra caudalis;

3,   4   Ala ossis ilii:
3   Dorsaler Rand – Dorsal border,
4   Ventraler Rand – Ventral border;
5   Spina alaris;
6   Tuber coxae, Spina iliaca ventralis cranialis;
7   Crista iliaca;
8,   9   Ala ossis ilii, Facies auricularis:
8   Ventraler Rand – Ventral border,
9   Dorsaler Rand – Dorsal border;
10,   11   Ala ossis sacri, Facies auricularis:
10   Ventraler Rand – Ventral border,
11   Dorsaler Rand – Dorsal border;
12   Corpus ossis ilii;

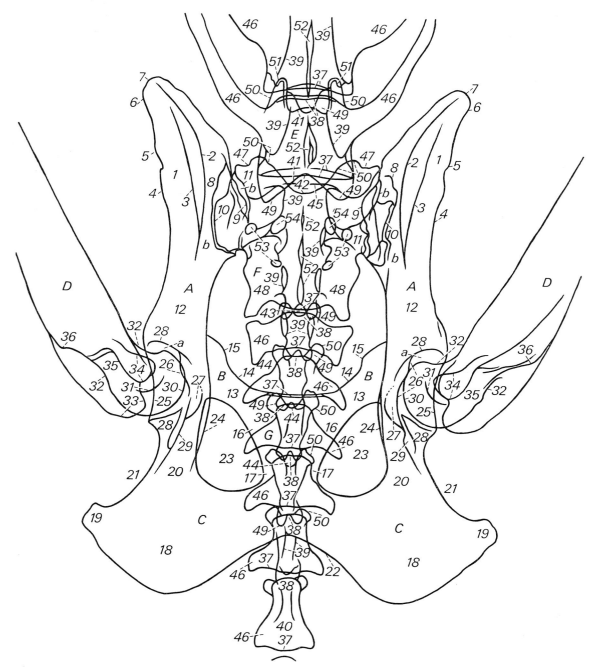

Abb. 121   Röntgenskizze zu Abb. 120   Fig. 121   X-ray sketch to fig. 120

13 Ramus cranialis ossis pubis;
14 Pecten ossis pubis;
15 Eminentia iliopubica;
16 Ramus caudalis ossis pubis;
17 Ramus ossis ischii;
18 Tabula ossis ischii;
19 Tuber ischiadicum;
20 Corpus ossis ischii;
21 Incisura ischiadica minor;
22 Arcus ischiadicus;
23 Foramen obturatum;
24 Spina ischiadica;
25, 26   Acetabulum:
25 Dorsaler Rand – Dorsal border,
26 Ventraler Rand – Ventral border;
27 Fossa acetabuli;
28 Verschattung, die sich aus der Konkavität der Facies lunata ergibt –
Shadow formed by the concavity of the lunate surface;
29 Incisura acetabuli;

Am Os femoris – On the femur:

30 Caput ossis femoris,
31 Ventrale Begrenzung seiner Facies articularis – Ventral border of its
articular surface;
32 Collum ossis femoris;
33 Trochanter major;
34 Orthograph getroffene Basis von 33 – Base of 33 struck orthographically;

35 Fossa trochanterica;
36 Trochanter minor;

An der Wirbelsäule – On the vertebral column:

37 Extremitas caudalis;
38 Extremitas cranialis;
39, 40   Canalis vertebralis:
39 Seitliche Begrenzung – Lateral border,
40 Öffnung und zugleich Ende – Combined entrance and termination;
41 Arcus vertebrae des 6. und 7. Lendenwirbels, Kaudalrand – Arcus
vertebrae of the 6th and 7th lumbar vertebrae, caudal border;
42 Spatium interarcuale lumbosacrale;
43 Spatium interarcuale sacrocaudale, von Arcus vertebrarum begrenzt
– Spatium interarcuale sacrocaudale limited by vertebral arches;
44 Spatia interarcualia zwischen Schwanzwirbeln – Spatia interarcualia
between caudal vertebrae;
45 Basis ossis sacri;
46 Processus transversus;
47 Ala ossis sacri, Kranialrand – Ala ossis sacri, cranial border;
48 Os sacrum, Pars lateralis;
49 Processus articularis caudalis;
50 Processus articularis cranialis;
51 Processus accessorius am 6. Lendenwirbel – Processus accessorius of
the 6th lumbar vertebra;
52 Processus spinosus;
53 Foramina sacralia dorsalia;
54 Foramina sacralia pelvina.

87

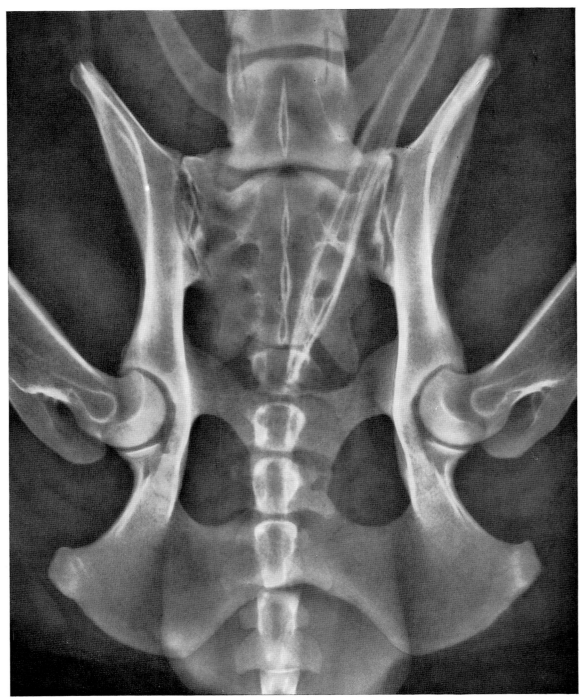

Abb. 122   Becken. Dorso-ventral. Deutscher Schäferhund, ♂, 7 Jahre.
Bucky-Blende – Feinzeichnende Folie – FFA 120 cm – 60 kV – 36 mAs
Verkleinerung von 24 × 30 cm
Lagerung Abb. 125

Fig. 122   Pelvis. Dorsoventral. German shepherd dog, ♂, 7 years old.
Bucky diaphragm – High definition screens – FFD 120 cm – 60 kV – 36 mAs
Diminution of 24 × 30 cm
Positioning fig. 125

A   Os ilium;
B   Os pubis;
C   Os ischii;
D   Os femoris;
E   6. Vertebra lumbalis;
F   Os sacrum;
G   3. Vertebra caudalis;
H   Os penis;

a   Articulatio coxae;

Am Becken – On the pelvis:

1 – 9 Ala ossis ilii:
2 Kompaktaschatten, der sich aus der Konkavität der Facies glutaea ergibt – Shadow of compacta formed by the concavity of the gluteal surface,
3 Dorsaler Rand – Dorsal border,
4 Ventraler Rand – Ventral border,
5 Spina alaris,
6 Tuber coxae, Spina iliaca ventralis cranialis,
7 Crista iliaca,
8, 9 Facies auricularis:
8 Ventraler Rand – Ventral border,
9 Dorsaler Rand – Dorsal border,
10, 11 Ala ossis sacri, Facies auricularis:
10 Ventraler Rand – Ventral border,
11 Dorsaler Rand – Dorsal border;
12 Corpus ossis ilii;
13 Ramus cranialis ossis pubis;
14 Pecten ossis pubis;
15 Eminentia iliopubica;
16 Ramus caudalis ossis pubis, links durch Schwanzwirbel überlagert – Ramus caudalis ossis pubis, left side superimposed by the caudal vertebrae;
17 Ramus ossis ischii;
18 Tabula ossis ischii;
19 Tuber ischiadicum;
20 Corpus ossis ischii;
21 Incisura ischiadica minor;
22 Arcus ischiadicus;
23 Foramen obturatum;
24 Spina ischiadica;
25, 26 Acetabulum:
25 Dorsaler Rand – Dorsal border,
26 Ventraler Rand – Ventral border;
27 Fossa et Incisura acetabuli;
28, 29 Verschattungen, die sich aus der Konkavität der Facies lunata ergeben – Shadows formed by the concavity of the lunate surface;

Am Os femoris – On the femur:

30 Caput ossis femoris;
31 Facies articularis, ventrale Begrenzung – Facies articularis, ventral border;
32 Collum ossis femoris;
33 Trochanter major;
34 Orthograph getroffene Basis von 33 – Base of 33 struck orthographically;
35 Fossa trochanterica;
36 Trochanter minor;

An der Wirbelsäule – On the vertebral column:

37 Extremitas caudalis;
38 Extremitas cranialis;
39, 40 Canalis vertebralis:
39 Seitliche Begrenzung – Lateral border,
40 Öffnung und zugleich Ende – Combined entrance and termination;
41 Arcus vertebrae des 6. und 7. Lendenwirbels, Kaudalrand – Arcus vertebrae of the 6th and 7th lumbar vertebrae, caudal border;
42 Spatium interarcuale lumbosacrale;
43 Spatium interarcuale sacrocaudale;
44 Spatia interarcualia zwischen 1. und 2. bzw. 2. und 3. bzw. 3. und 4. Schwanzwirbel – Spatia interarcualia between the 1st and 2nd, 2nd and 3rd, and 3rd and 4th caudal vertebrae respectively;
45 Basis ossis sacri, dorsaler Rand – Basis ossis sacri, dorsal border;

46 Processus transversus;
47 Ala ossis sacri, kranialer Rand – Ala ossis sacri, cranial border;
48 Os sacrum, Pars lateralis;
49 Processus articularis caudalis;
50 Processus articularis cranialis, am Os sacrum wegen Überlagerung nicht dargestellt – Processus articularis cranialis, because of superimposition not visible on the sacrum;
51 Processus accessorius;
52 Processus spinosus;
53 Foramina sacralia dorsalia;
54 Foramina sacralia pelvina;
55 Foramina intervertebralia.

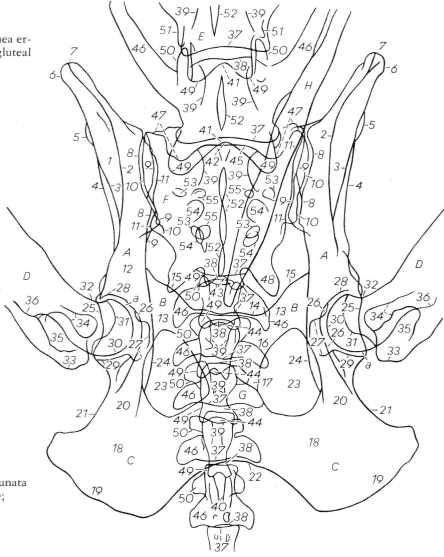

Abb. 123 Röntgenskizze zu Abb. 122
Fig. 123 X-ray sketch to fig. 122

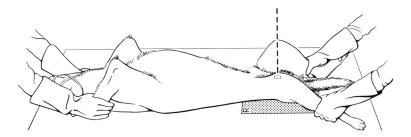

Abb. 124 Lagerung zur Aufnahme des Beckens mit gebeugten Beckengliedmaßen. Ventro-dorsal.

Diese Lagerung ist für den nicht narkotisierten Patienten bei Veränderungen im Bereich des Beckens weniger schmerzhaft und damit weniger schwierig als die in den Abbildungen 113 und 125 dargestellten Lagerungen. Die Gliedmaßen sollten in physiologischer „Standstellung" des Oberschenkels in Knie- und Sprunggelenk abgebeugt und dann angemessen abduziert werden. Bei richtiger Lagerung der Beckengliedmaßen liegt der Bereich „Hüfthöcker-Kreuzbein-Schwanzansatz" der Unterlage an. Um eine symmetrische Abbildung beider Beckenhälften zu erhalten, müssen die Medianebene des Rumpfes senkrecht zur Kassette stehen und die Hüfthöcker der Unterlage aufliegen.
Zur Reduzierung der Streustrahlung ist eine Bucky-Blende oder eine Kassette mit stehendem Raster erforderlich.
Der Zentralstrahl sollte die Mittellinie in Höhe des Schambeins treffen und im rechten Winkel auf die Kassette einfallen.

Fig. 124   Positioning of pelvis with flexed pelvic limbs. Ventrodorsal.

This positioning is less painful for the non-sedated patient with pathology in the pelvic region and therefore less difficult than those shown in figures 113 and 125. With the thighs placed in the physiological „standing position", the stifle and hock joints should be flexed and then abducted appropriately. With proper positioning of the pelvic limbs the region of the tuber coxae – sacrum – tail root lies on the cassette. In order to obtain a symmetrical radiograph of both halves of the pelvis, the median plane of the trunk must be perpendicular to the cassette and the coxal tuberosities must lie on the cassette.
To reduce scattered radiation, a Bucky diaphragm or a stationary grid should be used.
The central beam should strike the midline on a level of the pubis and fall at the right angle on the cassette.

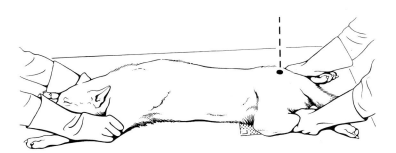

Abb. 125   Lagerung zur Aufnahme des Beckens. Dorso-ventral.

Bei Veränderungen im Bereich des Hüftgelenks ist die korrekte Lagerung wegen der notwendigen Abduktion der Gliedmaßen oft für den Patienten sehr schmerzhaft und damit schwierig. Bei einer Einschränkung der Beweglichkeit im Hüftgelenk, insbesondere der Abduktion bei Arthropathia deformans, kann außerdem in manchen Fällen das Becken nicht nahe genug auf der Kassette gelagert werden. Aus dem dann zu großen Objekt-Filmabstand kann eine unter Umständen erhebliche Verprojizierung resultieren.
Um eine symmetrische Abbildung beider Beckenhälften zu erhalten, muß die Medianebene des Rumpfes senkrecht zur Kassette stehen. Die Beckengliedmaßen sind so weit zu abduzieren, daß der Beckenboden der Kassette zumindest annähernd anliegt.
Zur Reduzierung der Streustrahlung ist eine Bucky-Blende oder eine Kassette mit stehendem Raster erforderlich.
Der Zentralstrahl sollte die Mittellinie in Höhe des 1. Schwanzwirbels treffen und im rechten Winkel auf die Kassette einfallen.

Fig. 125   Positioning of pelvis. Dorsoventral.

Because of the inevitable abduction of the limbs, the proper positioning of patients with pathology of the hip joints is frequently very painful and consequently difficult. In cases of limited mobility of the hip joints, especially abduction associated with arthropathia deformans, the pelvis can moreover not be aligned closely enough to the cassette. The resulting increased object to film distance can cause considerable distortion.
In order to obtain a symmetrical radiograph of both halves of the pelvis, the median plane of the trunk must be perpendicular to the cassette. The pelvic limbs should be abducted such that the pelvic floor at least is brought in close proximity to the cassette.
In order to reduce scattered radiation, a Bucky diaphragm or a cassette with a stationary grid should be used.
The central beam should strike the midline on a level of the 1st caudal vertebra and fall at the right angle on the cassette.

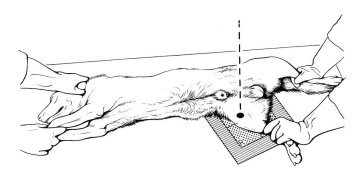

Abb. 126   Lagerung zur Aufnahme des Oberschenkels, des Kniegelenks und des Unterschenkels. Medio-lateral.

Die obenliegende Beckengliedmaße ist so weit abzuspreizen, daß das zu untersuchende Kniegelenk ohne Überlagerung dargestellt wird. Die die abgespreizte (obenliegende) Gliedmaße haltende Hand ist auf dem Tisch aufzustützen, gegebenenfalls fixiert sie zusätzlich die Rute.
Bei kräftiger Oberschenkelmuskulatur läßt sich die korrekte Lagerung der zu untersuchenden Gliedmaße durch Unterlegen eines Schaumgummikeils in Höhe der zu haltenden Hand erleichtern.
Os femoris: Zur Reduzierung der Streustrahlung ist eine Bucky-Blende oder eine Kassette mit stehendem Raster erforderlich.
Der Zentralstrahl sollte das Os femoris in Schaftmitte treffen und im rechten Winkel auf die Kassette einfallen.
Kniegelenk: Die Aufnahme ist mit einem folienlosen Film (Bleiunterlage nicht vergessen!) anzufertigen.
Der Zentralstrahl sollte den Gelenkspalt des Kniekehlgelenks treffen und im rechten Winkel auf den Film einfallen.
Tibia: Die Aufnahme ist mit einem folienlosen Film (Bleiunterlage nicht vergessen!) anzufertigen.
Der Zentralstrahl sollte die Tibia in Schaftmitte treffen und im rechten Winkel auf den Film einfallen.

Fig. 126   Positioning of thigh, stifle joint and tibia. Mediolateral.

The upper leg has to be abducted sufficiently to ensure radiography of the stifle joint free from overlaying. The hand holding the abducted leg should rest on the table and also secure the tail. In dogs with strong thigh muscles, the use of foam rubber pads will facilitate the desired positioning.
Femur: A Bucky diaphragm or a cassette with stationary grid is necessary to reduce scattered radiation.
The central beam should strike the mid-shaft of the femur and fall at the right angle on the cassette.
Stifle joint: The radiograph should be taken with a non-screen film (lead blocker not to be forgotten!).
The central beam should strike the articular space of the femorotibial joint and fall at the right angle on the film.
Tibia: The radiograph should be taken with a non-screen film (lead blocker not to be forgotten!).
The central beam should strike the mid-shaft of the tibia and fall at the right angle on the film.

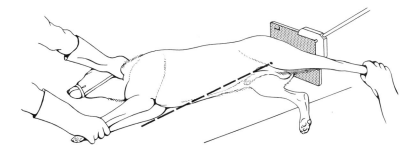

Abb. 127   Lagerung zur Aufnahme des Oberschenkels bei
Frakturverdacht. Kranio-kaudal.

Die obenliegende Beckengliedmaße des auf der gesunden Seite gelagerten
Hundes wird horizontal vorgezogen und so gehalten. Die Schultergliedmaßen sind nach kranial gezogen zu fixieren. Die Kassette ist kaudal am
Oberschenkel anzulegen.
Der Zentralstrahl sollte das Os femoris in Schaftmitte treffen und im
rechten Winkel auf die Kassette fallen.

Fig. 127   Positioning of thigh in case of suspected fracture of the femur.
Craniocaudal.

The dog is placed in lateral position, healthy side down. Thoracic limbs
are secured in cranial position, moderately stretched. The limb to be examined is pulled horizontally and held so. The cassette is placed caudally,
close to the thigh.
The central beam should strike the mid-shaft of the femur and fall at the
right angle on the cassette.

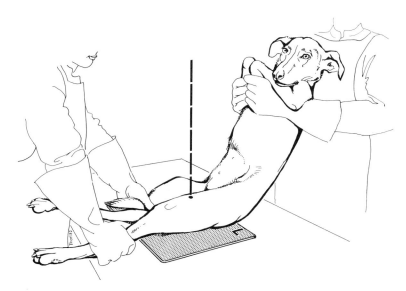

Abb. 128   Lagerung zur Aufnahme des Oberschenkels und des
Unterschenkels bei Frakturverdacht. Sitzend. Kranio-kaudal.

Der Hund wird auf die Oberschenkel gesetzt und aufrecht sitzend, gering
nach dorsal gekippt, gehalten. Die zu untersuchende Gliedmaße sollte in
Knie- und Sprunggelenk annähernd gestreckt gehalten werden.
Os femoris: Zur Reduzierung der Streustrahlung ist eine Bucky-Blende
oder eine Kassette mit stehendem Raster erforderlich.
Der Zentralstrahl sollte das Os femoris in Schaftmitte treffen und im
rechten Winkel auf die Kassette einfallen.
Tibia: Die Aufnahme ist mit einem folienlosen Film (Bleiunterlage nicht
vergessen!) anzufertigen.
Der Zentralstrahl sollte die Tibia in Schaftmitte treffen und im rechten
Winkel auf die Kassette einfallen.

Fig. 128   Positioning of thigh and lower leg in case of suspected fracture.
Sitting position. Craniocaudal.

The dog is placed in a vertical sitting position, tilted slightly backward.
The extremity to be examined should be held stretched moderately, both
in stifle and tarsal joints.
Femur: A Bucky diaphragm or a cassette with stationary grid is necessary
to reduce scattered radiation.
The central beam should strike the mid-shaft of the femur and fall at the
right angle on the cassette.
Tibia: The radiograph should be taken with a non-screen film (lead
blocker not to be forgotten!).
The central beam should strike the mid-shaft of the tibia and fall at the
right angle on the cassette.

Abb. 129   Lagerung zurAufnahme des Unterschenkels bei
Frakturverdacht. Kaudo-kranial.

Die obenliegende Beckengliedmaße des auf der gesunden Seite gelagerten
Hundes wird horizontal vorgezogen und so gehalten. Die Kassette ist
kranial am Unterschenkel anzulegen.
Der Zentralstrahl sollte die Tibia in Schaftmitte treffen und im rechten
Winkel auf die Kassette einfallen.

Fig. 129   Positioning of lower leg in case of suspected fracture
of the tibia. Caudocranial.

The dog is placed in lateral recumbency, healthy side down. The upper
pelvic limb (to be examined) is stretched moderately and held in horizontal position. The cassette is placed cranially, close to the tibia.
The central beam should strike the mid-shaft of the tibia and fall at the
right angle on the cassette.

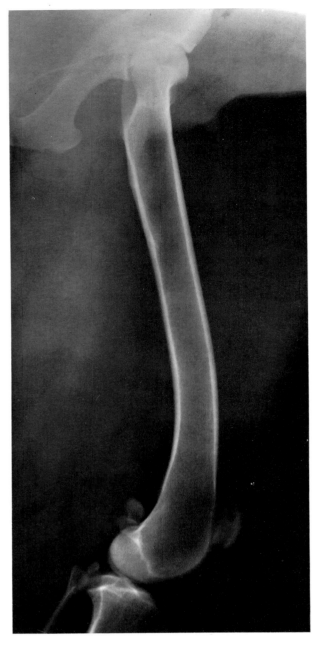

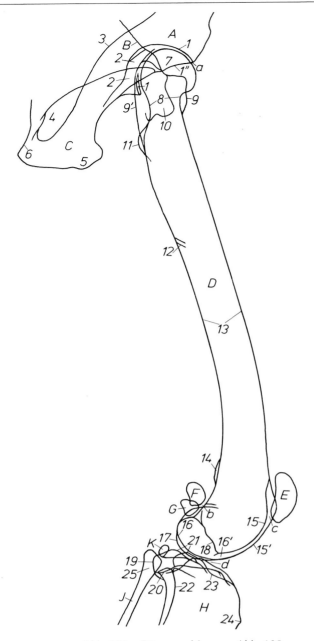

Abb. 130   Linker Oberschenkel. Medio-lateral.
Deutscher Schäferhund, 2 Jahre.
Bucky-Blende – Feinzeichnende Folie – FFA 120 cm – 66 kV – 36 mAs
Verkleinerung von 24 × 30 cm
Lagerung Abb. 126

Fig. 130   Left thigh. Mediolateral. German shepherd dog, 2 years old.
Bucky-diaphragm – High definition screens –
FFD 120 cm – 66 kV – 36 mAs
Diminution of 24 × 30 cm
Positioning fig. 126

Abb. 131   Röntgenskizze zu Abb. 130
Fig. 131   X-ray sketch to fig. 130

A	Os ilium;		F, G	Ossa sesamoidea m. gastrocnemii;
B	Os pubis;		H	Tibia;
C	Os ischii;		J	Fibula;
D	Os femoris;		K	Os sesamoideum m. poplitei;
E	Patella;			

a   Articulatio coxae;
b   Gelenk zwischen lateralem Os sesamoideum m. gastrocnemii und Condylus lateralis – Joint between the lateral sesamoid bone in the gastrocnemius muscle and the lateral condyle;
c,   d   Articulatio genus:
c   Articulatio femoropatellaris,
d   Articulatio femorotibialis;

Am Becken – On the pelvis:

1   Acetabulum, Facies lunata,
1″   Medialer Margo acetabuli – Medial margin of the acetabulum;
2   Fossa et Incisura acetabuli;
3   Spina ischiadica;
4   Foramen obturatum;

5   Tuber ischiadicum;
6   Arcus ischiadicus;

Am Os femoris – On the femur:

7   Caput ossis femoris;
8   Collum ossis femoris;
9, 9′   Trochanter major:
9   Kranialer Rand – Cranial border,
9′   Kaudaler Rand – Caudal border;
10   Fossa trochanterica;
11   Trochanter minor;
12   Foramen nutricium;
13   Corpus ossis femoris;
14   Tuberositas supracondylaris lateralis;
15, 15′   Trochlea ossis femoris:
15   Rollfurche – Groove,
15′   Medialer Rollkamm – Medial ridge;
16   Condylus lateralis,
16′   Fossa extensoria;
17   Condylus medialis;
18   Verschattung, die sich aus der Fossa intercondylaris ergibt – Shadow formed by the intercondylar fossa;

An der Tibia – On the tibia:

19   Condylus lateralis;
20   Condylus medialis;
21   Eminentia intercondylaris;
22   Verschattung, die sich aus der Konkavität der Incisura poplitea ergibt – Shadow formed by the concavity of the popliteal notch;
23   Area intercondylaris cranialis;
24   Tuberositas tibiae;

An der Fibula – On the fibula:

25   Caput fibulae.

Beckengliedmaße Pelvic limb

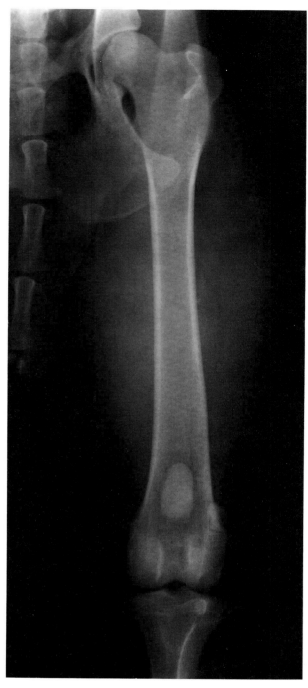

Abb. 132   Linker Oberschenkel. Kranio-kaudal.
Deutscher Schäferhund, 2 Jahre.
Bucky-Blende – Feinzeichnende Folie – FFA 120 cm – 70 kV – 40 mAs
Verkleinerung von 24 × 30 cm, Lagerung Abb. 127, 128

Fig. 132   Left thigh. Craniocaudal. German shepherd dog, 2 years old.
Bucky diaphragm – High definition screens – FFD 120 cm – 70 kV –
40 mAs, Diminution of 24 × 30 cm, Positioning fig. 127, 128

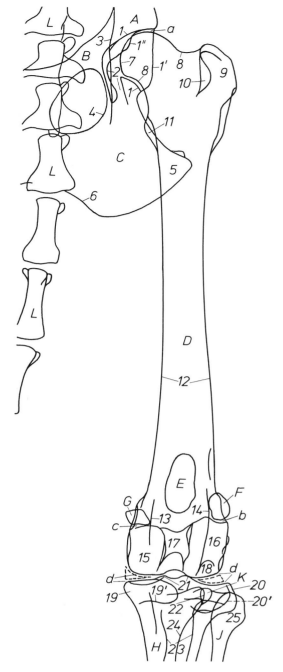

Abb. 133   Röntgenskizze zu Abb. 132
Fig. 133   X-ray sketch to fig. 132

A	Os ilium;	F, G	Ossa sesamoidea m. gastrocnemii;
B	Os pubis;	H	Tibia;
C	Os ischii;	J	Fibula;
D	Os femoris;	K	Os sesamoideum m. poplitei;
E	Patella;	L	Vertebrae caudales;

a   Articulatio coxae;
b,   c   Gelenke zwischen den Ossa sesamoidea m. gastrocnemii und
Condylus lateralis bzw. Condylus medialis – Joints between the sesamoid bones in the gastrocnemius muscle and the lateral condyle and
the medial condyle respectively;
d   Articulatio femorotibialis;

Am Becken – On the pelvis:

1	Acetabulum, Facies lunata,	3	Spina ischiadica;
1'	Dorsaler Rand – Dorsal border,	4	Foramen obturatum;
1"	Ventraler Rand – Ventral border;	5	Tuber ischiadicum;
2	Fossa et Incisura acetabuli;	6	Arcus ischiadicus;

Am Os femoris – On the femur:

7   Caput ossis femoris;
8   Collum ossis femoris;

9   Trochanter major;
10   Fossa trochanterica;
11   Trochanter minor;
12   Corpus ossis femoris;
13, 14   Trochlea ossis femoris, orthograph getroffene Corticalis –
Trochlea ossis femoris, orthographically struck corticalis:
13   Medialer Rollkamm – Medial ridge;
14   Lateraler Rollkamm – Lateral ridge;
15   Condylus medialis;
16   Condylus lateralis;
17   Fossa intercondylaris;
18   Fossa extensoria;

An der Tibia – On the tibia:

19   Condylus medialis,
19' Facies articularis proximalis, kaudale Begrenzung – Facies articularis
proximalis, caudal border;
20   Condylus lateralis,
20' Facies articularis proximalis, kaudale Begrenzung – Facies articularis
proximalis, caudal border;
21   Eminentia intercondylaris;
22   Sulcus extensorius;
23   Margo cranialis;
24   Verschattung, die sich aus der Konkavität der Facies lateralis ergibt –
Shadow formed by the concavity of the lateral surface;

An der Fibula – On the fibula:

25   Caput fibulae.

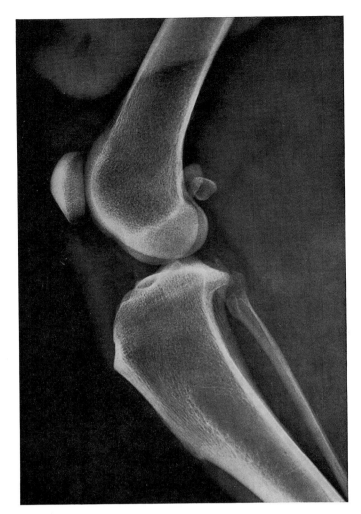

Abb. 134   Rechtes Kniegelenk. Medio-lateral.
Deutscher Schäferhund, ♂, 12 Jahre.
Folienloser Film – FFA 120 cm – 64 kV – 24 mAs
Originalgröße (Ausschnitt aus 13 × 18 cm)
Lagerung Abb. 126

Fig. 134   Right stifle joint. Mediolateral.
German shepherd dog, ♂, 12 years old.
Non-screen film – FFD 120 cm – 64 kV – 24 mAs
Original size (section of 13 × 18 cm)
Positioning fig. 126

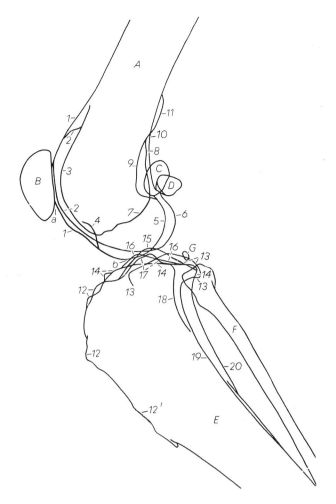

Abb. 135   Röntgenskizze zu Abb. 134
Fig. 135   X-ray sketch to fig. 134

A   Os femoris;
B   Patella;
C   Laterales Os sesamoideum m. gastrocnemii – Lateral sesamoid bone in the gastrocnemius muscle;
D   Mediales Os sesamoideum m. gastrocnemii – Medial sesamoid bone in the gastrocnemius muscle;
E   Tibia;
F   Fibula;
G   Os sesamoideum m. poplitei;

a   Articulatio femoropatellaris;
b   Articulatio femorotibialis;

Am Os femoris – On the femur:

1, 2, 3   Trochlea ossis femoris:
1   Lateraler Rollkamm – Lateral ridge,
2   Medialer Rollkamm – Medial ridge,
3   Rollfurche – Groove;
4   Fossa extensoria;
5   Condylus lateralis;
6   Condylus medialis;

7   Fossa intercondylaris;
8   9, 10   Facies poplitea:
9   Kaudolateraler Rand – Caudolateral border,
10   Kaudomedialer Rand – Caudomedial border;
11   Tuberositas supracondylaris lateralis;

An der Tibia – On the tibia:

12   Tuberositas tibiae;
12' Margo cranialis;
13   Condylus lateralis;
14   Condylus medialis;
15   Tuberculum intercondylare laterale;
16   Tuberculum intercondylare mediale;
17   Verschattung, die sich aus der Konkavität der Areae intercondylares cranialis, centralis et caudalis ergibt und in die der Incisura poplitea (18) übergeht – Shadow formed by the concavity of the cranial, central and caudal intercondylar areas, and merging into that of the popliteal notch (18);
19   Margo lateralis;
20   Margo medialis.

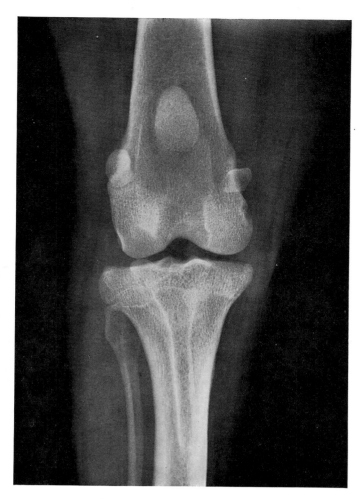

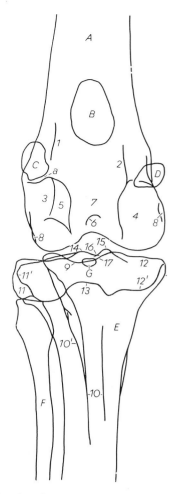

Abb. 136   Rechtes Kniegelenk, Kranio-kaudal.
Deutscher Schäferhund, 6 Jahre.
Folienloser Film – FFA 120 cm – 75 kV – 36 mAs
Originalgröße (Ausschnitt aus 13 × 18 cm)
Lagerung Abb. 140

Fig. 136   Right stifle joint. Craniocaudal.
German shepherd dog, 6 years old.
Non-screen film – FFD 120 cm – 75 kV – 36 mAs
Original size (section of 13 × 18 cm)
Positioning fig. 140

Abb. 137   Röntgenskizze zu Abb. 136
Fig. 137   X-ray sketch to fig. 136

A   Os femoris;
B   Patella;
C   Laterales Os sesamoideum m. gastrocnemii – Lateral sesamoid bone in the gastrocnemius muscle;
D   Mediales Os sesamoideum m. gastrocnemii – Medial sesamoid bone in the gastrocnemius muscle;
E   Tibia;
F   Fibula;
G   Os sesamoideum m. poplitei;

a   Gelenk zwischen lateralem Os sesamoideum m. gastrocnemii und Condylus lateralis – Joint between the lateral sesamoid bone in the gastrocnemius muscle and the lateral condyle;

Am Os femoris – On the femur:
1, 2   Trochlea ossis femoris:
1   Lateraler Rollkamm – Lateral ridge,
2   Medialer Rollkamm – Medial ridge;
3   Condylus lateralis;
4   Condylus medialis;
5   Verschattung an der medialen Fläche des Condylus lateralis: Bandgrube des Ligamentum cruciatum craniale – Shadow on the medial surface of the lateral condyle: area for attachment of the cranial cruciate ligament;
6   Kaudale Begrenzung einer flachen Grube am Übergang der Trochlea ossis femoris zur Fossa intercondylaris (7) – Caudal border of a shallow groove at the transition of the trochlea and intercondylar fossa (7);
8   Bandgruben – Depressions for ligamentous attachment;

An der Tibia – On the tibia:
9   Tuberositas tibiae;
10   Margo cranialis, Basis – Margo cranialis, base,
10'   Laterale Begrenzung – Lateral limit;
11   Condylus lateralis,
11'   Facies articularis proximalis, laterale Grenze – Facies articularis proximalis, lateral limit;
12   Condylus medialis,
12'   Facies articularis proximalis, mediale und kaudale Grenze – Facies articularis proximalis, medial and caudal limits;
13   Incisura poplitea;
14   Tuberculum intercondylare laterale;
15   Tuberculum intercondylare mediale;
16   Area intercondylaris centralis;
17   Area intercondylaris cranialis.

95

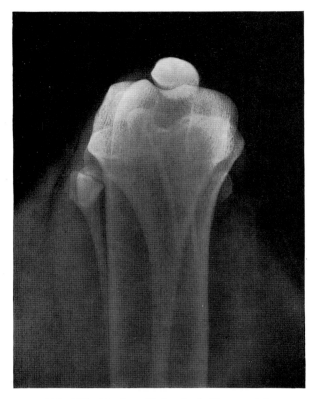

Abb. 138   Rechtes Kniegelenk. Tangential.
Deutscher Schäferhund, 6 Jahre.
Folienloser Film – FFA 120 cm – 65 kV – 24 mAs
Originalgröße (Ausschnitt aus 13 × 18 cm)
Lagerung Abb. 141

Fig. 138   Right stifle joint. Tangential.
German shepherd dog, 6 years old.
Non-screen film – FFD 120 cm – 65 kV – 24 mAs
Original size (section of 13 × 18 cm)
Positioning fig. 141

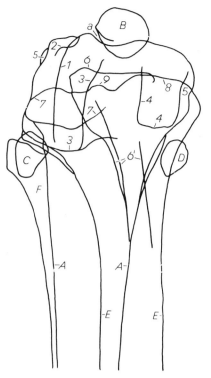

Abb. 139   Röntgenskizze zu Abb. 138
Fig. 139   X-ray sketch to fig. 138

A   Os femoris;
B   Patella;
C   Laterales Os sesamoideum m. gastrocnemii – Lateral sesamoid bone
in the gastrocnemius muscle;
D   Mediales Os sesamoideum m. gastrocnemii – Medial sesamoid bone
in the gastrocnemius muscle;
E   Tibia;
F   Fibula;

a   Articulatio femoropatellaris;

Am Os femoris – On the femur:

1   Trochlea ossis femoris, lateraler Rollkamm – Trochlea ossis femoris,
lateral ridge;
2   Fossa extensoria;
3   Condylus lateralis;
4   Condylus medialis;
5   Bandgrube – Depression for ligamentous attachment;

An der Tibia – On the tibia:

6   Tuberositas tibiae;
6' Margo cranialis;
7   Condylus lateralis;
8   Condylus medialis;
9   Tuberculum intercondylare laterale.

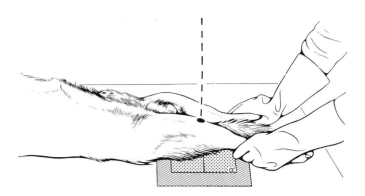

Abb. 140   Lagerung zur Aufnahme des Kniegelenks. Kranio-kaudal.

Die Beckengliedmaßen des auf dem Rücken liegenden Hundes sind so weit nach hinten zu strecken, daß sich die haltenden Hände auf den Tisch stützen können. Mit einem unter das Kniegelenk gelegten Schaumgummikeil läßt sich die Lagerung erleichtern. Während die Kniescheibe des zu untersuchenden Gelenks durch mäßige Innenrotation der Gliedmaße exakt über der Gelenkfläche liegen muß, kann die andere Beckengliedmaße in leichter Abduktion und Außenrotation fixiert werden.
Die Aufnahme ist mit einem folienlosen Film (Bleiunterlage nicht vergessen!) anzufertigen.
Der Zentralstrahl sollte die Mitte der Gliedmaßenvorderfläche dicht proximal der Tuberositas tibiae treffen und im rechten Winkel auf den Film einfallen.

Abb. 140   Positioning of stifle joint. Craniocaudal.

The hind legs of the dog lying on its back should be stretched far enough caudally to allow the hands to rest upon the table. The positioning can be facilitated by placing foam rubber pads under the knee joints. The patella of the leg to be examined must lie exactly over the articular surface. This can be accomplished by slight inward rotation of the limb. The other leg may be secured in slight abduction and outward rotation.
The radiograph should be taken with a non-screen film (lead blocker not to be forgotten!).
The central beam should strike the center of the cranial aspect of the limb immediately proximal to the tibial tuberosity and fall at the right angle on the film.

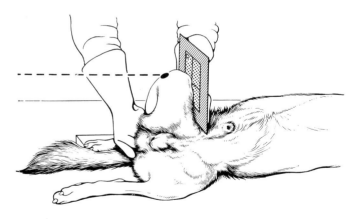

Abb. 141 Lagerung zur Aufnahme des Kniegelenks. Tangential.

Die zu untersuchende Beckengliedmaße des in halber Rückenlage befindlichen Hundes ist im Kniegelenk maximal abzubeugen und so weit aufzurichten, daß der an der Gliedmaßenvorderfläche angelegte Film senkrecht steht. Die Lagerung läßt sich durch Aufstützen der Pfote auf ein Schaumgummipolster erleichtern.
Die Aufnahme ist mit einem folienlosen Film (Bleiunterlage nicht vergessen!) anzufertigen.
Der Zentralstrahl sollte die proximale Kante der Tuberositas tibiae treffen und im rechten Winkel auf den Film einfallen.

Fig. 141   Positioning of stifle joint. Tangential.

The dog is placed in a dorsolateral position. The knee joint to be examined should be flexed completely and raised until the film placed against its cranial aspect is perpendicular to the table. A foam rubber pad supporting the paw will facilitate the positioning.
The radiograph should be taken with a non-screen film (lead blocker not to be forgotten!).
The central beam should strike the proximal edge of the tibial tuberosity and fall at the right angle on the film.

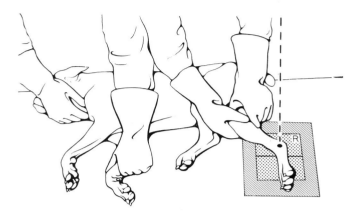

Abb. 142   Lagerung zur Aufnahme des Tarsalgelenks. Medio-lateral.

Person 1 hält mit einer Hand die untenliegende Schultergliedmaße und drückt dabei den Hund mit dem Unterarm auf den Tisch. Die andere Hand fixiert den Kopf. Die am Rücken des Hundes stehende 2. Person faßt mit beiden Händen die zu untersuchende Beckengliedmaße im Bereich des Kniegelenks und lagert das leicht gestreckte Sprunggelenk so auf dem Film, daß die Zehen senkrecht übereinander liegen. Dabei wird der Hund auch von der 2. Person mit beiden Unterarmen auf den Tisch gedrückt.
Die Aufnahme ist mit einem folienlosen Film (Bleiunterlage nicht vergessen!) anzufertigen.
Der Zentralstrahl sollte die Mitte der Beckengliedmaße in Höhe des Talus treffen und im rechten Winkel auf den Film einfallen.

Fig. 142   Positioning of hock joint. Mediolateral.

One assistant holds the lower limb with one hand, pressing the dog with the forearm against the table. The other hand secures the head. The second assistant, standing at the back of the dog, grasps the knee of the leg to be radiographed with both hands and places the slightly extended tarsal joint upon the film in such a manner that the digits are aligned perpendicularly to each other. The assistant also presses the dog with both forearms against the table.
The radiograph should be taken with a non-screen film (lead blocker not to be forgotten!).
The central beam should strike the middle of the pelvic limb at the level of the talus and fall at the right angle on the film.

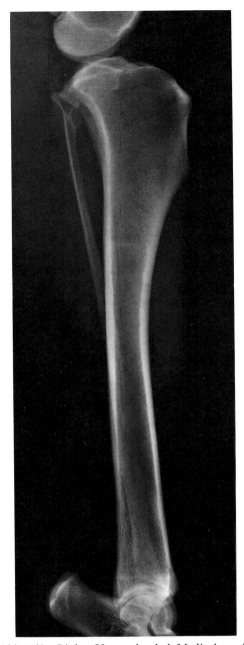

Abb. 143   Linker Unterschenkel. Medio-lateral.
Deutscher Schäferhund, 2 Jahre.
Bucky-Blende – Feinzeichnende Folie – FFA 120 cm – 66 kV – 25 mAs
Verkleinerung von 24 × 30 cm
Lagerung Abb. 126

Fig. 143   Left lower leg. Mediolateral.
German shepherd dog, 2 years old.
Bucky diaphragm – High definition screens –
FFD 120 cm – 66 kV – 25 mAs
Diminution of 24 × 30 cm
Positioning fig. 126

A   Os femoris;
B   Ossa sesamoidea m. gastrocnemii;
C   Tibia;
D   Fibula;
E   Os sesamoideum m. poplitei;
F   Talus;
G   Calcaneus;

a   Articulatio femorotibialis;
b   Articulatio tarsocruralis;

Am Os femoris – On the femur:

1   Condylus lateralis;
2   Condylus medialis;
3   Verschattung, die sich aus dem Grund der Fossa intercondylaris ergibt – Shadow formed by the base of the intercondylar fossa;
4   Fossa extensoria;

An der Tibia – On the tibia:

5   Condylus lateralis;
6   Condylus medialis;

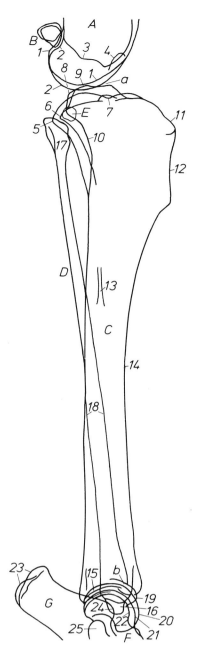

Abb. 144   Röntgenskizze zu Abb. 143
Fig. 144   X-ray sketch to fig. 143

7   Area intercondylaris;
8, 9   Eminentia intercondylaris:
8   Tuberculum intercondylare laterale,
9   Tuberculum intercondylare mediale;
10   Verschattung, die sich aus der Konkavität der Incisura poplitea ergibt – Shadow formed by the concavity of the popliteal notch;
11   Tuberositas tibiae;
12   Margo cranialis;
13   Foramen nutricium;
14   Corpus tibiae;
15   Cochlea tibiae;
16   Malleolus medialis;

An der Fibula – On the fibula:

17   Caput fibulae;
18   Corpus fibulae;
19   Malleolus lateralis;

Am Talus – On the talus:

20, 21, 22   Trochlea tali proximalis:
20   Lateraler Kamm – Lateral ridge,
21   Medialer Kamm – Medial ridge,
22   Verschattung, die sich aus der Furche zwischen den Rollkämmen ergibt – Shadow formed by the groove between the ridges;

Am Calcaneus – On the calcaneus:

23   Tuber calcanei;
24   Processus coracoideus;
25   Sustentaculum tali.

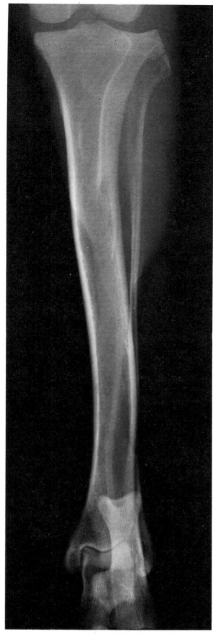

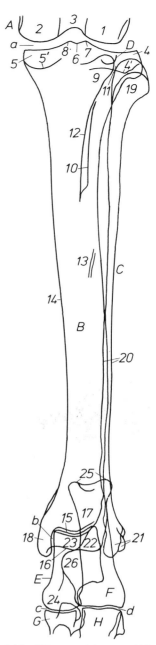

Abb. 145   Rechter Unterschenkel. Kaudo-kranial.
Deutscher Schäferhund, 6 Jahre.
Feinzeichnende Folie – FFA 120 cm – 66 kV – 25 mAs
Verkleinerung von 24 × 30 cm
Lagerung Abb. 129

Fig. 145   Right lower leg. Caudocranial.
German shepherd dog, 6 years old.
High definition screens – FFD 120 cm – 66 kV – 25 mAs
Diminution of 24 × 30 cm
Positioning fig. 129

Abb. 146   Röntgenskizze zu Abb. 145
Fig. 146   X-ray sketch to fig. 145

A   Os femoris;
B   Tibia;
C   Fibula;
D   Os sesamoideum m. poplitei;
E   Talus;
F   Calcaneus;
G   Os tarsi centrale;
H   Os tarsale IV;

a   Articulatio femorotibialis;
b   Articulatio tarsocruralis;
c   Articulatio talocalcaneocentralis;
d   Articulatio calcaneoquartalis;

Am Os femoris – On the femur:

1   Condylus lateralis;
2   Condylus medialis;
3   Fossa intercondylaris;

An der Tibia – On the tibia:

4   Condylus lateralis;
4'   Facies articularis proximalis, kaudaler Rand – Facies articularis proximalis, caudal border;

5   Condylus medialis;
5'   Facies articularis proximalis, kaudaler Rand – Facies articularis proximalis, caudal border;
6   Area intercondylaris;
7, 8   Eminentia intercondylaris:
7   Tuberculum intercondylare laterale,
8   Tuberculum intercondylare mediale;
9   Tuberositas tibiae;
10   Margo cranialis;
11   Sulcus extensorius;
12   Verschattung, die sich aus der Konkavität an der Basis der Facies lateralis ergibt – Shadow formed by the concavity on the base of the lateral surface;
13   Foramen nutricium;
14   Corpus tibiae;
15, 16, 17   Cochlea tibiae:
16   Kranialer Rand – Cranial border,
17   Kaudaler Rand – Caudal border;
18   Malleolus medialis;

An der Fibula – On the fibula:

19   Caput fibulae;
20   Corpus fibulae;
21   Malleolus lateralis;

Am Talus – On the talus:

22, 23   Trochlea tali proximalis:
22   Lateraler Kamm – Lateral ridge,
23   Medialer Kamm – Medial ridge;
24   Caput tali;

Am Calcaneus – On the calcaneus:

25   Tuber calcanei;
26   Sustentaculum tali.

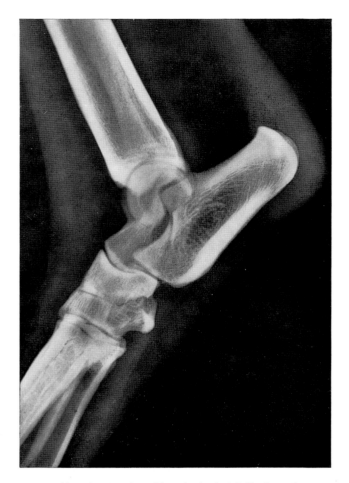

Abb. 147   Rechtes Tarsalgelenk. Medio-lateral.
Deutscher Schäferhund, 7 Jahre.
Folienloser Film – FFA 120 cm – 60 kV – 18 mAs
Originalgröße (Ausschnitt aus 13 × 18 cm)
Lagerung Abb. 142

Fig. 147   Right hock joint, Mediolateral.
German shepherd dog, 7 years old.
Non-screen film – FFD 120 cm – 60 kV – 18 mAs
Original size (section of 13 × 18 cm)
Positioning fig. 142

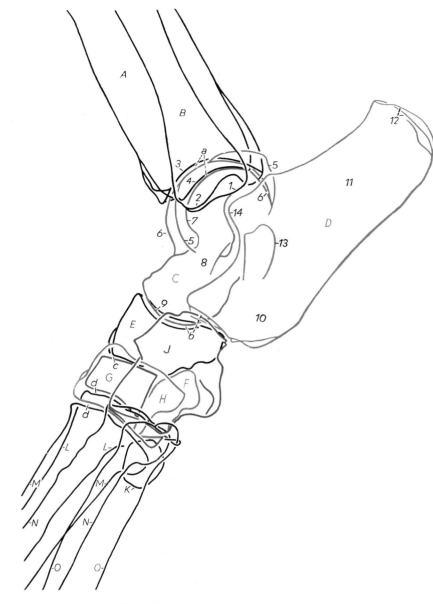

Abb. 148**   Röntgenskizze zu Abb. 147
Fig. 148**   X-ray sketch to fig. 147

A	Tibia;	1	Malleolus lateralis;
B	Fibula;	2	Malleolus medialis;
C	Talus;	3	Ineinander projizierte laterale und mediale Furche der Cochlea tibiae – Lateral and medial grooves of the tibial cochlea projected into one another;
D	Calcaneus;		
E	Os tarsi centrale;		
F	Os tarsale I;	4	Medianer Führungskamm der Cochlea tibiae – Median ridge of the tibial cochlea;
G	Os tarsale II;		
H	Os tarsale III;	5, 6, 7	Trochlea tali proximalis:
J	Os tarsale IV;	5	Lateraler Kamm – Lateral ridge,
K	Os metatarsale I;	6	Medialer Kamm – Medial ridge,
L	Os metatarsale II;	7	Furche – Groove;
M	Os metatarsale III;	8	Corpus tali;
N	Os metatarsale IV;	9	Caput tali;
O	Os metatarsale V;	10, 11	Calcaneus;
		12	Tuber calcanei;
a	Articulatio tarsocruralis;	13	Sustentaculum tali;
b	Articulationes talocalcaneocentralis et calcaneoquartalis;	14	Processus coracoideus.
c	Articulatio centrodistalis;		
d	Articulationes tarsometatarseae;		

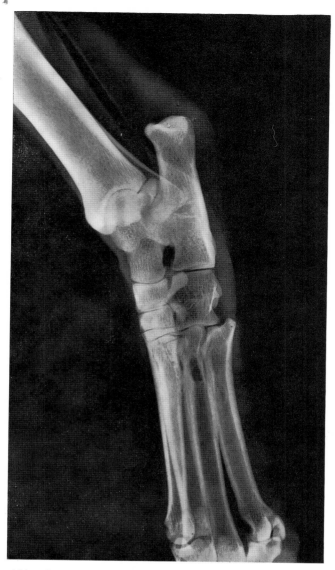

Abb. 149   Linkes Tarsalgelenk. Dorsolateral-medioplantar.
Deutscher Schäferhund, 8 Jahre.
Folienloser Film – FFA 120 cm – 65 kV – 24 mAs
Originalgröße (Ausschnitt aus 13 × 18 cm)
Lagerung Abb. 155

Fig. 149   Left hock joint. Dorsolateral-medioplantar.
German shepherd dog, 8 years old.
Non-screen film – FFD 120 cm – 65 kV – 24 mAs
Original size (section of 13 × 18 cm)
Positioning fig. 155

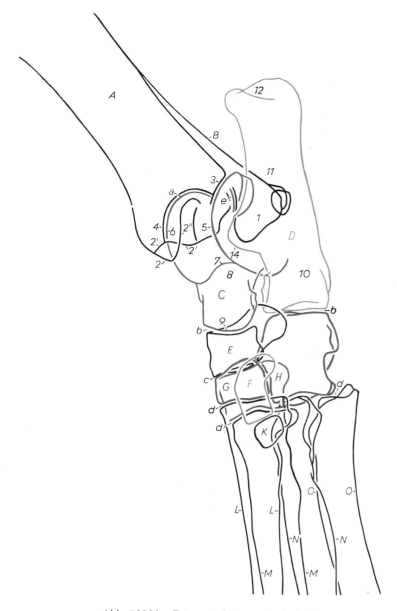

Abb. 150**   Röntgenskizze zu Abb. 149
Fig. 150**   X-ray sketch to fig. 149

A	Tibia;	
B	Fibula;	
C	Talus;	
D	Calcaneus;	
E	Os tarsi centrale;	
F	Os tarsale I;	
G	Os tarsale II;	
H	Os tarsale III;	
J	Os tarsale IV;	
K	Os metatarsale I;	
L	Os metatarsale II;	
M	Os metatarsale III;	
N	Os metatarsale IV;	
O	Os metatarsale V;	

a   Articulatio tarsocruralis;
b   Articulationes talocalcaneocentralis et calcaneoquartalis;
c   Articulatio centrodistalis;

d   Articulationes tarsometatarseae;
e   Articulatio tibiofibularis distalis;

1   Malleolus lateralis;
2   Malleolus medialis;
2', 2", 3, 4   Cochlea tibiae:
2'   Mediokaudaler Rand – Mediocaudal border,
2"   Kranialer Rand – Cranial border,
3   Laterale Furche – Lateral groove,
4   Mediale Furche – Medial groove;
5, 6, 7   Trochlea tali proximalis:
5   Lateraler Kamm – Lateral ridge,
6   Medialer Kamm – Medial ridge,
7   Furche – Groove;
8   Corpus tali;
9   Caput tali;
10, 11   Calcaneus;
12   Tuber calcanei;
14   Processus coracoideus.

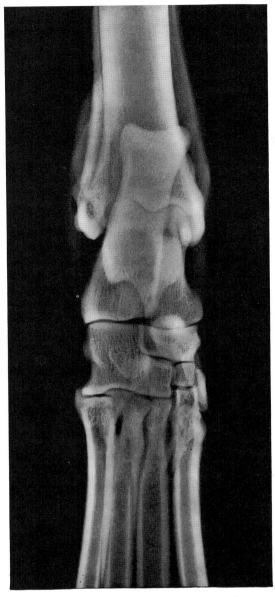

Abb. 151   Rechtes Tarsalgelenk. Dorso-plantar.
Deutscher Schäferhund, 10 Jahre.
Folienloser Film – FFA 120 cm – 65 kV – 24 mAs
Originalgröße (Ausschnitt aus 13 × 18 cm)
Lagerung Abb. 156

Fig. 151   Right hock joint. Dorsoplantar.
German shepherd dog, 10 years old.
Non-screen film – FFD 120 cm – 65 kV – 24 mAs
Original size (section of 13 × 18 cm)
Positioning fig. 156

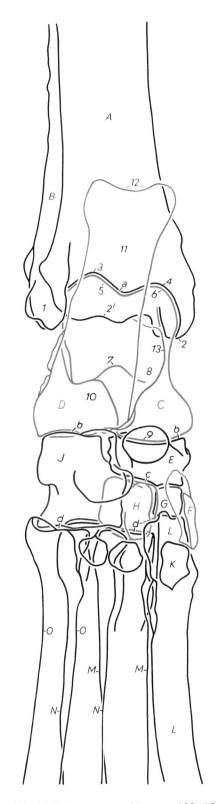

Abb. 152**   Röntgenskizze zu Abb. 151
Fig. 152**   X-ray sketch to fig. 151

A	Tibia;
B	Fibula;
C	Talus;
D	Calcaneus;
E	Os tarsi centrale;
F	Os tarsale I;
G	Os tarsale II;
H	Os tarsale III;
J	Os tarsale IV;
K	Os metatarsale I;
L	Os metatarsale II;
M	Os metatarsale III;
N	Os metatarsale IV;
O	Os metatarsale V;

a   Articulatio tarsocruralis;
b   Articulationes talocalcaneocentralis et calcaneoquartalis;

c   Articulatio centrodistalis;
d   Articulationes tarsometatarseae;

1   Malleolus lateralis;
2   Malleolus medialis;
2', 3, 4   Cochlea tibiae:
2'   Kranialer Rand – Cranial border,
3   Laterale Furche – Lateral groove,
4   Mediale Furche – Medial groove;
5, 6, 7   Trochlea tali proximalis:
5   Lateraler Kamm – Lateral ridge,
6   Medialer Kamm – Medial ridge,
7   Furche – Groove;
8   Corpus tali;
9   Caput tali;
10, 11   Calcaneus;
12   Tuber calcanei;
13   Sustentaculum tali.

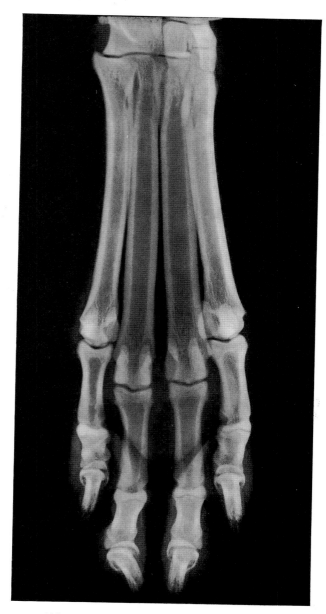

Abb. 153   Rechter Hinterfuß, Dorso-plantar.
Deutscher Schäferhund, 8 Jahre.
Folienloser Film – FFA 120 cm – 62 kV – 20 mAs
Originalgröße (Ausschnitt aus 13 × 18 cm)
Lagerung Abb. 157

Fig. 153   Right hindpaw. Dorsoplantar.
German shepherd dog, 8 years old.
Non-screen film – FFD 120 cm – 62 kV – 20 mAs
Original size (section of 13 × 18 cm)
Positioning fig. 157

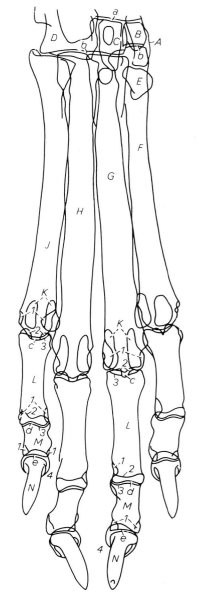

Abb. 154   Röntgenskizze zu Abb. 153
Fig. 154   X-ray sketch to fig. 153

A   Os tarsale I;
B   Os tarsale II;
C   Os tarsale III;
D   Os tarsale IV;
E   Os metatarsale I;
F   Os metatarsale II;
G   Os metatarsale III;
H   Os metatarsale IV;
J   Os metatarsale V;
K   Ossa sesamoidea proximalia;

An der 2. bis 5. Zehe – On the 2nd to 5th digits:

L   Phalanx proximalis;
M   Phalanx media;
N   Phalanx distalis;

a   Articulatio centrodistalis;
b   Articulationes tarsometatarseae;
c   Articulatio metatarsophalangea;
d   Articulatio interphalangea proximalis pedis;
e   Articulatio interphalangea distalis pedis;

1   Bandgruben – Depressions for ligamentous attachment;
2, 3   Fovea articularis des proximalen bzw. mittleren Zehenglieds – Fovea articularis of the proximal and middle phalanges respectively:
2   Plantare Begrenzung – Plantar limit,
3   Dorsale Begrenzung – Dorsal limit;
4   Crista unguicularis.

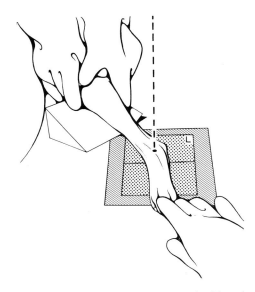

Abb. 155   Schräglagerung zur Aufnahme des Tarsalgelenks.
Dorsolateral-medioplantar.

Die im Sprunggelenk gestreckte Beckengliedmaße wird mit beiden Händen in Höhe des Kniegelenks umfaßt und so gelagert, daß der Kalkaneus und der Malleolus medialis dem Film anliegt. Dabei besteht zwischen der Unterlage und dem Unterschenkel ein Winkel von 45 Grad. Durch einen Schaumgummikeil unter dem proximalen Abschnitt des Unterschenkels läßt sich die Lagerung erleichtern. Eine 2. Person drückt die Zehen auf die Unterlage. Durch die Schräglagerung des Unterschenkels und die geringe Drehung im Tarsus werden der Talus, das Os tarsale IV und das Os metatarsale V weitgehend isoliert dargestellt.
Die Aufnahme ist mit einem folienlosen Film (Bleiunterlage nicht vergessen!) anzufertigen.
Der Zentralstrahl sollte die Mitte der Gliedmaße in Höhe des Talus treffen und im rechten Winkel auf den Film einfallen.

Fig. 155   Oblique position of hock joint.
Dorsolateral-medioplantar.

The limb with extended hock joint is grasped with both hands in the region of the stifle joint and placed with the calcaneus and medial malleolus adjacent to the film. The angle between the table and the crus is approximately 45°. A foam rubber pad placed under the proximal end of the tibia facilitates the positioning. The second assistant presses the digits against the table. Because of the oblique positioning of the tibia and a slight rotation of the tarsus, the talus, 4th tarsal, and 5th metatarsal bones are seen mostly separately.
The radiograph should be taken with a non-screen film (lead blocker not to be forgotten!).
The central beam should strike the middle of the limb on a level of the talus and fall at the right angle on the film.

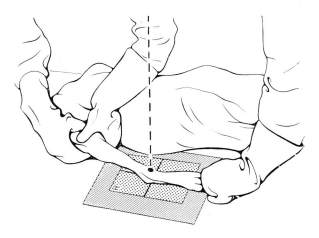

Abb. 156   Lagerung zur Aufnahme des Tarsalgelenks. Dorso-plantar.

Eine Person faßt die zu untersuchende Beckengliedmaße mit beiden Händen im Bereich des Oberschenkels und drückt mit einem Unterarm den auf Bauch und Brust liegenden Hund auf den Tisch. Sie lagert und fixiert die Gliedmaße in leichter Abduktion und mit gestrecktem Sprunggelenk so, daß der Tarsus im proximalen Bereich nur mit dem Kalkaneus dem Film aufliegt. Bei dieser Lagerung befinden sich die Malleoli lateralis und

medialis in gleicher Höhe. Eine Schrägaufnahme ergibt sich, wenn der Kalkaneus und entweder der Malleolus medialis oder der Malleolus lateralis dem Film anliegen. Die zweite Person fixiert mit einer Hand die Pfote und mit der anderen den Kopf des Hundes.
Die Aufnahme ist mit einem folienlosen Film (Bleiunterlage nicht vergessen!) anzufertigen.
Der Zentralstrahl sollte die Mitte der Gliedmaßenvorderfläche in Höhe des Talus treffen und im rechten Winkel auf den Film einfallen.

Fig. 156   Positioning of hock joint. Dorsoplantar.

The dog lies on its abdomen. One assistant holds the limb with both hands at the thigh and presses the animal with one forearm against the table. The limb must be secured in slight abduction with the tarsal joint extended. In the proximal region of the tarsus only the calcaneus should rest on the film. In this position, both the lateral and medial malleoli must lie on the same level. If the calcaneus and either the medial or lateral malleolus are in contact with the film an oblique radiograph will result. The second assistant secures the paw with one hand and with the other hand the head of the dog.
The radiograph should be taken with a non-screen film (lead blocker not to be forgotten!).
The central beam should strike the middle of the dorsal aspect of the limb on a level of the proximal intertarsal joint and fall at the right angle on the film.

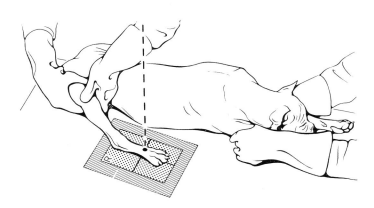

Abb. 157   Lagerung zur Aufnahme des Hinterfußes. Dorso-plantar.

Person 1 umfaßt die zu untersuchende Beckengliedmaße mit beiden Händen im Bereich des Oberschenkels und drückt mit einem Unterarm den auf Bauch und Brust liegenden Hund auf den Tisch. Sie lagert und fixiert die Gliedmaße in leichter Abduktion und mit gestrecktem Sprunggelenk so, daß die Zehenballen und der Sohlenballen dem Film aufliegen. Person 2 hält die Vordergliedmaßen in Höhe der Ellbogengelenke und stabilisiert so die Lage des Hundes.
Die Aufnahme ist mit einem folienlosen Film (Bleiunterlage nicht vergessen!) anzufertigen.
Der Zentralstrahl sollte die Mitte der Gliedmaße in Höhe des 1. Zehengelenks treffen und im rechten Winkel auf den Film einfallen.

Fig. 157   Positioning of hindpaw. Dorsoplantar.

The dog lies on its abdomen. One assistant holds the limb to be radiographed at the thigh and presses the animal with one forearm against the table. The limb must be secured in slight abduction and with an extended tarsal joint so that the digital and metatarsal pads are in contact with the film. The second assistant holds the thoracic limbs at the elbows, securing the position of the dog.
The radiograph should be taken with a non-screen film (lead blocker not to be forgotten!).
The central beam should strike the middle of the limb on a level of the 1st phalangeal joint and fall at the right angle on the film.

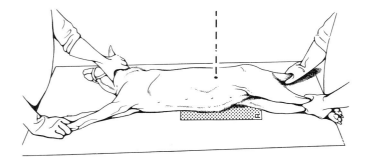

Abb. 158   Lagerung zur Aufnahme des Abdomens. Liegend.
Latero-lateral.

Fig. 158   Positioning of abdomen. Recumbent. Laterolateral.

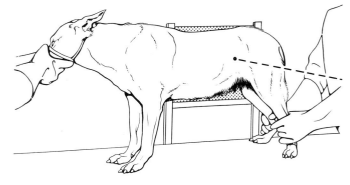

Abb. 159   Lagerung zur Aufnahme des Abdomens. Stehend.
Latero-lateral.

Fig. 159   Positioning of abdomen. Standing. Laterolateral.

Die Lage der Organe in der Bauchhöhle ist von der Lage des Tieres, von dem Füllungszustand der Organe und bei einem pathologischen Prozeß von dessen Art und Ausdehnung abhängig. Für eine schnelle und vor allem sichere Orientierung ist es zweckmäßig, die Aufnahmen in der gleichen Weise, also mit der der rechten oder der linken Bauchwand anliegenden Kassette, anzufertigen. Die Aufnahme mit der der rechten Bauchwand anliegenden Kassette hat sich bewährt. Davon sollte nur in den Fällen abgegangen werden, in denen die Veränderung in der linken Bauchhöhlenhälfte lokalisiert wurde.
Zur Verminderung der Streustrahlung sollte eine Bucky-Blende oder eine Kassette mit stehendem Raster verwendet werden. Nur bei kleinen, nicht fettleibigen Hunden kann auf einen Raster verzichtet werden.
Aus der Fragestellung ergibt sich, ob die Aufnahme am stehenden oder liegenden Hund anzufertigen ist. Zum Beispiel sind durch Flüssigkeit bedingte Spiegel am stehenden, nicht aber am liegenden Patienten nachweisbar.
Die Lagerung in Seitenlage (Abb. 158) ist einfacher. Durch Strecken der Gliedmaßen werden störende Überlagerungen durch die Oberschenkelmuskulatur vermieden. Der Patient ist so zu lagern, daß die Medianebene parallel zur Kassette liegt. Bei fettleibigen Hunden ist einer Verkantung durch Unterlegen von Schaumgummikeilen unter die auf der Tischplatte liegenden Gliedmaßen in Höhe der haltenden Hände und durch eine Fixation der Rute auf der Tischplatte unter leichtem Zug vorzubeugen.
Bei der Lagerung des stehenden Hundes vor der Kassette (Abb. 159) ist darauf zu achten, daß er der Kassette anliegt. Hunde der großen Rassen sollten selbst stehen. Kleine Hunde sind einfacher vor der Kassette zu halten, wobei durch leichtes Strecken die Überlagerung durch die Oberschenkelmuskulatur vermieden werden kann. Es ist darauf zu achten, daß die Medianebene parallel zur Kassette liegt.
Der Zentralstrahl sollte im rechten Winkel auf die Kassette einfallen und die seitliche Bauchwand so treffen, daß das zu untersuchende Gebiet in das Zentrum des Filmes zu liegen kommt.

Einstellungen des Zentralstrahls:
Übersichtaufnahme und Darm: Mitte der seitlichen Bauchwand in Höhe des 3. Lendenwirbels.
Magen: Mitte der seitlichen Bauchwand in Höhe der 12. Rippe.
Gallenblase: Seitliche Brustwand am Übergang vom ventralen zum mittleren Drittel in Höhe des 7. Interkostalraums.
Nieren: Seitliche Bauchwand am Übergang vom mittleren zum dorsalen Drittel in Höhe des 2. Lendenwirbels.
Eierstöcke: Seitliche Bauchwand am Übergang vom mittleren zum dorsalen Drittel in Höhe des 3. Lendenwirbels.
Gebärmutter: Mitte der seitlichen Bauchwand in Höhe des 5. Lendenwirbels.
Harnblase: Seitliche Bauchwand am Übergang vom ventralen zum mittleren Drittel in Höhe des 7. Lendenwirbels.

The location of the organs in the abdominal cavity depends on the position of the animal, the fullness of the organs themselves, and, in case of a pathological condition, upon its nature and extent. For rapid and reliable orientation radiographs should preferably be taken in a uniform manner, i.e. either with the left or the right abdominal wall against the cassette. The latter proved to be most succesful.
A Bucky diaphragm or a cassette with a stationary grid should be used to reduce scattered radiation. In small lean dogs a grid is not necessary.
Correct positioning of the dog depends on the merits of the individual case presented. For example, fluid levels are demonstrated in the standing position, not in recumbent patients.
The positioning in lateral recumbency (fig. 158) is less complicated. By stretching the extremities troublesome overlaying on the thigh musculature can be avoided. The patient is to be aligned with the median plane parallel to the cassette. In obese dogs, tilting can be eliminated with the aid of foam rubber pads placed at the proper level under the extremities resting on the table. In addition the tail can be fixed by slight pull.
In positioning a standing dog (fig. 159), it is necessary to place in direct contact with the cassette. Dogs of large breeds should stand on their own. Small dogs are easier to hold in front of the cassette. Gentle stretching eliminates overlaying of the thigh musculature. The median plane must be parallel to the cassette.
The central beam should fall at the right angle on the cassette and strike the lateral abdominal wall in such a way, that the region to be examined coincides with the center of the film.

Directing the central beam:
Routine radiograph and intestines: Center of the lateral abdominal wall on a level of the 3rd lumbar vertebra.
Stomach: Center of the lateral abdominal wall on a level of the 12th rib.
Gall-bladder: Lateral thoracic wall at the junction of the ventral and middle thirds on a level of the 7th intercostal space.
Kidneys: Lateral abdominal wall at the junction of the middle and dorsal thirds on a level of the 2nd lumbar vertebra.
Ovaries: Lateral abdominal wall at the junction of the middle and dorsal thirds on a level of the 2nd lumbar vertebra.
Uterus: Center of the lateral abdominal wall on a level of the 5th lumbar vertebra.
Urinary bladder: Lateral abdominal wall at the junction of the ventral and middle thirds on a level of the 7th lumbar vertebra.

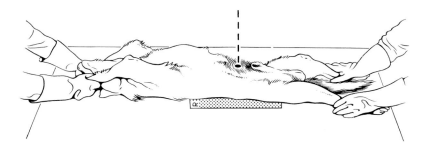

Abb. 160   Lagerung zur Aufnahme des Abdomens. Ventro-dorsal.

Bei der Lagerung in Rückenlage ist darauf zu achten, daß die Medianebene senkrecht zur Kassette steht. Es ist zweckmäßig, die Beckengliedmaßen in leichter Abduktion so weit nach hinten zu strecken, daß sich die Hände, die die Gliedmaßen im Bereich der Sprunggelenke halten, auf die Tischplatte stützen können.
Zur Verminderung der Streustrahlung ist eine Bucky-Blende oder eine Kassette mit stehendem Raster notwendig. Nur bei kleinen, nicht fettleibigen Hunden kann auf den Raster verzichtet werden.
Der Zentralstrahl sollte im rechten Winkel auf die Kassette auftreffen.

Einstellung des Zentralstrahls:
Übersichtsaufnahme, Darm und Nieren: Mittellinie in Höhe des 2. Lendenwirbels.
Magen: Mittellinie in Höhe des 12. Brustwirbels.
Harnblase: Mittellinie in Höhe des 7. Lendenwirbels.

Fig. 160   Positioning of abdomen. Ventrodorsal.

When positioning the animal on its back, care must be exercised to align the median plane perpendicularly to the cassette. The hind legs should be stretched and slightly abducted to allow the hands holding the hocks to rest on the table.
To reduce scattered radiation, a Bucky diaphragm or a cassette with a stationary grid should be used. A grid is not necessary for small lean dogs.
The central beam should fall at the right angle on the cassette.

Directing the central beam:
Routine radiograph, intestines and kidneys: Midline on a level of the 2nd lumbar vertebra.
Stomach: Midline on a level of the 12th thoracic vertebra.
Urinary bladder: Midline on a level of the 7th lumbar vertebra.

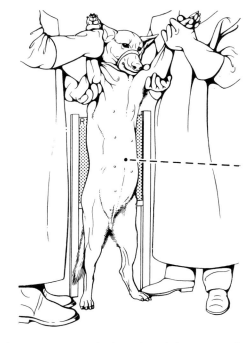

Abb. 161   Lagerung zur Aufnahme des Abdomens. Aufrecht stehend. Ventro-dorsal.

Bei der Lagerung des auf den Beckengliedmaßen aufrecht stehenden oder in aufrechter Stellung vor der Kassette gehaltenen Hundes ist darauf zu achten, daß sich die Medianebene im rechten Winkel zur Kassette befindet und daß der Hund mit dem Rücken der Kassette anliegt.
Zur Verminderung der Streustrahlung ist eine Bucky-Blende oder eine Kassette mit stehendem Raster erforderlich.
Der Zentralstrahl sollte die Mittellinie in Höhe des 12. Brustwirbels treffen und im rechten Winkel auf die Kassette einfallen.

Fig. 161   Positioning of abdomen. Standing on hind legs. Ventrodorsal.

In positioning the dog standing on its hind legs or holding it in this position, the median plane should be aligned at the right angle to the cassette with the back of the animal against the latter.
To reduce scattered radiation, a Bucky diaphragm or a cassette with a stationary grid is required.
The central beam should strike the midline on a level of the 12th thoracic vertebra and fall at the right angle on the cassette.

Magen-Darm-Kontrastuntersuchung

Vorbereitung: Abgesehen von dringlichen Untersuchungen (Ileusverdacht), sollte eine Magen-Darmpassage im nüchternen Zustand nach zwölfstündigem Hungern und Dursten vorgenommen werden. Ein Kontaktlaxans oder ein Reinigungseinlauf ist zu verordnen, wenn der bei der Untersuchung (Durchleuchtung, Leeraufnahme) festgestellte Darminhalt die Darstellung in einem bestimmten Abschnitt und damit die Beurteilung erschweren oder verhindern kann.

Kontrastmittel: Barium sulfuricum purissimum in einer der handelsüblichen Abfüllungen in einer wässerigen homogenen Aufschwemmung von rahmiger Konsistenz. Die Suspension (je nach Größe des Hundes 75–150 ml) ist schluckweise in die Backentasche zu applizieren.
Trijodierte Kontrastmittel in wässeriger Lösung. Es sind hochprozentige Lösungen mit Geschmackskorrigentien im Handel. Wegen der guten Verträglichkeit sind diese Präparate im Fall einer Aspiration und bei perforierenden Verletzungen der Speiseröhre, des Magens oder Darmes dem Barium sulfuricum überlegen. Im Vergleich zu den Barium sulfuricum enthaltenden Präparaten passieren die trijodierten Kontrastmittel in wässeriger Lösung den Darm schneller. Außerdem sind sie weniger schattengebend und werden zum Teil resorbiert. Sie können beim gesunden Hund das Caecum und Colon ascendens schon nach ca. 30 bis 40 Minuten erreichen. Dem sind die Abstände zwischen den Untersuchungen anzupassen.
Mit der Magensonde sollte das Kontrastmittel nur im Ausnahmefall gegeben werden, weil dabei Luft in den Magen gelangen, also eine atypische Situation entstehen kann. Außerdem kann bei Anwendung der Magensonde das Einfließen in den Magen nicht beobachtet werden.

Untersuchungstechnik: Bei der Passageuntersuchung liegt das Gewicht in der Erfassung von Bewegungsabläufen. Aufschluß darüber ist nur bei der Durchleuchtung zu erhalten. Bei jeder Unregelmäßigkeit im Passageablauf ist der betreffende Abschnitt eingehend zu untersuchen. Aufnahmen sind zur Objektivierung bestimmter Zustände „gezielt" anzufertigen. Die Zahl der notwendigen Aufnahmen und die Lagerung hängen deshalb von der jeweiligen Situation ab.

Folgendes Untersuchungsschema hat sich bewährt:
a) Durchleuchtung stehend seitlich.
b) Eingabe von 2-3 Schlucken Kontrastmittel, Untersuchung stehend seitlich, anschließend liegend, seitlich und in Rückenlage (dorso-ventraler Strahlengang).
Für die Beurteilung der Magenschleimhaut (Durchleuchtung, Aufnahme) ist die Lagerung des Hundes auf dem Rücken mit einer geringen Neigung des Thorax (etwa 15°) nach der rechten Seite hin zweckmäßig.
c) Schluckweise Eingabe des Kontrastmittels bis zur „mäßigen Füllung" des Magens.
Beim gesunden Hund beginnt die Entleerung des Kontrastmittels 5-10 Minuten nach der Applikation und ist nach 2 1/2 - 3 Stunden beendet. Zweckmäßig ist die Untersuchung (Durchleuchtung, Aufnahme) stehend und liegend seitlich sowie in Rückenlage mit einer geringen Neigung des Thorax (etwa 15°) nach der rechten Seite hin.
d) Das Duodenum ist beim gesunden Hund etwa 2–3 Minuten nach Beginn der Magenentleerung meist in ganzer Länge dargestellt. Zweckmäßig ist die Untersuchung (Durchleuchtung, Aufnahme) in Rückenlage mit geringer Neigung des Thorax (15°) nach der rechten Seite hin und in rechter Seitenlage.
e) Barium sulfuricum purissimum (rahmige Konsistenz) erreicht beim gesunden Hund das Caecum und Colon ascendens etwa 3 1/2–4 Stunden, das Colon descendens etwa 6 Stunden und das Rectum etwa 6–8 Stunden nach der Kontrastmitteleingabe.
Trijodierte Kontrastmittel in wässeriger Lösung können das Caecum und Colon ascendens in 30-40 Minuten erreichen.
Zweckmäßig ist die Untersuchung (Durchleuchtung, Aufnahme) in Rückenlage und rechter Seitenlage.

Contrast studies of the gastro-intestinal tract

Preparation of the patient: Except in emergencies (e.g. ileus), a gastrointestinal passage should be undertaken after withholding food and water for at least 12 hours. A contact laxative or an evacuation enema should be prescribed if intestinal contents can still be demonstrated by fluoroscopy or radiography prior to administration of the contrast medium.

Contrast medium: Barium sulphate in a watery, homogeneous suspension and of creamy consistency. It is administered into the corner of the mouth in doses of 75-150 ml depending on the size of the animal.
Tri-iodized contrast media in watery solution. High percentage solutions with taste-correcting components are available. Since they are well tolerated these drugs are superior to Barium sulphate if aspirated, and in perforating lesions of the esophagus, stomach or intestine. In comparison with Barium sulphate containing drugs the tri-iodized watery soluble contrast media pass more rapidly through the intestine. Moreover, they produce less shadows and are partly resorbed. In normal dogs they may reach the cecum and colon ascendens already within approximately 30-40 minutes. For the latter the intervals between the examinations should be adapted.
Only in exceptional cases should the contrast medium be administered with a stomach tube, since air could enter the stomach and create an atypical picture. Moreover, passage of the contrast medium into the stomach cannot be observed under these circumstances.

Technique: In studying the passage of the contrast medium it is important to determine the motility-sequences. Only fluoroscopy can furnish the desired information. In case of any irregularity in the passage the particular segment should be closely examined. In order to register particular conditions, specific radiographs should be aimed at. The number of radiographs necessary and the positioning, therefore, depend on the individual case.

The following routine procedure proved to be successful:
a) Fluoroscopy – standing position, lateral;
b) Administration of 2-3 swallows of contrast medium followed by fluoroscopy in the standing, lateral recumbent and dorsal recumbent position. In order to interpret the mucosa of the stomach (fluoroscopy, radiography) the dog should be placed in a dorsal recumbent position with the thorax tilted slightly (approximately 15°) toward the right side.
c) Swallow by swallow administration of the contrast medium until the stomach is well filled. In a normal dog emptying of the stomach begins 5-10 minutes after the medium has been swallowed and is completed within 2 1/2-3 hours. The examination (fluoroscopy, radiography) should be performed in the standing, lateral recumbent and dorsal recumbent positions. In the latter, the chest should be tilted approximately 15° to the right.
d) In an healthy dog the duodenum is outlined in the majority of cases in its full length approximately 2-3 minutes after emptying of the stomach has commenced. The examination (fluoroscopy, radiography) is the most effective in dorsal recumbency with slight tilting of the thorax (15°) to the right side and in the right lateral recumbent position.
e) In the healthy dog pure Barium sulphate (creamy consistency) reaches the cecum and ascending colon approximately 3 1/2-4 hours after administration, the descending colon approximately after 6 hours and the rectum approximately after 6-8 hours.
Tri-iodized contrast media in watery solution can reach the cecum and ascending colon in 30-40 minutes.
The examination (fluoroscopy, radiography) is most effective in dorsal recumbency and in right lateral recumbency.

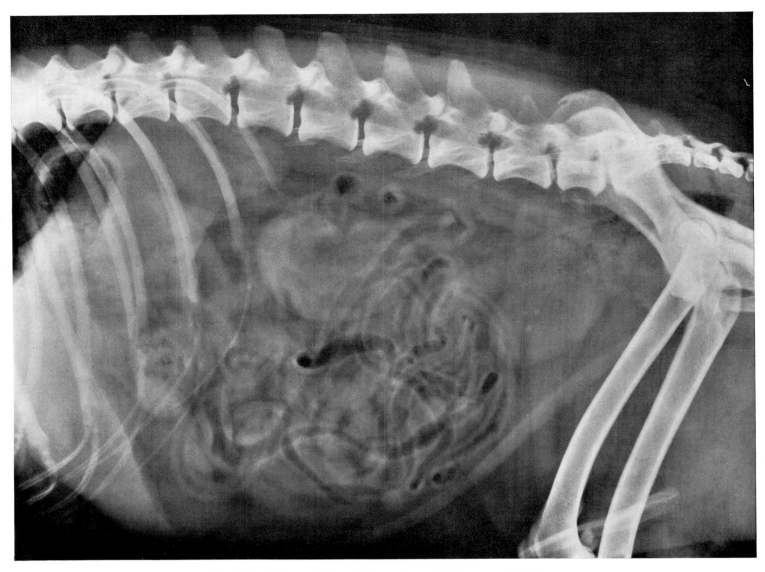

Abb. 162   Abdomen. Liegend. Latero-lateral. Schnauzer, ♂, 9 Jahre.
Bucky-Blende – Mittelverstärkende Folie – FFA 120 cm – 65 kV – 35 mAs
Verkleinerung von 30 × 40 cm
Lagerung Abb. 158

Fig. 162   Abdomen. Recumbent. Laterolateral. Schnauzer, ♂, 9 years old.
Bucky diaphragm – Standard screens – FFD 120 cm – 65 kV – 35 mAs
Diminution of 30 × 40 cm
Positioning fig. 158

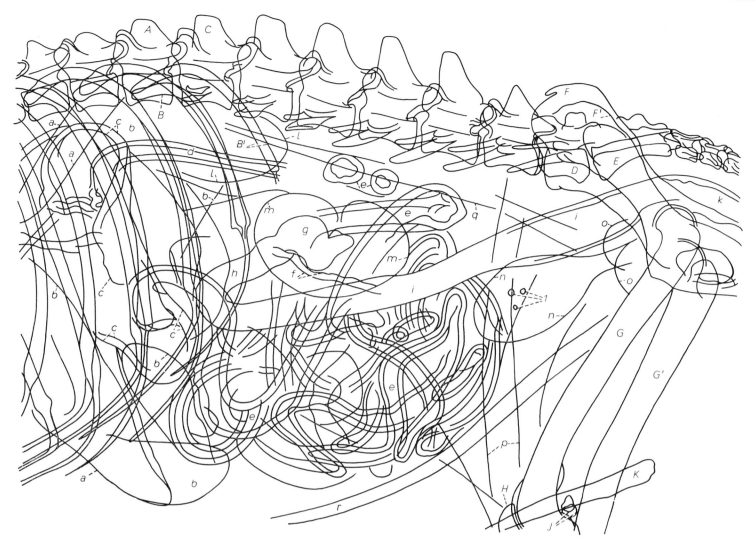

Abb. 163** Röntgenskizze zu Abb. 162   Fig. 163** X-ray sketch to fig. 162

A	13. Vertebra thoracica;		f	Ileum;
B	13. Os costale;		g	Caecum;
C	1. Vertebra lumbalis;		h	Colon ascendens;
D	7. Vertebra lumbalis;		i	Colon descendens;
E	Os sacrum;		k	Rectum;
F	Ala ossis ilii;		l	Ren dexter;
G	Os sacrum;		m	Ren sinister;
H	Patella;		n	Vesica urinaria;
J	Ossa sesamoidea m. gastrocnemii;		o	Lobi prostatae, vergrößert – Lobi prostatae, enlarged;
K	Os penis;		p	Oberschenkelmuskulatur, kraniale Begrenzung – Cranial limit of the thigh musculature;
			q	Lendenmuskulatur, ventrale Begrenzung – Ventral limit of the lumbar muscles;
a	Diaphragma;			
b	Hepar;		r	Ventrale Bauchwand – Ventral abdominal wall;
c	Ventriculus;			
d	Duodenum;		1	Blasensteine – Vesical calculi.
e	Jejunum, zum Teil gashaltig – Jejunum, partly filled with gas;			

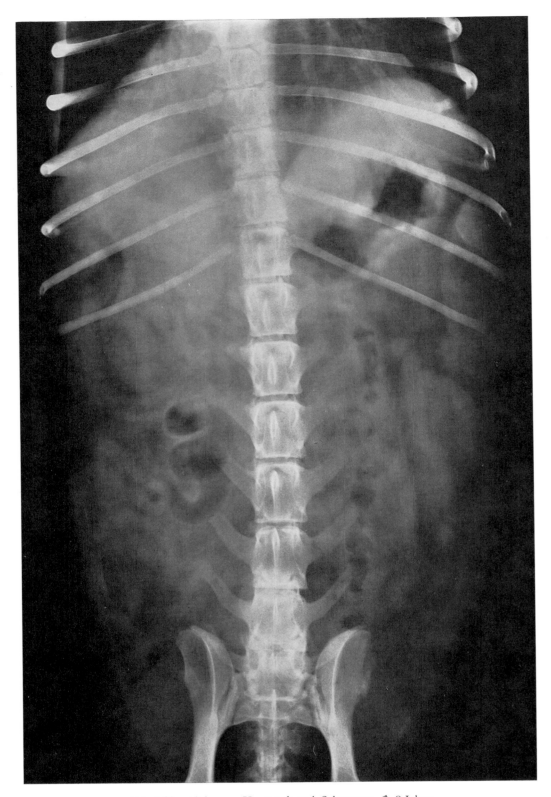

Abb. 164   Abdomen. Ventro-dorsal. Schnauzer, ♂, 9 Jahre.
Bucky-Blende – Mittelverstärkende Folie – FFA 120 cm – 65 kV – 48 mAs
Verkleinerung von 30 × 40 cm
Lagerung Abb. 160

Fig. 164   Abdomen. Ventrodorsal. Schnauzer, ♂, 9 years old.
Bucky diaphragm – Standard screens – FFD 120 cm – 65 kV – 48 mAs
Diminution of 30 × 40 cm
Positioning fig. 160

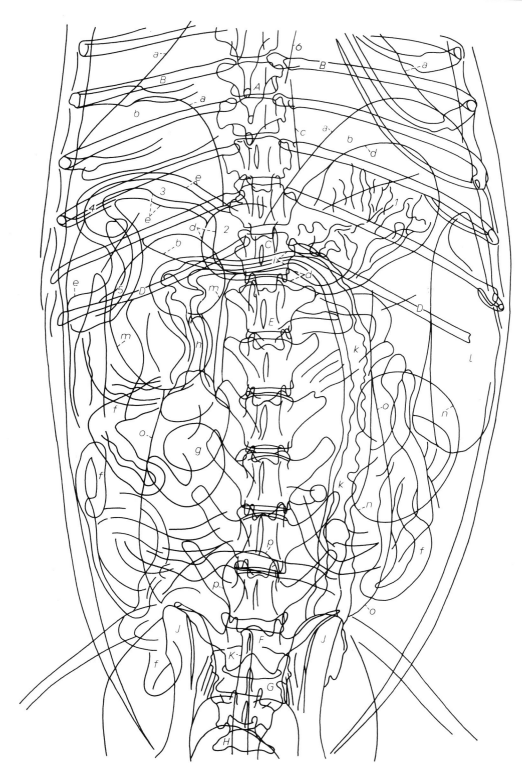

Abb. 165   Röntgenskizze zu Abb. 164   Fig. 165   X-ray sketch to fig. 164

A	9. Vertebra thoracica;	i	Colon transversum;
B	9. Os costale;	k	Colon descendens;
C	13. Vertebra thoracica;	l	Lien;
D	13. Os costale;	m	Ren dexter;
E	1. Vertebra lumbalis;	n	Ren sinister;
F	7. Vertebra lumbalis;	o	Stammuskulatur, laterale Begrenzung – Lateral limit of the trunk musculature;
G	Os sacrum;		
H	1. Vertebra caudalis;	p	Praeputium;
J	Ala ossis ilii;		
K	Os penis;		

Am Magen – On the stomach:

1   Fundus ventriculi, Plicae villosae;
2   Pars pylorica;

a	Diaphragma;
b	Hepar;
c	Oesophagus;
d	Ventriculus;
e	Duodenum;
f	Jejunum;
g	Caecum;
h	Colon ascendens;

Am Duodenum – On the duodenum:

3   Pars cranialis;
4   Flexura duodeni cranialis;
5   Pars descendens;
6   Aorta thoracica.

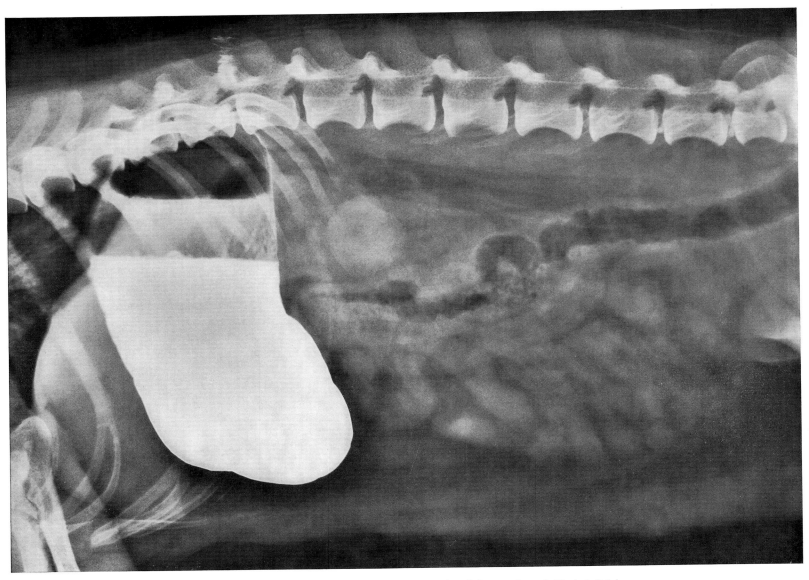

Abb. 166   Magen-Kontrast (Barium sulfuricum). Stehend. Latero-lateral. Dackel, 3 Jahre.
Bucky-Blende – Feinzeichnende Folie – FFA 120 cm – 65 kV – 30 mAs
Verkleinerung von 24 × 30 cm
Lagerung Abb. 159

Fig. 166   Stomach-contrast (Barium sulphate). Standing. Laterolateral. Dachshund, 3 years old.
Bucky diaphragm – High definition screens – FFD 120 cm – 65 kV – 30 mAs
Diminution of 24 × 30 cm
Positioning fig. 159

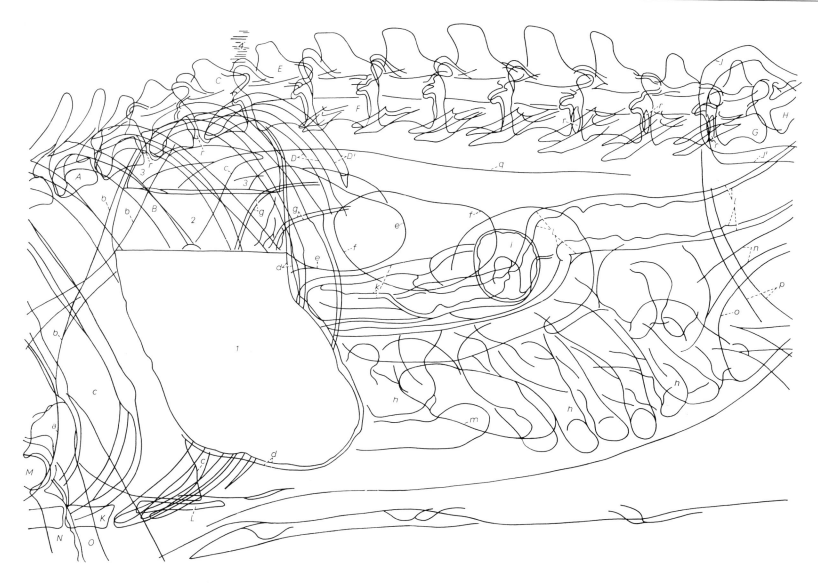

Abb. 167*   Röntgenskizze zu Abb. 166   Fig. 167*   X-ray sketch to fig. 166

A	9. Vertebra thoracica;		e	Ren dexter;
B	9. Os costale;		f	Ren sinister;
C	12. Vertebra thoracica;		g	Duodenum;
D	12. Os costale;		h	Intestinum tenue;
E	13. Vertebra thoracica, 13. Os costale fehlt – 13th thoracic vertebra, 13th rib is absent;		i	Caecum;
F	1. Vertebra lumbalis;		k	Colon ascendens;
G	7. Vertebra lumbalis;		l	Colon descendens;
H	Os sacrum;		m	Lien;
J	Ala ossis ilii;		n	Uterus;
K	Sternum, 6. Sternebra;		o	Vesica urinaria;
L	Processus xiphoideus;		p	Oberschenkelmuskulatur, kraniale Begrenzung – Cranial limit of the thigh musculature;
M	Humerus;			
N	Radius;		q	Lendenmuskulatur, ventrale Begrenzung – Ventral limit of the lumbar muscles;
O	Ulna;			
			r	Nucleus pulposus, verkalkt! – Nucleus pulposus, calcified;
a	Cor;		1	Kontrastmittel, Barium sulfuricum – Contrast medium, Barium sulphate;
b	Diaphragma;			
c	Hepar;		2	Magensaft – Gastric juice;
d	Ventriculus;		3	Luftblase – Air bubble;
			4	Kontrastmittel auf dem Fell – Contrast medium on the skin.

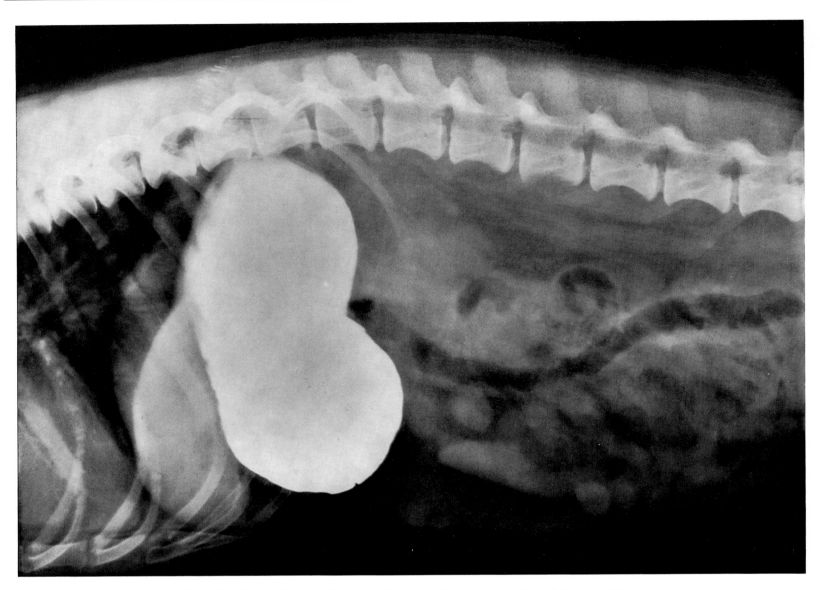

Abb. 168   Magen-Kontrast (Barium sulfuricum). Liegend. Latero-lateral. Dackel, 3 Jahre.
Bucky-Blende – Feinzeichnende Folie – FFA 120 cm – 65 kV – 30 mAs
Verkleinerung von 24 × 30 cm
Lagerung Abb. 158

Fig. 168   Stomach-contrast (Barium sulphate). Recumbent. Laterolateral. Dachshund, 3 years old.
Bucky diaphragm – High definition screens – FFD 120 cm – 65 kV – 30 mAs
Diminution of 24 × 30 cm
Positioning fig. 158

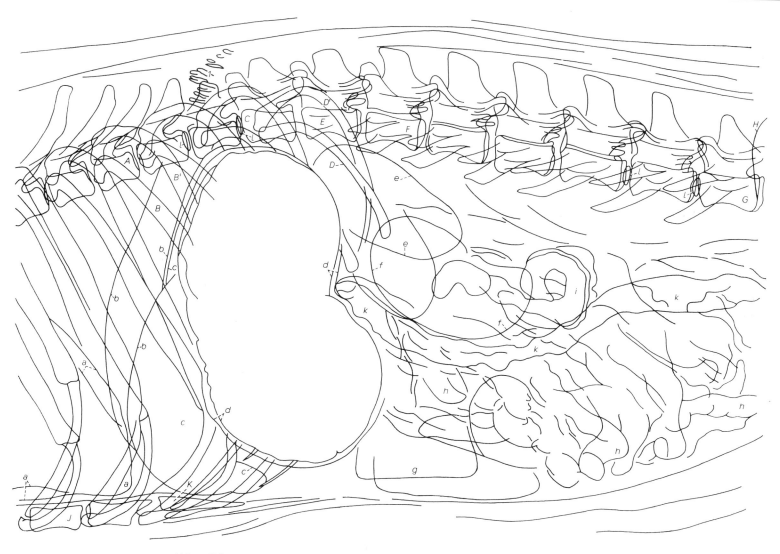

Abb. 169*   Röntgenskizze zu Abb. 168   Fig. 169*   X-ray sketch to fig. 168

A   9. Vertebra thoracica;
B   9. Os costale;
C   12. Vertebra thoracica;
D   12. Os costale;
E   13. Vertebra thoracica, 13. Os costale fehlt! – 13th thoracic vertebra, 13th rib is absent;
F   1. Vertebra lumbalis;
G   6. Vertebra lumbalis;
H   Ala ossis ilii;
J   Sternum, 5. Sternebra;
K   Processus xiphoideus;

a   Cor;
b   Diaphragma;
c   Hepar;
d   Ventriculus;
e   Ren dexter;
f   Ren sinister;
g   Lien;
h   Intestinum tenue;
i   Caecum;
k   Colon descendens;
l   Nucleus pulposus, verkalkt – Nucleus pulposus, calcified;

l   Kontrastmittel auf dem Fell – Contrast medium on the skin.

Legende zu den Abbildungen 173 – 175 (S. 117) – Legend to figs. 173 – 175 (p. 117):

A   10. Vertebra thoracica;
B   10. Os costale;
C   13. Vertebra thoracica;
D   13. Os costale;
E   Processus xiphoideus;
F   1. Vertebra lumbalis;

a   Apex cordis;
b   V. cava caudalis;
c   Diaphragma;
d   Oesophagus;
e   Ventriculus;
f   Duodenum, Tunica muscularis;
g   Lien;
h   Jejunum;

Am Magen – On the stomach:

1   Curvatura ventriculi major;
2   Curvatura ventriculi minor;
3   Pars cardiaca;
4   Fundus ventriculi;
5   Corpus ventriculi mit Plicae villosae, teilweise kontrastiert – Corpus ventriculi with gastric folds partly contrasted;
6   Antrum pyloricum;
7   Canalis pyloricus;

Am Duodenum – On the duodenum:

8   Pars cranialis;
9   Pars descendens.

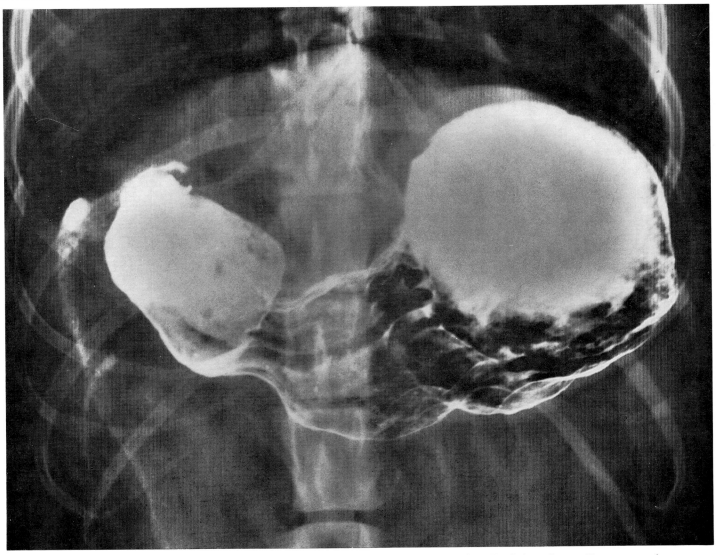

Abb. 170, 171, 172   Magen-Kontrast (Barium sulfuricum). Pylorus - Verschiedene Funktionsphasen. Dorso-ventral.
Aufnahme aus der Durchleuchtung. Pudel, 6 Jahre. Bucky-Blende – Mittelverstärkende Folie – FFA 80 cm – 70 kV – 7 mAs
Originalgröße (Ausschnitt aus 18 × 24 cm bzw. 13 × 18 cm)

Figs. 170, 171, 172   Stomach-contrast (Barium sulphate). Pylorus - Various functional phases. Dorsoventral.
Radiograph of fluoroscopic image. Poodle, 6 years old. Bucky diaphragm – Standard screens – FFD 80 cm – 70 kV – 7 mAs
Original size (section of 18 × 24 cm and 13 × 18 cm respectively)

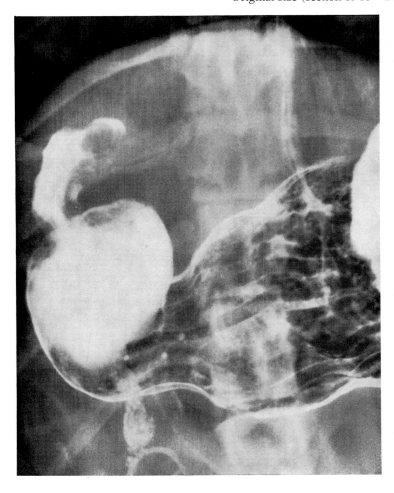
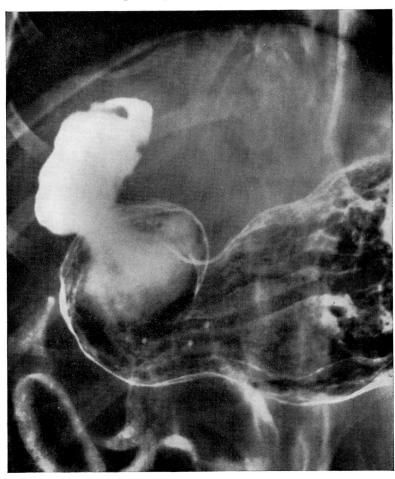

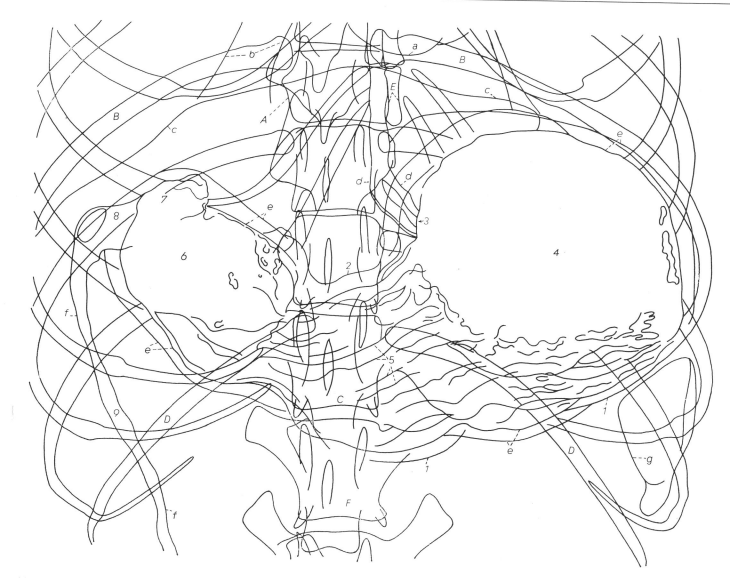

Abb. 173, 174, 175   Röntgenskizzen zu Abb. 170, 171, 172   Figs. 173, 174, 175   X-ray sketches to figs. 170, 171, 172

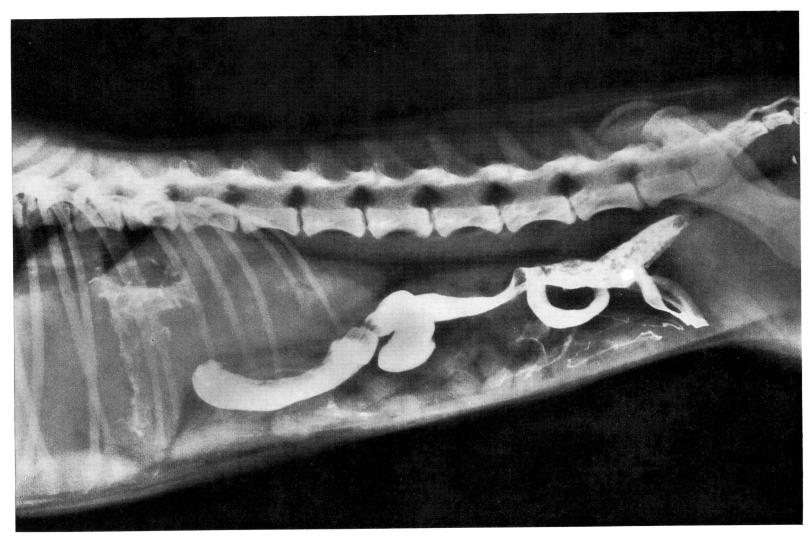

Abb. 176   Darm-Kontrast (Barium sulfuricum). Stehend. Latero-lateral. Pudel, 1 Jahr.
Bucky-Blende – Hochverstärkende Folie – FFA 100 cm – 45 kV – 13 mAs
Verkleinerung von 18 × 24 cm
Lagerung Abb. 159

Fig. 176   Intestine-contrast (Barium sulphate). Standing. Laterolateral. Poodle, 1 year old.
Bucky diaphragm – High speed screens – FFD 100 cm – 45 kV – 13 mAs
Diminution of 18 × 24 cm
Positioning fig. 159

Abb. 177*   Röntgenskizze zu Abb. 176   Fig. 177*   X-ray sketch to fig. 176

A   12. Vertebra thoracica;
B   13. Os costale;
C   1. Vertebra lumbalis;
D   4. Vertebra lumbalis;
E   Os sacrum;
F   1. Vertebra caudalis;
G   Ala ossis ilii;

a   Ventriculus;
b   Jejunum;
c   Ileum;
d   Caecum;
e   Colon ascendens;
f   Colon transversum;
g   Colon descendens, nur im Endabschnitt kontrastmittelhaltig – Colon descendens, containing contrast medium only in its caudal part;

h   Rectum;
i   Diaphragma;
k   Hepar;
k″ Processus caudatus;
l   Lien;
m   Ren;
n   Vesica urinaria;
o   Lendenmuskulatur, ventraler Rand — Ventral limit of the lumbar muscles;

1   Plicae villosae mit Kontrastmittelresten — Gastric folds with the residual contrast medium;
2   Gasblase im Fundus ventriculi — Gas bubble in fundus of the stomach;
3   Ostium ileale.

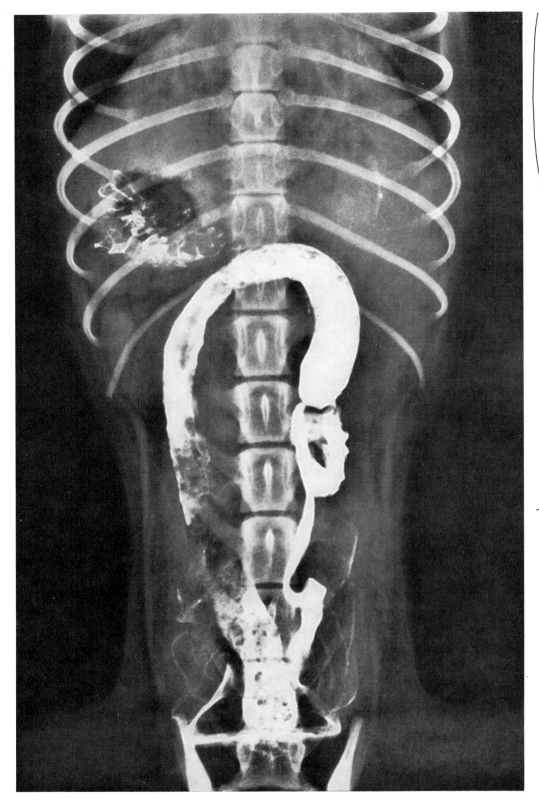

Abb. 178   Darm-Kontrast (Barium sulfuricum). Dorso-ventral. Pudel, 1 Jahr.
Bucky-Blende – Hochverstärkende Folie – FFA 120 cm – 45 kV – 12 mAs
Verkleinerung von 18 × 24 cm

Fig. 178 Intestine-contrast (Barium sulphate). Abdominal recumbency. Dorsoventral.
Poodle, 1 year old.
Bucky diaphragm – High speed screens – FFD 120 cm – 45 kV – 12 mAs
Diminution of 18 × 24 cm

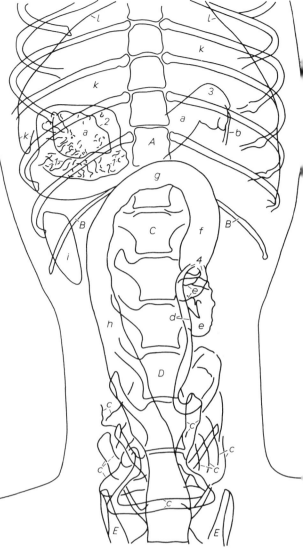

Abb. 179   Röntgenskizze zu Abb. 178
Fig. 179   X-ray sketch to fig. 178

A   12. Vertebra thoracica;
B   13. Os costale;
C   1. Vertebra lumbalis;
D   4. Vertebra lumbalis;
E   Ala ossis ilii;

a   Ventriculus;
b   Duodenum, Kontrastmittelrest im Anfangsteil –
    Duodenum, residual contrast medium in its cra-
    nial part;
c   Jejunum;
d   Ileum;
e   Caecum;
f   Colon ascendens;
g   Colon transversum;
h   Colon descendens;
i   Lien;
k   Leberschatten, Grenzen der Leberlappen nicht zu differenzieren –
    Liver shadows, borders of lobes cannot be differentiated;
l   Diaphragma;

1   Kontrastmittelrest auf der Schleimhaut des Fundus ventriculi – Resi-
    dual contrast medium on mucosa of the fundus of the stomach;
2   Gasblase im Fundus ventriculi – Gas bubble in fundus of the stom-
    ach;
3   Gasblase im Canalis pyloricus – Gas bubble in pyloric canal;
4   Ostium ileale.

Retrograde Kontrastdarstellung des Dickdarms (Irrigoskopie)

Vorbereitung: Die Untersuchung sollte nüchtern nach 12stündigem Hungern (nicht Dursten!) erfolgen. Ist Darminhalt (fest oder gasförmig) nachweisbar (Durchleuchtung, Leeraufnahme), so wird zunächst ein Reinigungseinlauf vorgenommen und dem Hund Bewegungsmöglichkeit gegeben.

Kontrastmittel: Barium sulfuricum purissimum (eine der handelsüblichen Abfüllungen) in einer wässerigen homogenen Aufschwemmung von rahmiger Konsistenz.
Je nach Größe des Hundes sind 75 – 250 ml der Suspension erforderlich.

Untersuchungstechnik:

a) Leeraufnahme in rechter Seitenlage, latero-lateraler Strahlengang. Die Aufnahme kann Aufschluß über Veränderungen geben, die eine Erkrankung des Dickdarms vortäuschen können. Da die Veränderungen oft nur geringgradig sind, evtl. nur einen zarten Schatten geben, können sie bei einer Kontrastaufnahme verlorengehen.

b) Applikation des Kontrastmittels unter Schirmkontrolle. Der Kontrastmitteleinlauf erfolgt mit einem bis in das Colon descendens vorgeschobenen Schlauch, an dem entweder ein Irrigator oder eine größere Spritze angeschlossen wird. Um ein Abfließen des Kontrastmittels zu verhindern, ist der Anus durch Andrücken an den Schlauch zu verschließen. Einfacher und zuverlässiger ist die Applikation mit einem Darmtamponator oder der Straus'schen Sonde.
Das Kontrastmittel muß langsam einfließen, da bei einer übermäßigen Dehnung des Rektums der Hund mit starkem Pressen (Stuhldrang) reagiert.
Nach praller Füllung des Dickdarms sind 2 Aufnahmen (latero-lateraler Strahlengang am stehenden oder liegenden Hund und ventrodorsaler Strahlengang in Rückenlage) anzufertigen.

c) Möglichst gute Entleerung des Kontrastmittels.

d) Luftinsufflation unter Schirmkontrolle. Der Grad der erforderlichen Füllung ist von der Fragestellung und der Situation abhängig zu machen. Nach ausreichender Füllung sind 2 Aufnahmen (latero-lateraler Strahlengang am stehenden oder liegenden Hund und ventro-dorsaler Strahlengang in Rückenlage) anzufertigen.
Durch die dosierte Luftapplikation wird die Schleimhaut entfaltet, so daß das Schleimhautrelief sichtbar wird.

Retrograde contrast radiography of the colon (irrigoradioscopy)

Preparation of the patient: Withhold food but not water for 12 hours. Should the intestine not be quite empty (fluoroscopy, preliminary radiograph), the animal must be given an evacuation enema and allowed some exercise.

Contrast medium: Barium sulphate enema (homogeneous watery suspension of creamy consistency).
75 – 200 ml of the suspension will be needed, depending on the size of the dog.

Technique:

a) A preliminary radiograph should be taken in the right lateral position with laterolateral direction of the beam.
This radiograph can give information on changes resembling diseases of the colon. Because these changes are often very slight and may produce only faint shadows, they may disappear in an contrast radiograph.

b) Administration of the contrast medium should be performed with the help of fluoroscopy. A rubber tube, connected either to an irrigator or a large syringe should be introduced into the colon descendens. In order to avoid back flow of the contrast medium, the anus should be closed by pressing it against the tube. Administration of the contrast medium with a „Bardex" or a „Straus" catheter is easier and more reliable.
The contrast medium must be introduced slowly, since a rapid distention of the rectum could induce tenesmus.
When the colon has been completely filled two radiographs should be taken: in the lateral position (standing or lying) with laterolateral direction of the beam and in dorsal recumbency with ventrodorsal direction of the beam.

c) Evacuation of the contrast medium should be as complete as possible.

d) Inflation of air – with the help of fluoroscopy. The desired degree of filling depends on the circumstances. Two radiographs should be taken as described under b. The proper introduction of air into the colon allows unfolding and visibility of the mucous membrane.

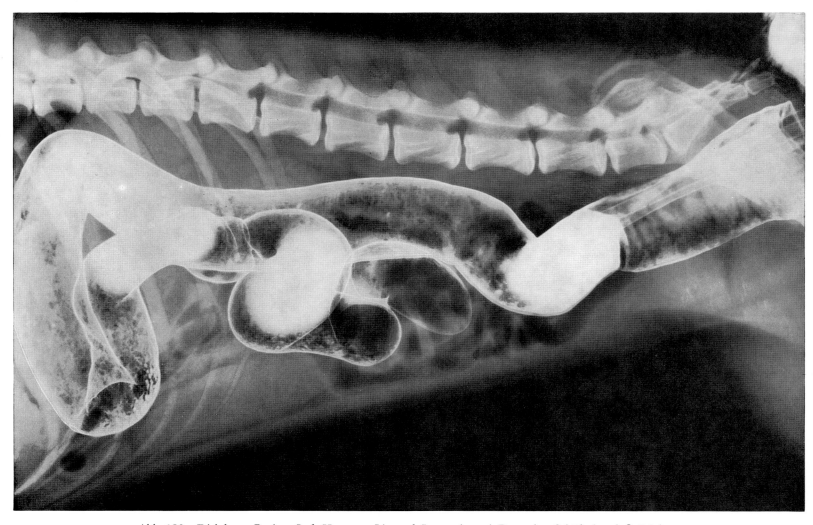

Abb. 180 Dickdarm. Barium-Luft-Kontrast. Liegend. Latero-lateral. Deutscher Schäferhund, ♀, 3 Jahre.
Bucky-Blende – Mittelverstärkende Folie – FFA 120 cm – 60 kV – 30 mAs
Verkleinerung von 30 × 40 cm
Lagerung Abb. 158

Fig. 180 Large intestine. Barium-air-contrast. Recumbent. Laterolateral. German shepherd dog, ♀, 3 years old.
Bucky diaphragm – Standard screens – FFD 120 cm – 60 kV – 30 mAs
Diminution of 30 × 40 cm
Positioning fig. 158

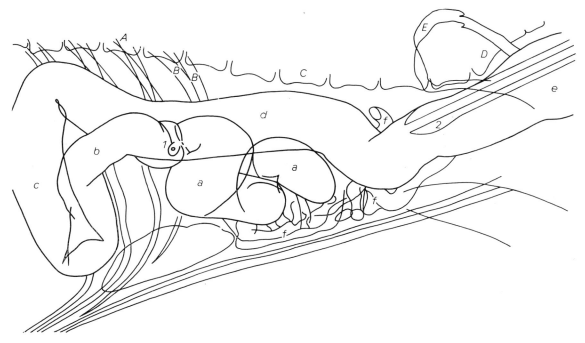

Abb. 181   Röntgenskizze zu Abb. 180   Fig. 181   X-ray sketch to fig. 180

A   13. Vertebra thoracica;
B   13. Os costale;
C   4. Vertebra lumbalis;
D   Os sacrum;
E   Os ilium;

a   Caecum;
b   Colon ascendens;
c   Colon transversum;

d   Colon descendens;
e   Rectum;
f   Jejunumschlingen – Intestinal loops;

1   Ostium ileale;
2   Sonde – Probe.

Streifenförmige Schatten im Oberschenkelweichteilschatten: äußerliche Verunreinigung mit Kontrastmittel – Striped shadows in the thigh region represent external soiling by contrast medium.

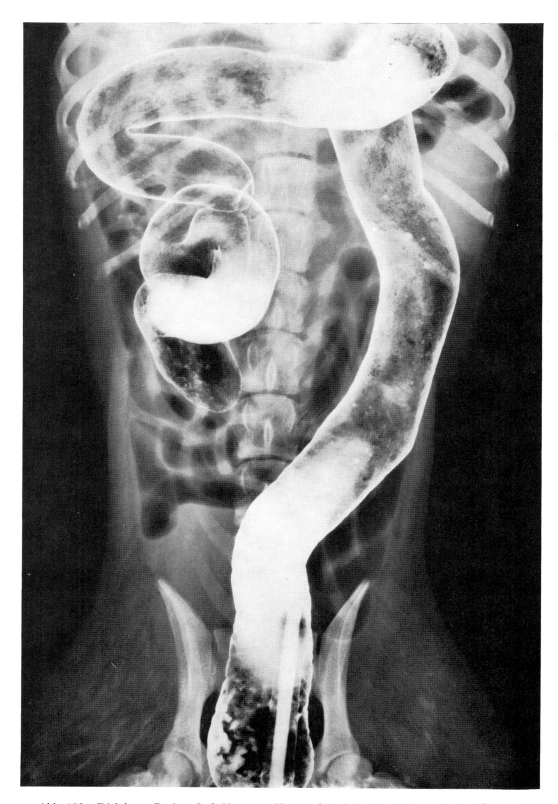

Abb. 182 Dickdarm. Barium-Luft-Kontrast. Ventro-dorsal. Deutscher Schäferhund, ♀, 3 Jahre.
Bucky-Blende – Mittelverstärkende Folie – FFA 120 cm – 65 kV – 35 mAs
Verkleinerung von 30 × 40 cm
Lagerung Abb. 160

Fig. 182 Large intestine. Barium-air-contrast. Ventrodorsal. German shepherd dog, ♀, 3 years old.
Bucky diaphragm – Standard screens – FFD 120 cm – 65 kV – 35 mAs
Diminution of 30 × 40 cm
Positioning fig. 160

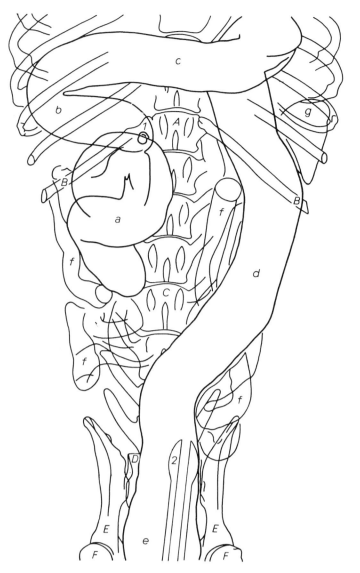

Abb. 183   Röntgenskizze zu Abb. 182   Fig. 183   X-ray sketch to fig. 182

A   13. Vertebra thoracica;
B   13. Os costale;
C   4. Vertebra lumbalis;
D   Os sacrum;
E   Os ilium;
F   Caput ossis femoris;

a   Caecum;
b   Colon ascendens;
c   Colon transversum;
d   Colon descendens;

e   Rectum;
f   Jejunumschlingen – Intestinal loops;
g   Lien;

1   Ostium ileale;
2   Sonde – Probe.

Homogene Verschattung im Bereich der Regio inguinalis dextra: Zustand nach Hysterektomie – Homogeneous shading in the right inguinal region: result of hysterectomy.
Streifenförmige Schatten im Oberschenkelweichteilschatten: äußerliche Verunreinigung mit Kontrastmittel – Striped shadows in the thigh region represent external soiling by contrast medium.

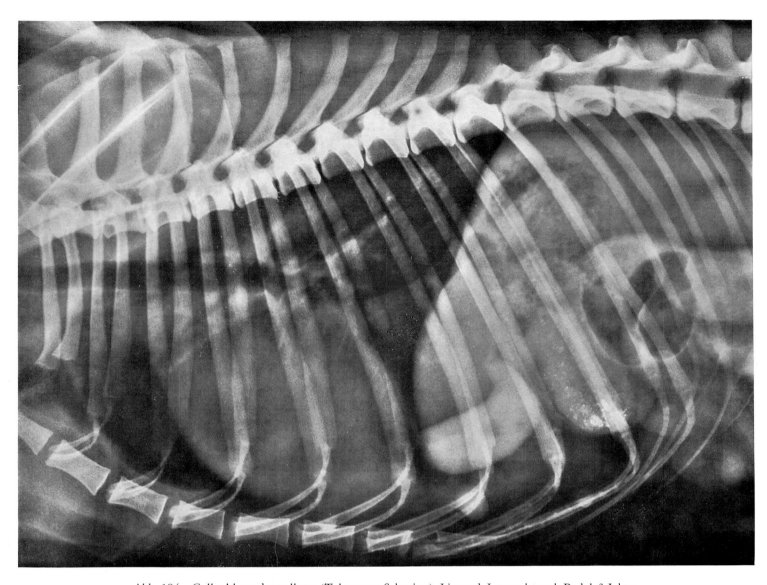

Abb. 184   Gallenblasendarstellung (Telepaque, Schering). Liegend. Latero-lateral. Pudel, 3 Jahre.
Bucky-Blende – Feinzeichnende Folie – FFA 120 cm – 65 kV – 45 mAs
Verkleinerung von 24 × 30 cm
Lagerung Abb. 158

Fig. 184   Gall-bladder (Telepaque, Schering). Recumbent. Laterolateral. Poodle, 3 years old.
Bucky diaphragm – High definition screens – FFD 120 cm – 65 kV – 45 mAs
Diminution of 24 × 30 cm
Positioning fig. 158

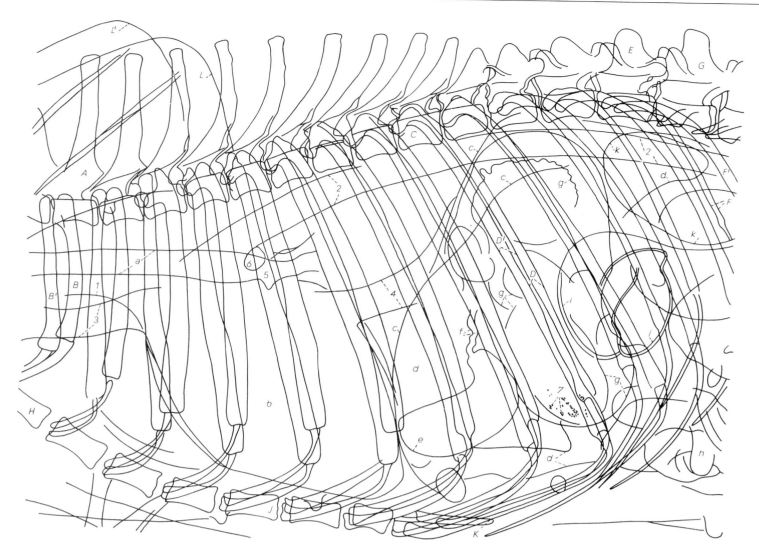

Abb. 185*   Röntgenskizze zu Abb. 184   Fig. 185*   X-ray sketch to fig. 184

A	1. Vertebra thoracica;	e	Vesica fellea;
B	1. Os costale;	f	Ductus cysticus;
C	9. Vertebra thoracica;	g	Ventriculus;
D	9. Os costale;	h	Intestinum tenue;
E	13. Vertebra thoracica;	i	Caecum;
F	13. Os costale;	k	Ren dexter;
G	1. Vertebra lumbalis;		
H	Manubrium sterni;		
J	Corpus sterni, 4. Sternebra;	1	Truncus brachiocephalicus;
K	Processus xiphoideus;	2	Aorta;
L	Scapula;	3	V. cava cranialis;
		4	V. cava caudalis;
a	Trachea;	5	Bifurcatio tracheae;
b	Cor;	6	Bronchus principalis;
c	Diaphragma;	7	Kontrastmittelrest im Magen – Residual contrast medium in stomach.
d	Hepar;		

Hund Dog

Orale Cholezystographie

Vorbereitung: Am Tag vor der Untersuchung ist nur eine schlackenarme breiige Kost anzubieten. Etwa 12 Stunden vor der Untersuchung wird der Füllungszustand des Darmes überprüft (Durchleuchtung, Leeraufnahme). Falls notwendig, Reinigungseinlauf vornehmen und dem Hund ausreichend Bewegung geben. Die Applikation des Kontrastmittels erfolgt per os, danach keine Futter- und Flüssigkeitsaufnahme mehr.

Kontrastmittel: Di- bzw. trijodierte Präparate, deren gebundenes Jod über die Gallenwege ausgeschieden wird. Perorale Applikation12 Stunden vor der Untersuchung.

Untersuchungstechnik:

a) 12 Stunden nach der peroralen Applikation erste Aufnahme, laterolateraler Strahlengang.
Die Gallenblase ist dargestellt, wenn das Kontrastmittel den Magen passieren und im Dünndarm ausreichend resorbiert werden konnte.

b) Reizfütterung, wenn die Gallenblase dargestellt ist. Reizfutter: 2 Eigelb mit 1 Teelöffel Zucker und 1 Teelöffel Speiseöl anrühren. Geeignet ist auch eine handelsübliche Reizmahlzeit in entsprechender Dosierung. Applikation mit der Magensonde, falls das Futter nicht aufgenommen wird.

c) 15 und 30, ggf. auch 60 Minuten nach der Reizfütterung 2. und 3., evtl. auch 4. Aufnahme, latero-lateraler Strahlengang.
Die Aufnahmen geben Aufschluß über die Entleerungsfähigkeit der Gallenblase. Da Kontrastmittelreste an Konkrementen länger haften können, ist es notwendig, 10 – 15 Minuten nach der Entleerung bzw. nach der weitgehenden Entleerung des Kontrastmittels noch eine Aufnahme anzufertigen.

Oral cholecystography

Preparation of the patient: On the day prior to the examination the dog is fed semiliquid food. Approximately 12 hours prior to radiography, the degree of filling of the intestinal tract is checked (fluoroscopy, preliminary radiography). If necessary, the dog is given an evacuation enema and allowed exercise. The contrast medium is administered orally; subsequently the animal is not allowed any food or liquids.

Contrast media: Iodized organic compounds, excreted through the liver and concentrated in the gall-bladder. Oral administration 12 hours prior to examination.

Technique:

a) 12 hours after oral administration of the contrast medium, the first radiograph is taken with laterolateral direction of the beam.
The gall-bladder ist outlined, when the contrast medium has passed the stomach and has been absorbed satisfactorily in the small intestine.

b) Challenge feeding – when the gall-bladder ist outlined. The challenge food: 2 egg yolks mixed with 1 teaspoonful of sugar and 1 teaspoonful of oil. A challenge food available in the trade is also suitable in appropriate doses. If the animal refuses the food it should be administered with a stomach tube.

c) 15, 30, and if necessary 60 minutes after administration of the challenge food, the second, third, and fourth radiographs are taken respectively with laterolateral direction of the beam.
These radiographs furnish information on the emptying ability of the gall-bladder. Since contrast medium residues may adhere longer to the concrements, a radiograph is necessary 10 – 15 minutes after the contrast medium has passed into the small intestine.

Intravenöse Cholezystographie

Indikation: Nach erfolgloser oraler Cholezystographie (Erbrechen, Pylorusspasmus, fehlende oder unzureichende Resorption im Dünndarm) und zur Bestimmung des Beginns und der Dauer der Kontrastmittelausscheidung.

Vorbereitung: 12 Stunden vor der Untersuchung wird der Füllungszustand des Darmes geprüft (Durchleuchtung, Leeraufnahme). Falls notwendig, Reinigungseinlauf vornehmen und dem Hund Bewegungsmöglichkeit geben.

Kontrastmittel: Jodreiche organische Verbindungen in wässeriger Lösung, deren Jod über die Gallenwege ausgeschieden wird. Die Dosis beträgt je nach Gewicht des Hundes 5 – 15 ml, die Applikation der körperwarmen Lösung erfolgt langsam (mindestens 4 – 5 Minuten!) intravenös. Bei zu rascher Injektion ist mit Unverträglichkeitsreaktionen zu rechnen. Das Kontrastmittel wird zum Teil auch über die Nieren ausgeschieden.

Untersuchungstechnik:

a) 1. Aufnahme 45 Minuten post injectionem, latero-lateraler Strahlengang.
Die kontrastmittelhaltigen Gallengänge sind zumindest abschnittsweise, die Gallenblase ist noch nicht, nur wenig oder nur abschnittsweise dargestellt. Man achte auf Stauungen oder Aussparungen!

b) 2. Aufnahme 2 Stunden post injectionem, latero-lateraler Strahlengang. Bei ungestörter Leberfunktion und unbehindertem Gallenabfluß ist die Ausscheidung abgeschlossen, also die Gallenblase deutlich dargestellt.
Stellt sich die Gallenblase nicht dar, sind Untersuchungen (Durchleuchtung, Aufnahme) 3, 6 und ggf. 24 Stunden post injectionem zur Erfassung des evtl. eingedickten und dann nachweisbaren Kontrastmittels in der Gallenblase zweckmäßig.

c) Reizfütterung, wenn die Gallenblase dargestellt ist. Reizfutter: 2 Eigelb mit 1 Teelöffel Zucker und 1 Teelöffel Speiseöl. Geeignet ist auch eine handelsübliche Reizmahlzeit in entsprechender Dosierung. Falls das Futter nicht aufgenommen wird, Applikation mit der Magensonde.

d) 3., 4. und ggf. 5. Aufnahme 15, 20 und ggf. 30 Minuten nach der Reizfütterung, latero-lateraler Strahlengang. Die Aufnahmen geben Aufschluß über die Entleerungsfähigkeit der Gallenblase. Da Kontrastmittelreste an Konkrementen länger haften können, ist es notwendig, 10 – 15 Minuten nach der Entleerung bzw. nach der weitgehenden Entleerung des Kontrastmittels in den Dünndarm noch eine Aufnahme anzufertigen.

Intravenous cholecystography

Indications: Failure of oral cholecystography (vomiting, spasm of the pylorus, insufficient or no resorption in the small intestine); to determine the commencement and the duration of contrast medium elimination.

Preparation of the patient: 12 hours before the examination the degree of filling of the intestine is checked (fluoroscopy, preliminary radiograph). If necessary, an evacuation enema is administered and the animal allowed exercise.

Contrast media: Iodized organic water soluble compounds. They are excreted through the biliary system and to some extent through the kidneys. The dose is 5 – 15 ml, depending on the weight of the dog. The solution must be injected intravenously at body temperature and slowly (4 – 5 minutes!). If injected too rapidly the patient may show symptoms of intolerance.

Technique:

a) First radiograph to be taken 45 minutes after injection with laterolateral direction of the beam. The biliary ducts are reproduced at least partially. The gall-bladder ist not delineated or only slightly or in part. Observe possible accumulations.

b) Second radiograph is taken 2 hours after injection with laterolateral direction of the beam. With normal liver function and flow of gall, the excretion of the contrast medium is completed and the gallbladder is clearly outlined.
If the gall-bladder is not outlined, examination (fluoroscopy, radiograph) are useful at 3, 6, and if necessary 24 hours after injection to locate the possibly thickened and the demonstrable contrast medium in the gall-bladder.

c) Challenge feeding after the gall-bladder is outlined. Challenge food: 2 egg yolks mixed with 1 teaspoonful of sugar and 1 teaspoonful of oil. A challenge food available in the trade is also suitable in appropriate doses. If the animal refuses the food it should be administered with a stomach tube.

d) A third, fourth, and if necessary a fifth radiograph should be taken 15, 20, and 30 minutes respectively after administration of the challenge food with laterolateral direction of the beam. The radiographs furnish information on the emptying ability of the gall-bladder. Since the residues of the contrast medium may adhere longer to the concrements, it is essential to make a radiograph 10 – 15 minutes after passage of the contrast medium into the small intestine.

Pneumoperitonaeum

Vorbereitung: Die Untersuchung sollte nüchtern nach 12stündigem Hungern (nicht Dursten!) erfolgen. Falls im Dickdarm reichlich fester Darminhalt oder eine pralle Füllung der Harnblase nachweisbar ist (Durchleuchtung, Leeraufnahme), ist ein Reinigungseinlauf zu veranlassen bzw. die Harnblase zu entleeren. Widersetzliche Hunde sollten sediert oder, falls es erforderlich ist und es der Zustand des Patienten erlaubt, in Narkose untersucht werden.

Untersuchungstechnik: Rasur und Desinfektion der Haut. Punktion der Bauchhöhle in der Mitte zwischen Nabel und Schambeinrand 1 – 2 fingerbreit neben der Linea alba. Die Luftinsufflation kann mit einer genügend großen Injektionsspritze (100 – 200 ml), mit einem kleinen Blasebalg oder über ein Reduzierventil aus einer Sauerstoff-Flasche unter Zwischenschaltung eines Luftfilters vorgenommen werden. Je nach Größe des Hundes sind 700 – 2000 ml Luft zu instillieren. Eine ausreichende Auffüllung des Abdomens ist am tympanischen Klopfschall zu erkennen. Die Insufflation muß langsam erfolgen und sollte bei irgendwelchen Reaktionen des Hundes (z. B. Schwierigkeit bei der Atmung) sofort gestoppt werden. Die Kanüle ist vor dem Röntgen herauszuziehen.
Lagerung und Anzahl der erforderlichen Aufnahmen hängen von der Fragestellung ab.

Nachbehandlung: Bei richtiger Ausführung wird das Pneumoperitonaeum gut vertragen. Die insufflierte Luft wird in 2 – 3 Wochen resorbiert, so daß der Perkussionsschall in der Regel nach 8 – 10 Tagen nicht mehr tympanisch ist.
Bei medikamentell ruhiggestellten Patienten und bei Hunden, denen reichlich Luft (straffe Spannung der Bauchdecken!) insuffliert wurde, sollte zur Verkürzung der Resorptionszeit das Abdomen nach der Röntgenuntersuchung nochmals punktiert werden, um einen Teil der instillierten Luft abströmen zu lassen.

Pneumoperitoneum

Preparation of the patient: The animal is starved for 12 hours but allowed water. Should the colon or urinary bladder be full (fluoroscopy, preliminary radiograph), the patient is given an enema and the bladder is emptied by catheterization. Restless dogs should be sedated or examined under anesthesia, their condition permitting.

Technique: Shaving and desinfection of the skin. The needle should be inserted into the abdominal cavity midway between the umbilicus and the os pubis, 1 – 2 fingerwidths lateral to the linea alba. Inflation of air can be done with a large syringe (100 – 200 ml), a rubber tube and a tri-valve, or a Baumanometer bulb or any other air insufflator. An oxygen cylinder with a manometer (reduction valve) can also be used. Depending on the size of the dog, 700 – 2000 ml of gas (air, oxygen) will be necessary. Adequate filling of the abdomen can be determined by percussion. The inflation must be performed slowly and should be stopped immediately in case of any untoward reaction of the patient (e.g., respiratory difficulty!). The needle is withdrawn before X-raying.
The positioning of the patient for radiography and the number of the radiographs to be taken depend on the merits of the case.

After care: With proper gas inflation pneumoperitoneum is well tolerated. The instilled air will be resorbed within 2 – 3 weeks. The tympanic percussion sound usually disappears after 8 – 10 days.
In order to shorten the time of resorption, at least part of the air should be withdrawn, especially in anesthetized animals and those with rather distended abdomens.

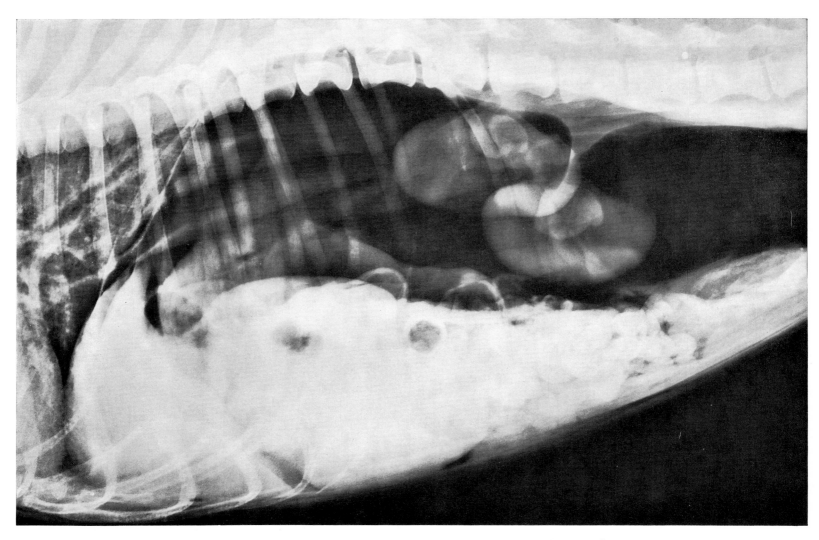

Abb. 186   Pneumoperitonaeum. Stehend. Latero-lateral. Deutscher Schäferhund, ♀, 4 Jahre.
Bucky-Blende – Mittelverstärkende Folie – FFA 120 cm – 65 kV – 18 mAs
Verkleinerung von 30 × 40 cm
Lagerung Abb. 159

Fig. 186   Pneumoperitoneum. Standing. Laterolateral. German shepherd dog, ♀, 4 years old.
Bucky diaphragm – Standard screens – FFD 120 cm – 65 kV – 18 mAs
Diminution of 30 × 40 cm
Positioning fig. 159

Abb. 187* Röntgenskizze zu Abb. 186   Fig. 187* X-ray sketch to fig. 186

A	7. Vertebra thoracica;	o	Ren dexter;
B	12. Vertebra thoracica;	p	Ren sinister;
C	1. Vertebra lumbalis;	q	Vesica urinaria;
D	5. Vertebra lumbalis;	r	Ovarium dextrum et Bursa ovarica;
E	12. Os costale;	s	Ovarium sinistrum et Bursa ovarica;
F	Processus xiphoideus;	t	Cornu uteri dextrum;
		u	Cornu uteri sinistrum;
a	Cor;	v	Corpus uteri;
b	Aorta;	w	Glandula suprarenalis dextra;
c	V. cava caudalis;	x	Glandula suprarenalis sinistra;
d	Bronchus lobaris caudalis;	y	Lendenmuskulatur, ventrale Begrenzung – Ventral limit of the lumbar muscles;
e	Oesophagus;		
f	Diaphragma;	1	A. coeliaca;
g	Crus dextrum;	2	A. mesenterica cranialis;
h	Crus sinistrum;	3	A. mesenterica caudalis;
i	Hepar;	4	Ligamentum triangulare;
k	Ventriculus;	5	Nierengekröse – Renal mesentery;
l	Lien;	6	Kontrastmittelrest von einer Myelographie – Residual contrast medium of myelography.
m	Intestinum;		
n	Colon descendens;		

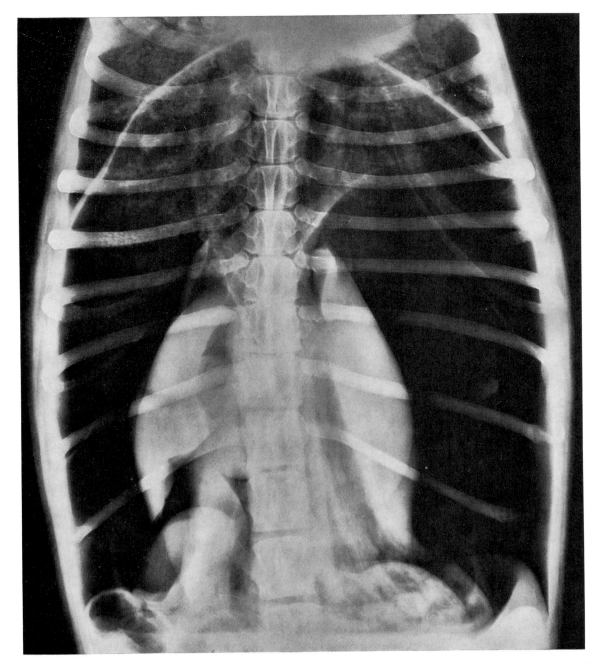

Abb. 188   Pneumoperitonaeum. Auf den Beckengliedmaßen stehend. Ventro-dorsal. Deutscher Schäferhund, 4 Jahre.
Bucky-Blende – Mittelverstärkende Folie – FFA 120 cm – 65 kV – 18 mAs
Verkleinerung von 30 × 40 cm
Lagerung Abb. 161

Fig. 188   Pneumoperitoneum. Standing on hind legs. Ventrodorsal. German shepherd dog, 4 years old.
Bucky diaphragm – Standard screens – FFD 120 cm – 65 kV – 18 mAs
Diminution of 30 × 40 cm
Positioning fig. 161

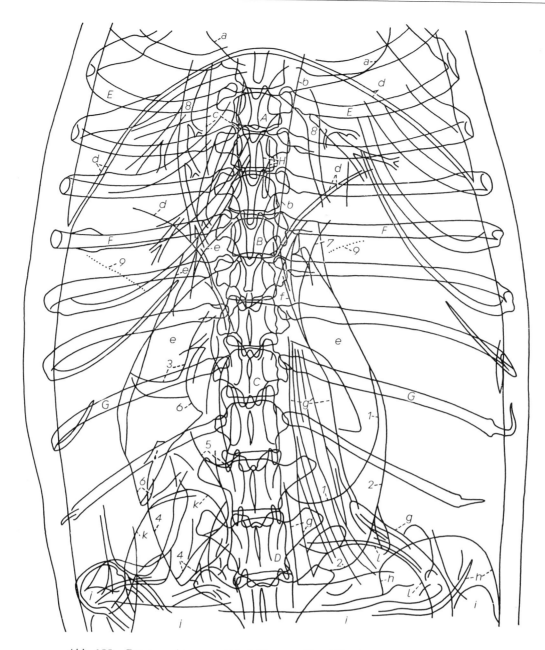

Abb. 189   Röntgenskizze zu Abb. 188   Fig. 189   X-ray sketch to fig. 188

A   6. Vertebra thoracica;
B   9. Vertebra thoracica;
C   12. Vertebra thoracica;
D   2. Vertebra lumbalis;
E   6. Os costale;
F   9. Os costale;
G   12. Os costale;
H   Processus xiphoideus;

a   Cor;
b   Aorta thoracica;
c   V. cava caudalis;
d   Diaphragma;
e   Hepar;
f   Oesophagus;
g   Ventriculus;

h   Lien;
i   Intestinum;
k   Ren dexter;
l   Ren sinister;

1 – 7   Hepar:
1   Lobus hepatis sinister lateralis,
2   Lobus hepatis sinister medialis,
3   Lobus hepatis dexter lateralis,
4   Lobus hepatis dexter medialis,
5   Lobus quadratus,
6   Lobus caudatus,
7   Ligamentum triangulare sinistrum;
8   Bronchus lobaris caudalis;
9   Lobus caudalis, kaudaler Rand — Lobus caudalis, caudal border.

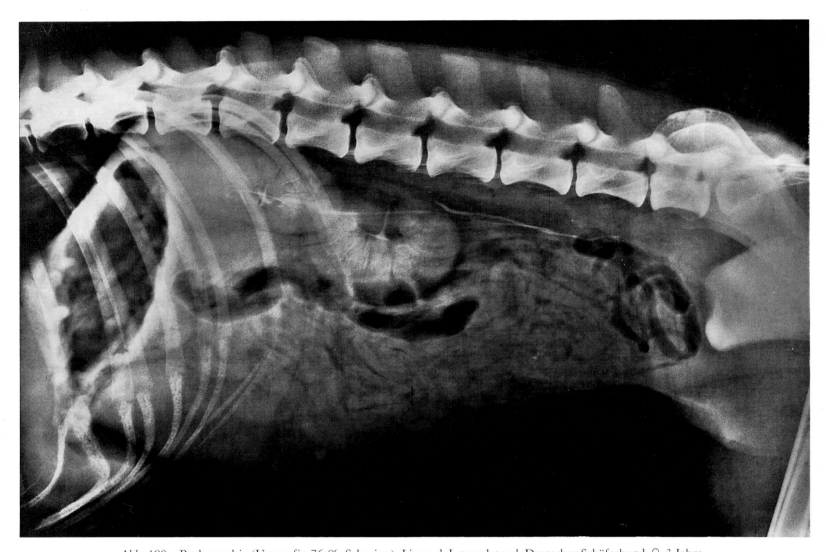

Abb. 190   Pyelographie (Urografin 76 %, Schering). Liegend. Latero-lateral. Deutscher Schäferhund, ♀, 3 Jahre.
Bucky-Blende – Mittelverstärkende Folie – FFA 120 – 65 kV – 24 mAs
Verkleinerung von 24 × 30 cm
Lagerung Abb. 158

Fig. 190   Pyelography (Urografin 76 %, Schering). Recumbent. Laterolateral. German shepherd dog, ♀, 3 years old.
Bucky diaphragm – Standard screens – FFD 120 cm – 65 kV – 24 mAs
Diminution of 24 × 30 cm
Positioning fig. 158

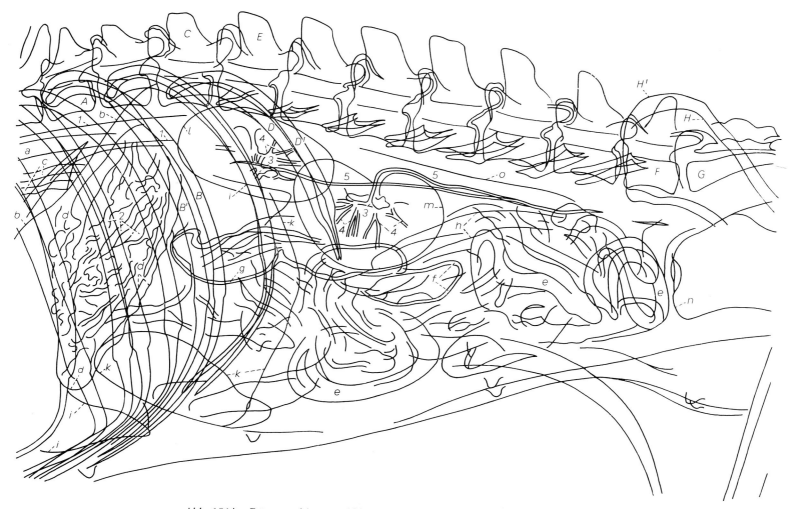

Abb. 191*   Röntgenskizze zu Abb. 190   Fig. 191*   X-ray sketch to fig. 190

A	11. Vertebra thoracica;		g	Colon ascendens;
B	1. Os costale;		h	Colon descendens;
C	13. Vertebra thoracica;		i	Hepar;
D	13. Os costale;		k	Lien;
E	1. Vertebra lumbalis;		l	Ren dexter;
F	7. Vertebra lumbalis;		m	Ren sinister;
G	Os sacrum;		n	Vesica urinaria;
H	Ala ossis ilii;		o	Lendenmuskulatur, ventrale Begrenzung – Ventral limit of the lumbar muscles;
a	Pulmo;			
b	Diaphragma;		1	Aorta;
c	Oesophagus;		2	Plicae villosae;
d	Ventriculus;		3	Pelvis renalis;
e	Intestinum tenue;		4	Recessus pelvis;
f	Caecum;		5	Ureter.

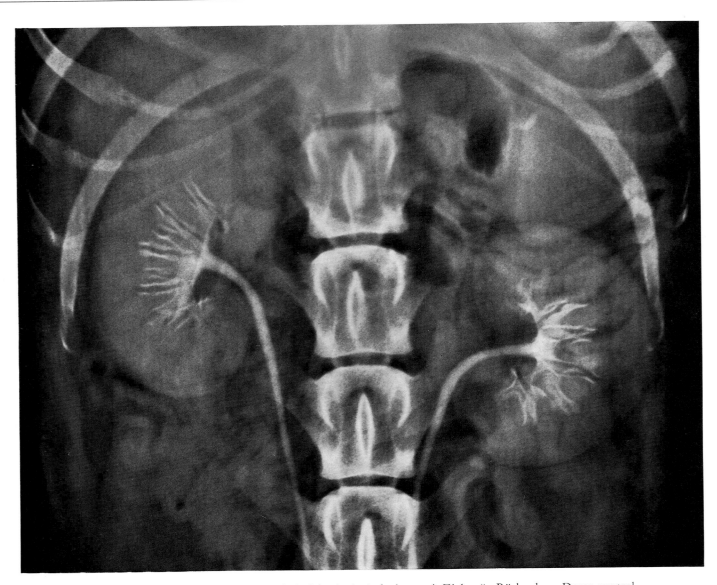

Abb. 192   Pyelographie (Urografin 76 %, Schering). Aufnahme mit Zielgerät. Rückenlage. Dorso-ventral.
Deutscher Schäferhund, 3 Jahre.
Bucky-Blende – Feinzeichnende Folie – FFA 60 cm – 65 kV – 30 mAs
Verkleinerung von 24 × 30 cm

Fig. 192   Pyelography (Urografin 76 %, Schering). Radiograph with spotfilm device. Dorsal recumbency. Dorsoventral.
German shepherd dog, 3 years old.
Bucky diaphragm – High definition screens – FFD 60 cm – 65 kV – 30 mAs
Diminution of 24 × 30 cm

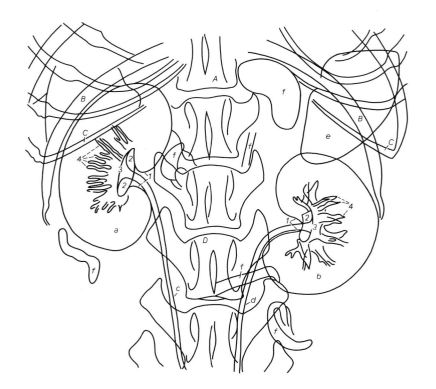

Abb. 193   Röntgenskizze zu Abb. 192

Fig. 193   X-ray sketch to fig. 192

A	13. Vertebra thoracica;
B	13. Os costale;
C	12. Cartilago costalis;
D	3. Vertebra lumbalis;

a	Ren dexter;
b	Ren sinister;
c	Ureter dexter;
d	Ureter sinister;
e	Lien;
f	Gashaltige Darmschlingen – Gas containing intestinal loops;

1	Hilus renalis, bei b nur abschnittsweise dargestellt – Hilus renalis, only partly shown at b;
2	Sinus renalis;
3	Pelvis renalis;
4	Recessus pelvis.

Ausscheidungsurographie

Vorbereitung: Es ist zweckmäßig, den Patienten 12 Stunden vor der Untersuchung hungern (nicht dursten!) zu lassen und den Füllungszustand des Darmes vor der Nierenuntersuchung zu überprüfen (Durchleuchtung oder Leeraufnahme). Wenn die Vorbereitung nicht möglich oder nicht ausreichend ist, werden ein Reinigungseinlauf ggf. unter Beigabe eines Kontaktlaxans vorgenommen und dem Hund ausreichend Bewegungsmöglichkeit zum Kot- und Harnabsatz gegeben.

Kontrastmittel: Gut verträglich und kontrastreich sind trijodierte wasserlösliche Präparate. Die Dosis beträgt je nach Größe des Hundes 5 - 15 ml. Die Applikation der körperwarmen Lösung erfolgt intravenös in 2 - 3 Minuten.

Untersuchungstechnik:

a)   3 Minuten post injectionem 1. Aufnahme, ventro-dorsaler Strahlengang.
Die Aufnahme ist zur Beurteilung der Nierenfunktion und der Durchgängigkeit der Harnleiter zweckmäßig.

b)   Kompression der Harnleiter durch Fixation einer Zellstoffrolle, eines Sandsacks oder ähnlichem mit Hilfe eines Bandkompressoriums.

c)   5 und 10 Minuten nach Anlegen der Ureterenkompression 2. und 3. Aufnahme, ventro-dorsaler Strahlengang.
Die Aufnahmen sollen eine einwandfreie Nierenbeckendarstellung ergeben.

d)   Unmittelbar nach Abnahme des Kompressoriums 4. Aufnahme, ventro-dorsaler Strahlengang.
Da der gestaute kontrasthaltige Harn abfließt, müssen die Harnleiter in ganzer Länge, zumindest aber in einem großen Abschnitt dargestellt sein.

e)   Umlagerung des Patienten in rechte Seitenlage und 5. Aufnahme, latero-lateraler Strahlengang.
Die Aufnahme ist zur Beurteilung der Harnblase und der Nieren erforderlich.

Das Kontrastmittel wird in etwa 30 Minuten ausgeschieden. Bei einer Störung der Nierenfunktion oder des Harntransports verzögert sich die Ausscheidung.

Excretory urography

Preparation of the patient: A full intestine could cause troublesome overlaying. Therefore it is necessary to withhold food, but not water, for 12 hours prior to the examination and check the condition of the intestine (fluoroscopy, preliminary radiograph). Should preliminary preparation not be possible, it is advisable to administer an evacuation enema and, if necessary, apply a contact laxative prior to the examination. The animal should be allowed freedom to defecate and urinate.

Contrast media: Well tolerated and rich in contrast are tri-iodized, water soluble, organic compounds. The dose ist 5 - 15 ml, depending on the size of a dog. The medium should be injected intravenously within 2 - 3 minutes at body temperature.

Technique:

a)   First radiograph to be taken 3 minutes after injection with ventrodorsal direction of the beam. This radiograph is suitable for the study of kidney function and the patency of the ureters.

b)   Compression of the ureters by means of a compression block.

c)   5 - 10 minutes after applying the compression block, the second and third radiographs are taken with ventrodorsal direction of the beam. These radiographs should produce a diagnostic pyelogram.

d)   Immediately after removal of the compression block, the fourth radiograph is taken with ventrodorsal direction of the beam.
Since the previously retained (compression!) urine now flows freely, both ureters or at least the greater part of them should be outlined.

e)   Rotating of the patient into the right lateral position. Fifth radiograph is taken with laterolateral direction of the beam. This radiograph is necessary for radiologic evaluation of the urinary bladder and kidneys.

The contrast medium is normally excreted in approximately 30 minutes. In case of disturbed kidney function of urine flow the excretion of the contrast medium is slowed down.

Retrograde Zystographie

Vorbereitung: Eine Ruhigstellung (Anästhesie) ist notwendig. Es sollten ein Reinigungseinlauf vorgenommen und die Harnblase entleert werden.

Kontrastmittel: Luft für die Pneumozystographie und einen Doppelkontrast; diatrizoierte Kontrastmittel für die einfache Zystographie oder den Doppelkontrast.

Dosis: Zu beachten ist eine ausreichende Dehnung der Harnblase; dafür werden 5 ml Flüssigkeit pro kg KGW empfohlen; für ein Pneumozystogramm oder einen Doppelkontrast sind 10 - 150 ml Luft erforderlich. Bei Doppelkontrast ist das flüssige Kontrastmittel zuerst zu applizieren; die Harnblase sollte massiert werden, um das Kontrastmittel in der Harnblase zu verteilen; überschüssiges Kontrastmittel ist so vollständig wie möglich zu entfernen. Wenn die Luft eingeblasen ist, werden der Katheter herausgezogen und 2 Röntgenaufnahmen gemacht, eine mit latero-lateralem Strahlengang und eine mit ventro-dorsalem Strahlengang.

Retrograde Cystography

Preparation of the patient: Sedation is necessary, so are cleansing enema and emptying of the urinary bladder.

Contrast media: Air for pneumocystography and double contrast; diatrizoate contrast media for simple cystography or double contrast.

Dosis: Care must be taken to distend the urinary bladder moderately, this will suggest a dosis of about 5 ml of liquid per kg of body weight. For pneumocystogram or double contrast 10 - 150 ml of air may be required. In double contrast, the liquid contrast medium is instilled first, the bladder massaged in order to spread the medium all over. Afterwards medium is to remove as completely as possible. Air is insufflated, catheter withdrawn, and 2 radiographs are taken: one in laterolateral direction, one in ventrodorsal direction of the beam.

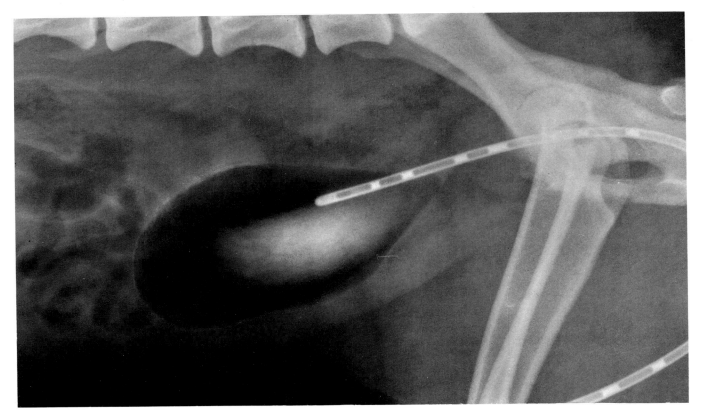

Abb. 194   Harnblase. Negativ- und Positiv-Kontrast (Luft und Kaliumjodid, 10 %ig). Liegend. Latero-lateral.
Spaniel, ♂, 4 Jahre.
Bucky-Blende – Feinzeichnende Folie – FFA 120 cm – 66 kV – 50 mAs
Originalgröße (Ausschnitt aus 24 × 30 cm)
Lagerung Abb. 158

Fig. 194   Urinary bladder. Negative and positive contrast (air and 10 % potassium iodide). Recumbent. Laterolateral.
Spaniel, ♂, 4 years old.
Bucky diaphragm – High definition screens – FFD 120 cm – 66 kV – 50 mAs
Original size (section of 24 × 30 cm)
Positioning fig. 158

A	5. Vertebra lumbalis;
B	7. Vertebra lumbalis;
C	Os sacrum;
D	Os ilium;
E	Os pubis;
F	Os ischii;
G	Os femoris;
a	Caecum;
b	Colon descendens;
c	Jejunumschlingen – Intestinal loops;
d	Vesica urinaria;
e	Urethra mit Katheter – Urethra with catheter;
f	Kontur der Muskulatur der Stammzone – Outline of the musculature of the stem zona;
g	Ventrale Bauchwand – Ventral abdominal wall;
h	Kraniale Kontur des Oberschenkels – Cranial outline of the thigh;
i	Kniefalte – Fold of the flank;
k	Articulatio coxae;
1	Blasenwand – Wall of the urinary bladder;
2	Vertex vesicae;
3	Collum vesicae.

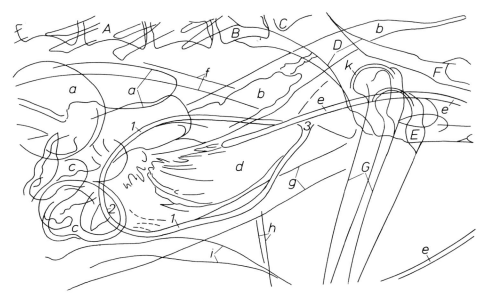

Abb. 195   Röntgenskizze zu Abb. 194

Fig. 195   X-ray sketch to fig. 194

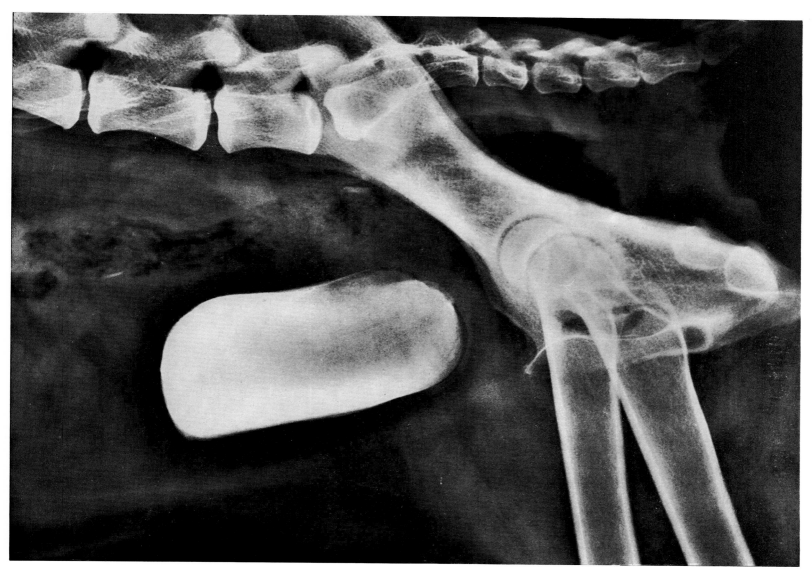

Abb. 196 Harnblase. Positiv-Kontrast (Kaliumjodid, 10 %ig).
Liegend. Latero-lateral. Deutscher Schäferhund, ♀, 8 Jahre.
Bucky-Blende – Feinzeichnende Folie – FFA 120 cm – 74 kV – 45 mAs
Verkleinerung von 24 × 30 cm
Lagerung Abb. 158

Fig. 196 Urinary bladder. Positive contrast (10 % potassium iodide).
Recumbent. Laterolateral. German shepherd dog, ♀, 8 years old.
Bucky diaphragm – High definition screens – FFD 120 cm – 74 kV – 45 mAs
Diminution of 24 × 30 cm
Positioning fig. 158

A   6. Vertebra lumbalis;
B   Os sacrum;
C   3. Vertebra caudalis;
D   Os ilium;
E   Os femoris;

a   Vesica urinaria, mit Kontrastmittel gefüllt –
     Vesica urinaria filled with contrast medium;
a″   Cervix vesicae;
b   Colon descendens mit Darminhalt –
     Colon descendens with contents;
c   Ampulla recti, gashaltig – Ampulla recti,
     filled with gas;
d   Canalis analis;
e   Jejunumschlingen – Intestinal loops;
f   Ventrale Bauchwand – Ventral abdominal wall;

1   Harnblasenwand – Wall of the urinary bladder;
2   Harnblasenschleimhaut – Mucous membrane
     of the urinary bladder;
3   Darmwand – Wall of the intestine.

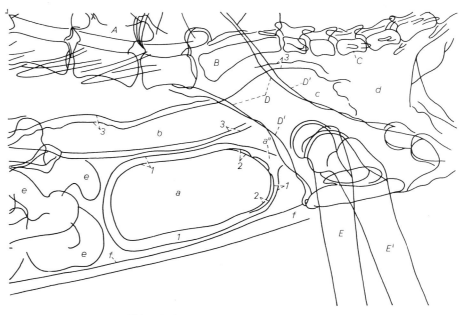

Abb. 197* Röntgenskizze zu Abb. 196
Fig. 197* X-ray sketch to fig. 196

139

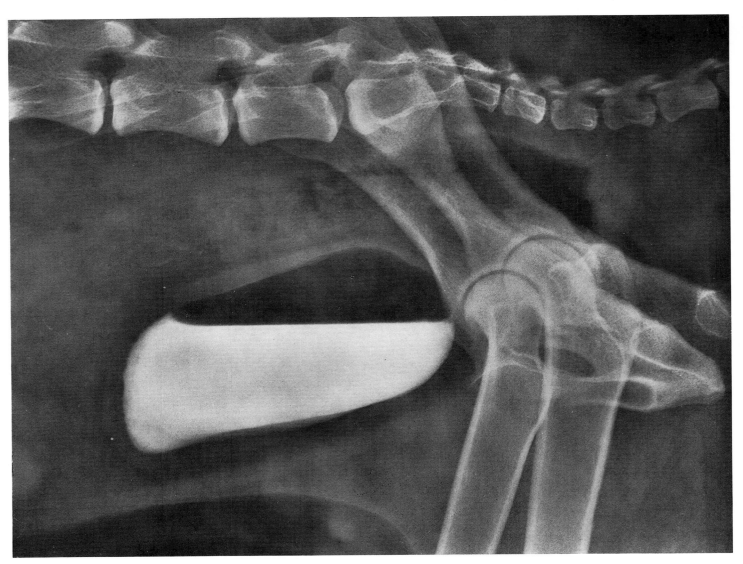

Abb. 198   Harnblase. Positiv- und Negativ-Kontrast (Kaliumjodid,
10 %ig, und Luft). Stehend. Latero-lateral.
Deutscher Schäferhund, ♀, 8 Jahre.
Bucky-Blende – Feinzeichnende Folie – FFA 120 cm – 74 kV – 45 mAs
Verkleinerung von 24 × 30 cm
Lagerung Abb. 159

Fig. 198   Urinary bladder. Positive and negative contrast
(10 % potassium iodide and air). Standing. Laterolateral.
German shepherd dog, ♀, 8 years old.
Bucky diaphragm – High definition screens – FFD 120 cm – 74 kV – 45 mAs
Diminution of 24 × 30 cm
Positioning fig. 159

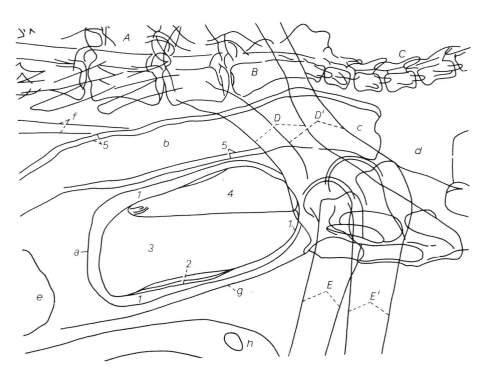

Abb. 199*   Röntgenskizze zu Abb. 198
Fig. 199*   X-ray sketch to fig. 198

A   6. Vertebra lumbalis;
B   Os sacrum;
C   3.Vertebra caudalis;
D   Os ilium;
E   Os femoris;

a   Vesica urinaria;
b   Colon descendens mit Darminhalt –
     Colon descendens with contents;
c   Rectum, gashaltig – Rectum, filled with gas;
d   Canalis analis;
e   Jejunumschlingen – Intestinal loops;
f   Lendenmuskulatur, Konturen –
     Contours of the lumbar muscles;
g   Ventrale Bauchwand – Ventral abdominal wall;
h   Papilla mammae;

1   Harnblasenwand, beachte unterschiedliche Wandstärke –
     Wall of the urinary bladder, notice varying thickness
     of the wall;
2   Harnblasenschleimhaut – Mucous membrane
     of the urinary bladder;
3   Kontrastmittel – Contrast medium;
4   Luft – Air;
5   Darmwand – Wall of the intestine.

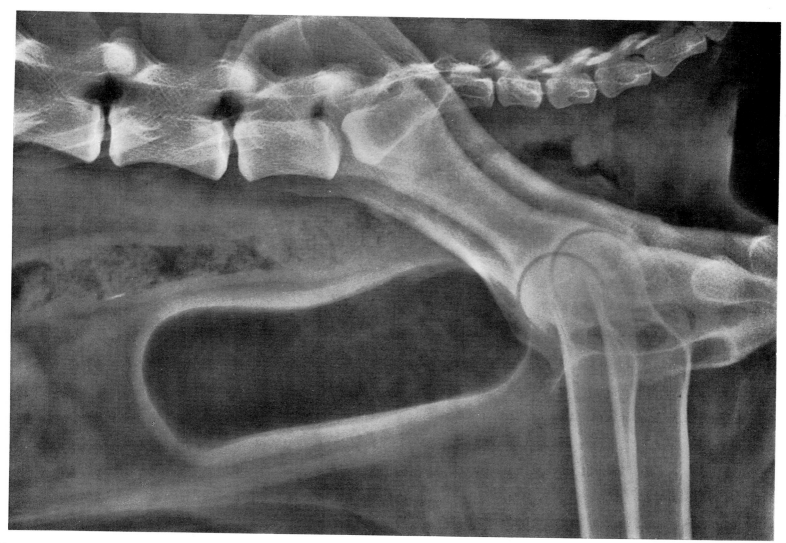

Abb. 200 Harnblase. Negativ-Kontrast (Luft). Liegend. Latero-lateral.
Deutscher Schäferhund, ♀, 8 Jahre.
Bucky-Blende – Feinzeichnende Folie – FFA 120 cm – 70 kV – 36 mAs
Verkleinerung von 24 × 30 cm
Lagerung Abb. 158

Fig. 200 Urinary bladder. Negative contrast (air). Recumbent.
Laterolateral. German shepherd dog, ♀, 8 years old.
Bucky diaphragm – High definition screens – FFD 120 cm – 70 kV – 36 mAs
Diminution of 24 × 30 cm
Positioning fig. 158

A   6. Vertebra lumbalis;
B   Os sacrum;
C   3. Vertebra caudalis;
D   Os ilium;
E   Os femoris;

a   Vesica urinaria;
b   Colon descendens mit Darminhalt –
    Colon descendens with contents;
c   Rectum, wenig gashaltig – Rectum,
    slightly filled with gas;
d   Canalis analis;
e   Jejunumschlingen – Intestinal loops;
f   Fornix vaginae;
g   Cervix uteri;
h   Corpus uteri;
i   Cornua uteri;
k   Ventrale Bauchwand – Ventral abdominal wall;
l   Subperitonäales Fettgewebe – Subperitoneal
    adipose tissue;

1   Harnblasenwand – Wall of the urinary bladder;
2   Scheidenwand – Wall of the vagina;
3   Darmwand – Wall of the intestine;

Punktierte Linien = Verunreinigung auf dem Bucky-Tisch –
Dotted lines represent dirt on the X-ray table.

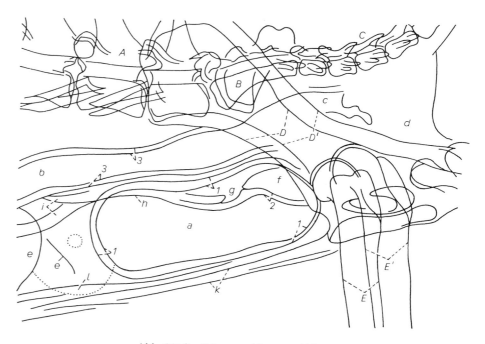

Abb. 201* Röntgenskizze zu Abb. 200
Fig. 201* X-ray sketch to fig. 200

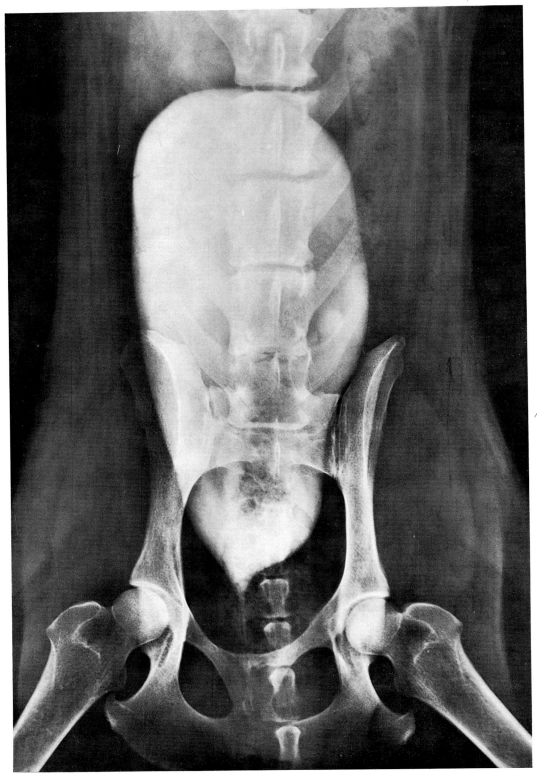

Abb. 203   Röntgenskizze zu Abb. 202
Fig. 203   X-ray sketch to fig. 202

Abb. 202   Harnblase. Positiv-Kontrast (Ausscheidungsurographie, Urografin 76 %, Schering).
Ventro-dorsal. Deutscher Schäferhund, ♀, 8 Jahre.
Bucky-Blende – Feinzeichnende Folie – FFA 120 cm – 74 kV – 45 mAs
Verkleinerung von 24 × 30 cm
Lagerung Abb. 160

Fig. 202   Urinary bladder. Positive contrast (excretory urography, 76 % Urografin, Schering).
Ventrodorsal. German shepherd dog, ♀, 8 years old.
Bucky diaphragm – High definition screens – FFD 120 cm – 74 kV – 45 mAs
Diminution of 24 × 30 cm
Positioning fig. 160

A   6. Vertebra lumbalis;
B   Os sacrum;
C   3. Vertebra caudalis;
D   Os ilium;
E   Os pubis;
F   Os ischii;
G   Os femoris;

a   Vesica urinaria;
b   Urethra;
c   Ampulla recti;
d   Colon descendens;
e   Peritonäalhöhle, Begrenzung – Border of the peritoneal cavity;
f   Papilla mammae;

1   Darmwand – Wall of the intestine.

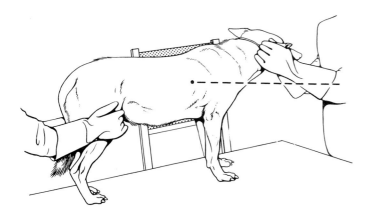

Abb. 204 Lagerung zur Aufnahme des Brustkorbs. Stehend. Latero-lateral.

Fig. 204 Positioning of thorax. Standing. Laterolateral.

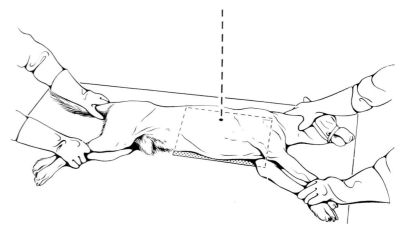

Abb. 205 Lagerung zur Aufnahme des Brustkorbs. Liegend. Latero-lateral.

Fig. 205 Positioning of thorax. Recumbent. Laterolateral.

Um miliare Herde und feine Strukturen nachweisen zu können, ist eine feinzeichnende Folie zu verwenden. Daraus resultiert eine längere Belichtungszeit und damit die Möglichkeit einer Unschärfe durch die Atembewegung. Falls die zur Vermeidung der Bewegungsunschärfe erforderliche kurze Belichtungszeit wegen der Leistung des Röntgengeräts nicht erreicht werden kann, sollte die Atembewegung durch kurzes Zuhalten der Nase, möglichst am Ende der Inspiration ausgeschaltet werden.
Zur Verringerung der Streustrahlung, die nur bei kleinen, nicht fettleibigen Hunden (Rehpinscher, Zwergpudel) unbedeutend ist, sollte eine Thoraxaufnahme mit einer Bucky-Blende angefertigt werden.
Die am stehenden Hund mit latero-lateralem Strahlengang (Abb. 204) angefertigte Aufnahme ist nützlich, aber die Lagerung nicht selten schwierig. Deshalb werden – von Ausnahmen abgesehen – die Aufnahmen mit latero-lateralem Strahlengang am in Seitenlage befindlichen Hund (Abb. 205) angefertigt.
Zur Beurteilung des Herzens und bei der Routineuntersuchung sollte die Aufnahme am in linker Seitenlage befindlichen Hund, d. h. links anliegender Kassette, angefertigt werden. Werden feine strukturelle Veränderungen in der linken Lungenhälfte vermutet, ist die Aufnahme mit latero-lateralem Strahlengang am in rechter Seitenlage befindlichen Patienten anzufertigen.
Es ist darauf zu achten, daß die Medianebene parallel zur Kassette liegt. Bei fettleibigen Hunden läßt sich die korrekte Lagerung in Seitenlage mit Hilfe von Schaumgummikeilen, die in Höhe der Hände unter die Gliedmaßen zu legen sind, erleichtern.
Der Zentralstrahl sollte die 5. Rippe in halber Höhe treffen und im rechten Winkel auf die Kassette einfallen.
Ventro-dorsaler Strahlengang (Abb. 206). Der Hund ist so zu lagern, daß die Medianebene senkrecht zur Kassette steht, also das Sternum senkrecht über den Brustwirbeln liegt.
Bei der in dieser Lagerung angefertigten Aufnahme erscheint das Lungenfeld größer, so daß feinstrukturelle Veränderungen im Bereich des Anhangslappens und im Bereich des kaudalen Abschnitts des Mediastinums besser zu erkennen sind.
Der Zentralstrahl sollte das Brustbein in der Mittellinie in Höhe des 6. Brustwirbels treffen und im rechten Winkel auf die Kassette einfallen.

Dorso-ventraler Strahlengang (Abb. 207). Der Hund ist so zu lagern, daß die Medianebene senkrecht zur Kassette steht, also die Brustwirbel senkrecht über dem Sternum liegen. Um eine stabile Lagerung zu erreichen, sollte sich der Hund beiderseits auf dem etwas abduzierten Ellbogen stützen. Die Beckengliedmaßen sollten abduziert und abgebeugt nur wenig nach kaudal gezogen fixiert werden.

Auf der in dieser Lagerung angefertigten Aufnahme sind die Gefäße der Zwerchfellappen besser zu erkennen und die Lagerung zur Beurteilung des Herzens günstiger.
Der Zentralstrahl sollte in der Mittellinie den 6. Brustwirbel treffen und im rechten Winkel auf die Kassette einfallen.

High definition screens must be used if miliary foci and fine structures are to be depicted. With longer exposures fogging through respiratory movements could result. If the capacity of the X-ray machine does not allow sufficient shortening of the exposure time to avoid fogging, the respiratory movement should be eliminated by briefly closing to the nostrils at the end of inspiration.
To reduce scattered radiation, insignificant in small lean dogs (toy terrier, toy poodle) only, the radiography should be taken with a Bucky diaphragm.
Taking a thoracic radiograph in standing laterolateral position (fig. 204) is often desirable, but also often difficult. Therefore a recumbent laterolateral positioning (fig. 205) is preferred, with exceptions.
For a standard thoracic radiography, including heart examination, the patient should be placed in left lateral position. A radiograph in right lateral recumbency is to take, if fine structural changes in the left lung are suspected.
The median plane should be parallel to the cassette. In obese animals, the proper alignment in a lateral position can be facilitated with the help of foam rubber pads placed under the extremities resting on the table.
The central beam should strike the 5th rib in the middle of its length and fall at the right angle on the cassette.
Ventrodorsal positioning (fig. 206). The sagittal plane of the dog must lie perpendicularly to the cassette and the sternum must rest exactly above and perpendicularly to the thoracic spine. In radiographs taken in this position, the lungs appear larger, this facilitating identification of fine structural changes in caudal mediastinum and accessory lobe.
The central beam should strike the sternum in the midline above the 6th thoracic vertebra and fall at the right angle on the cassette.
Dorsoventral positioning (fig. 207). The dog is be positioned with its sagittal plane perpendicular to the cassette, and thoracic spine situated exactly above the sternum. For better balance, the thorax should be supported by slightly abducted elbows of the patient, resting on the table. Hind limbs should be moderately abducted, flexed and held slightly stretched caudally. Radiographs taken in this position allow better visualization of the diaphragmatic lobes and heart.
The central beam should pass through the midpoint of the 6th thoracic limb and fall at the right angle on the cassette.

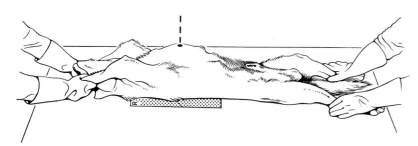

Abb. 206 Lagerung zur Aufnahme des Brustkorbs. Ventro-dorsal.

Fig. 206 Positioning of thorax. Ventrodorsal.

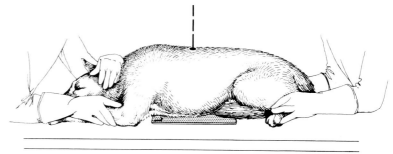

Abb. 207 Lagerung zur Aufnahme des Brustkorbs. Dorso-ventral.

Fig. 207 Positioning of thorax. Dorsoventral.

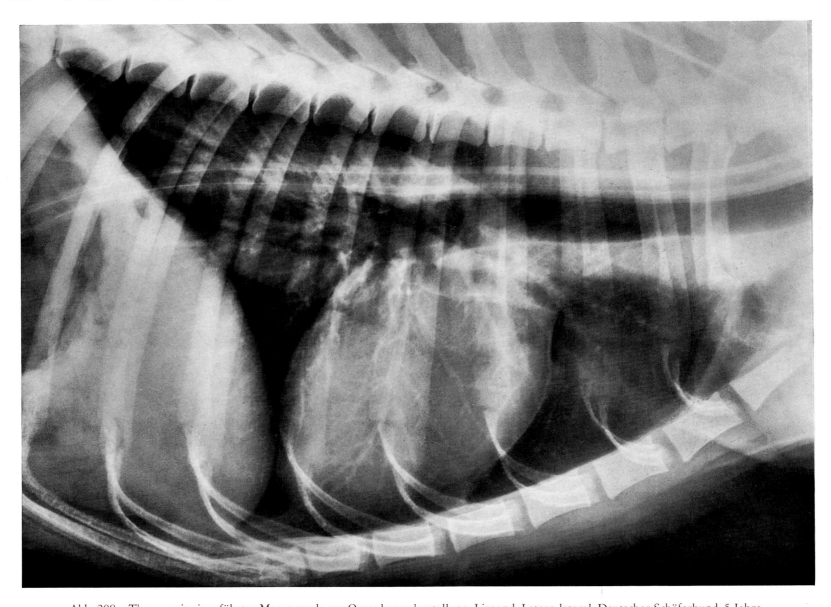

Abb. 208   Thorax mit eingeführter Magensonde zur Oesophagusdarstellung. Liegend. Latero-lateral. Deutscher Schäferhund, 5 Jahre.
Bucky-Blende – Feinzeichnende Folie – FFA 120 cm – 65 kV – 22 mAs
Verkleinerung von 30 × 40 cm
Lagerung Abb. 205

Fig. 208   Thorax with stomach tube introduced into esophagus. Recumbent. Laterolateral. German shepherd dog, 5 years old.
Bucky diaphragm – High definition screens – FFD 120 cm – 65 kV – 22 mAs
Diminution of 30 × 40 cm
Positioning fig. 205

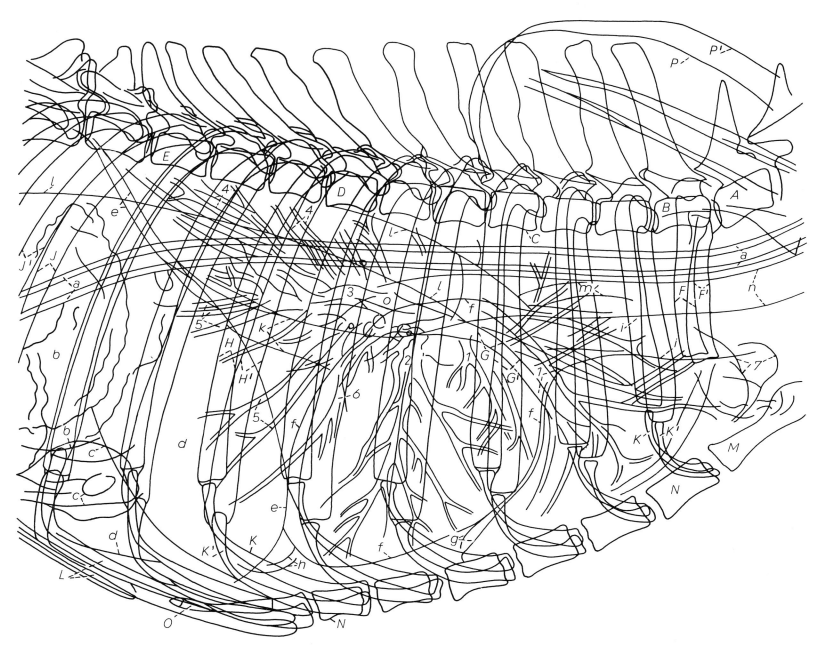

Abb. 209* Röntgenskizze zu Abb. 208   Fig. 209 * X-ray sketch to fig. 208

A   7. Vertebra cervicalis;
B   1. Vertebra thoracica;
C   4. Vertebra thoracica;
D   7. Vertebra thoracica;
E   10. Vertebra thoracica;
F   1. Os costale;
G   4. Os costale;
H   7. Os costale;
J   10. Os costale;
K   Cartilago costalis;
L   Arcus costalis;
M   Manubrium sterni;
N   Corpus sterni, 1. et 6. Sternebra;
O   Processus xiphoideus;
P   Scapula;

a   Magensonde im Oesophagus – Stomach tube in esophagus;
b   Ventriculus mit Gasblase im Fundus-Corpusbereich – Stomach with gas bubble in the fundus-body region;
c   Intestinum tenue;

d   Hepar;
e   Diaphragma;
f   Cor;
g   Pericardium;
h   Ligamentum phrenicopericardiacum;
i   V. cava cranialis;
k   V. cava caudalis;
l   Aorta thoracica;
m   A. subclavia sinistra;
n   Trachea;
o   Bifurcatio tracheae;

1 – 6   Aa. pulmonales:
1   Rami loborum cranialium dextri et sinistri,
2   Ramus lobi medii,
3   Rami loborum caudalium dextri et sinistri,
4   Rami dorsales von 3 – Dorsal branches of 3,
5   Rami ventrales von 3 – Ventral branches of 3,
6   Ramus lobi accessorii;
7   Cupula pleurae.

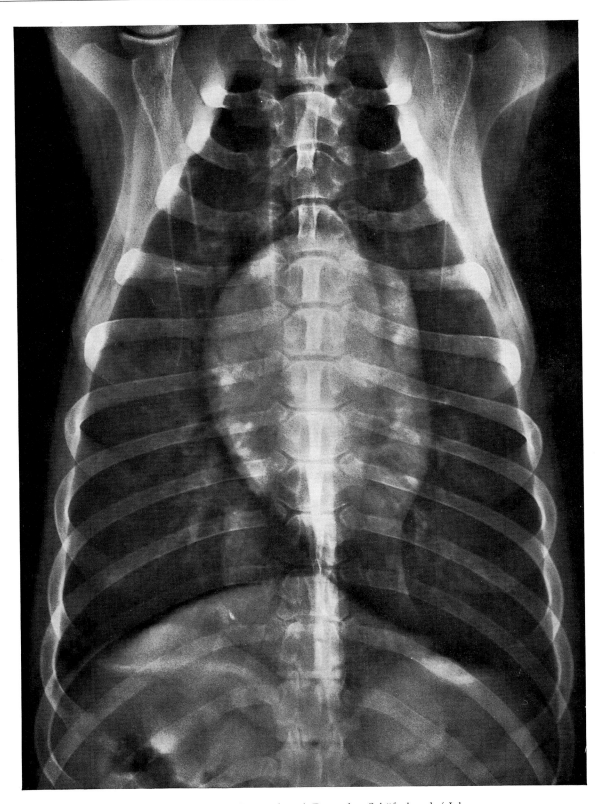

Abb. 210   Thorax. Ventro-dorsal. Deutscher Schäferhund, 4 Jahre.
Bucky-Blende – Feinzeichnende Folie – FFA 120 cm – 75 kV – 30 mAs
Verkleinerung von 30 × 40 cm
Lagerung Abb. 206

Fig. 210   Thorax. Ventrodorsal. German shepherd dog, 4 years old.
Bucky diaphragm – High definition screens – FFD 120 cm – 75 kV – 30 mAs
Diminution of 30 × 40 cm
Positioning fig. 206

A 7. Vertebra cervicalis;
B 1. Vertebra thoracica;
C 4. Vertebra thoracica;
D 7. Vertebra thoracica;
E 10. Vertebra thoracica;
F 13. Vertebra thoracica;
G 1. Os costale;
H 4. Os costale;
J 7. Os costale;
K 10. Os costale;
L 13. Os costale;
M Corpus sterni, Sternebrae;
N Scapula;
O Caput humeri;

a Trachea, punktiert – Trachea, dotted;
b Cor;
c Aorta thoracica;
d V. cava cranialis;
e V. cava caudalis;
f Oesophagus;
g Ligamentum phrenicopericardiacum;
h Pulmo dexter;
i Pulmo sinister mit Bronchenschatten, die in der Skizze nicht eingezeichnet sind – Pulmo sinister; shadows of the bronchi not illustrated;
k Cupula pleurae;
l Diaphragma;
m Hepar;
n Ventriculus;

An der 7. Vertebra cervicalis bis zur 13. Vertebra thoracica – On the 7th cervical to the 13th thoracic vertebrae:

1 Extremitas cranialis;
2 Extremitas caudalis;
3, 4 Arcus vertebrae:
3 Kranialer Rand – Cranial border,
4 Kaudaler Rand – Caudal border;
5 Processus articularis cranialis;
6 Processus articularis caudalis;
7 Processus spinosus;
8 Processus transversus;
9 Pediculus arcus vertebrae, zugleich Begrenzung des Foramen vertebrale – Pediculus arcus vertebrae forming the lateral margin of the vertebral foramen;
10 Articulatio costovertebralis: Foveae costales cranialis et caudalis;
11 Articulatio costotransversaria: Fovea costalis transversalis;
12 Processus accessorius;

An den Rippen – On the ribs:

13 Os costale;
14 Caput costae;
15 Articulatio costotransversaria: Tuberculum costae et Facies articularis;
16 Cartilago costalis;

An der Scapula – On the scapula:

17 Spina scapulae;
18 Basis spinae scapulae;
19 Acromion;
20 Orthograph getroffener Abschnitt im Bereich der Fossa supraspinata – Orthographically struck part in the region of the supraspinous fossa;
21 Margo dorsalis;
22 Margo caudalis;
23 Margo cranialis;
24 Cavitas glenoidalis.

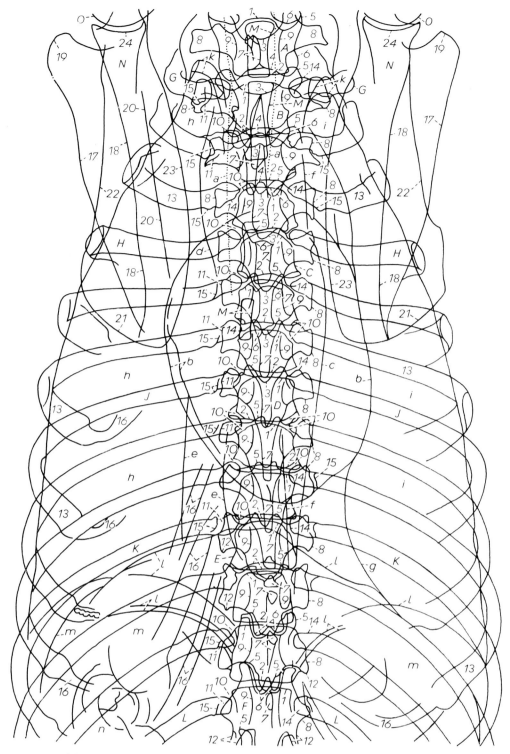

Abb. 211 Röntgenskizze zu Abb. 210   Fig. 211 X-ray sketch to fig. 210

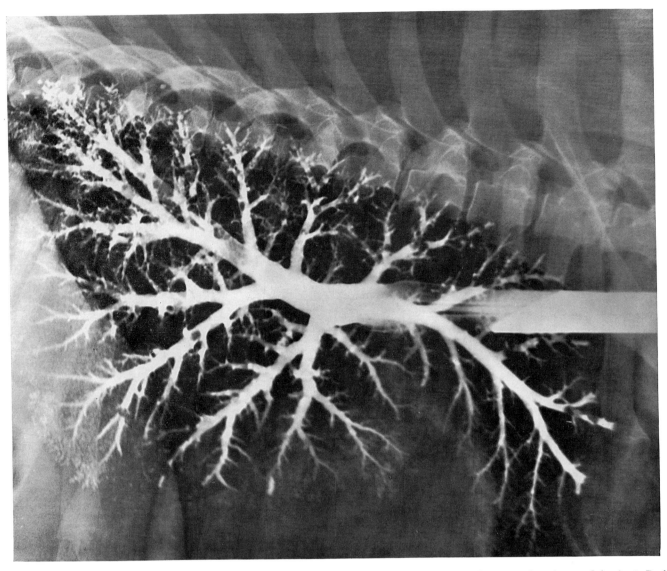

<table>
<tr>
<td>

Abb. 212   Bronchographie (Bronchoselectan, Schering). Aufnahme mit
Zielgerät. Latero-lateral. Deutscher Schäferhund, 4 Jahre.
Feinzeichnende Folie – FFA 60 cm – 65 kV – 15 mAs
Verkleinerung von 24 × 30 cm

</td>
<td>

Fig. 212   Bronchography (Bronchoselectan, Schering). Radiograph with
spotfilm device. Laterolateral. German shepherd dog, 4 years old.
High definition screens – FFD 60 cm – 65 kV – 15 mAs
Diminution of 24 × 30 cm

</td>
</tr>
</table>

Bronchographie

Vorbereitung: Prämedikation mit Atropinum sulfuricum, um die
Schleimsekretion und die Hustenbereitschaft herabzusetzen. Zur Erleichterung der Manipulationen und zur Vermeidung von Zwischenfällen in-
folge Überempfindlichkeit ist eine Narkose notwendig. Falls der Patient
nur sediert werden kann, ist zusätzlich eine Schleimhautanästhesie des
Pharynx und Larynx durch Versprayen eines Oberflächenanästhetikums
(Pantocain 0,5 %ig mit Adrenalinzusatz) zweckmäßig.

Kontrastmittel: Die gebräuchlichen Präparate enthalten wasserlösliches
Jod als Wirkstoff in Kombination mit einer kolloidalen Lösung als Verdickungsmittel.
Der Wirkstoff wird im Verlauf von 1–2 Stunden über die Nieren ausgeschieden. Das Verdickungsmittel wird zum Teil expektoriert und zum Teil
resorbiert.
Das Kontrastmittel ist körperwarm unter Kontrolle in den Hauptbronchus der zu untersuchenden Seite zu instillieren. Die Dosis für die einsei-
tige Bronchographie beträgt 3–15 ml.

Untersuchungstechnik: Der Hund wird auf Bauch-Brust oder auf die zu
untersuchende Seite gelegt und fixiert. Nach Einführen des Bronchoskops
bis an die Bifurkation wird der Katheter (Metras-Katheter, dünner Harn-
röhrenkatheter) unter Sicht in den Hauptbronchus der zu untersuchenden
Seite eingeführt. Die Instillation des Kontrastmittels erfolgt langsam und
unter Durchleuchtungskontrolle, um einer zu schnellen bzw. um einer
übermäßigen oder einer zu geringen Füllung des Bronchalbaums vorzubeugen. Nach der Applikation des Kontrastmittels sind 2 Aufnahmen (latero-lateraler und ventro-dorsaler Strahlengang) anzufertigen.

Nachbehandlung: Bis zum Erwachen aus der Narkose ist der Hund auf
der untersuchten Seite zu lagern und der Situation entsprechend zu ver-
sorgen.

Bronchography

Preparation of the patient: Premedication with Atropine sulphate is necessary to diminish mucus secretion and suppress coughing. To facilitate
manipulation and avoid incidents due to hypersensitivity the patient
should be anesthetized. If the dog can only be sedated spray anesthesia of
the pharynx and larynx is indicated.

Contrast media: Commonly used preparation contain water soluble iodized organic compounds in combination with a colloidal solution as
thickening agent. The active component is excreted via the kidneys within
1–2 hours. The thickening agent is partly expectorated and partly resorbed.
The contrast medium should be introduced at body temperature into the
main bronchus of the side to be examined.
The dose for unilateral bronchography is 3–15 ml.

Technique: The dog is placed in abdominal recumbency or on the side to
be examined. The bronchoscope is introduced to the bifurcation of the trachea; then a plastic catheter is pushed through into the main bronchus of
the lung under scrutiny. The installation of the contrast medium should be
done slowly and with the help of fluoroscopy in order to assure a homogeneous spreading of the medium in the lung. Two pictures should be tak-
en: one with laterolateral and one with ventrodorsal direction of the
beam.

After care: Until such time as it recovers from the narcosis the dog should
be placed on the examined side and treated symptomatically.

Abb. 213*   Röntgenskizze zu Abb. 212   Fig. 213*   X-ray sketch to fig. 212

A   1. Vertebra thoracica;
B   4. Vertebra thoracica;
C   7. Vertebra thoracica;
D   10. Vertebra thoracica;
E   1. Os costale;
F   8. Os costale;
G   Scapula;

a   Cor;
b   Diaphragma;
c   Bronchoskop – Bronchoscope;
d   Bronchalkatheter – Bronchal catheter;

e   Trachea, kontrastmittelhaltig – Trachea, containing contrast medium;
f, g, h   Bronchus lobaris cranialis:
f   Bronchus segmentalis cranialis,
g   Bronchi segmentales dorsales,
h   Bronchus segmentalis caudalis;
i, k, l   Bronchus lobaris caudalis:
k   Bronchi segmentales ventrales,
l   Bronchi segmentales dorsales;
m   Kontrastmittelhaltige Bronchuli und Alveolen, abschnittsweise eingezeichnet – Bronchuli and alveoli containing contrast medium, partly illustrated.

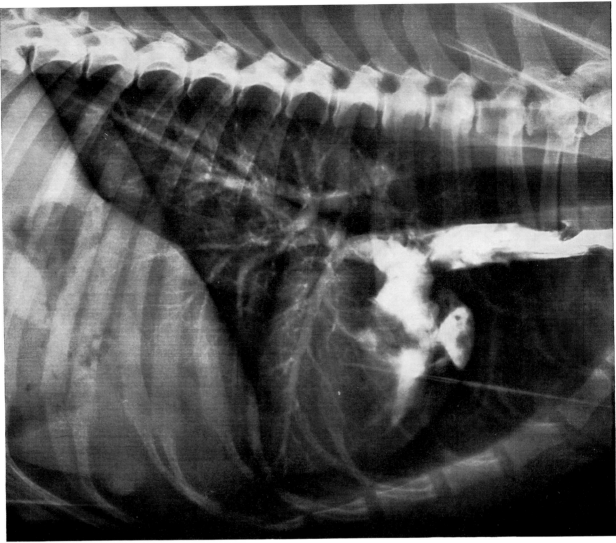

Abb. 214   Angiokardiographie. Aufnahme aus Bildserie, aufgenommen mit Angio-Kardio-Seriograph nach BUCHTALA (Urografin 76 %, Schering). Injektion in die V. jugularis externa. Venöse Seite, Endphase der Systole. Liegend. Latero-lateral. Deutscher Schäferhund, 4 Jahre.
Mittelverstärkende Folie – FFA 200 cm – 75 kV – 36 mAs
Verkleinerung von 24 × 30 cm
Lagerung Abb. 205

Fig. 214   Angiocardiography. Radiograph out of series, taken by angio-cardio-seriograph after BUCHTALA (76 % Urografin, Schering). Injection into external jugular vein. Venous side, end phase of systole. Recumbent.
Laterolateral. German shepherd dog, 4 years old.
Standard screens – FFD 200 cm – 75 kV – 36 mAs
Diminution of 24 × 30 cm
Positioning fig. 205

Angiokardiographie

Allgemeines: Für die Herz- und Gefäßdiagnostik sind Aufnahmen in schneller Bildfolge erforderlich. Die Voraussetzung dafür ist eine besondere apparative Ausrüstung.

Die Passage des Kontrastmittels kann reproduzierbar festgehalten werden, wenn eine Bildverstärkeranlage mit einer kinematographischen Einrichtung oder mit einem Bildbandspeichergerät vorhanden ist. Eine Serie von Röntgenaufnahmen, also einzelne Phasen der Kontrastmittelpassage, können mit Hilfe eines Seriographen aufgenommen werden.

Die Abbildungen 214 und 216 stammen aus einer mit dem Angio-Kardio-Seriographen nach BUCHTALA aufgenommenen Bildserie. Das Gerät arbeitet mit 2 Filmtransportgeschwindigkeiten (0,4 und 0,8 Sekunden pro Aufnahme einschließlich Filmtransport). In 6 bzw. 12 Sekunden können 16 Röntgenaufnahmen im Format 24 × 30 cm angefertigt werden. Da das Gerät außerdem einen exakt einstellbaren verlängerten Ablauf der Bildgeschwindigkeit hat, können Spätfüllungen oder ein verzögerter Kontrastmitteldurchlauf erfaßt werden.

Um die Injektion des Kontrastmittels rasch durchführen zu können, wurden Apparate konstruiert, die das Kontrastmittel unter hohem Druck zu injizieren gestatten.

Bei Verwendung eines Angio-Seriographen ist eine Kontaktkupplung zwischen Injektionsspritze und dem Gerät zweckmäßig. Die erste Aufnahme wird dadurch nach der Applikation einer bestimmten Kontrastmittelmenge automatisch ausgelöst.

Kontrastmittel: Trijodierte organische Verbindungen, je nach Größe des Tieres 3 – 15 ml, intravenöse Applikation.

Untersuchungstechnik: Um Bewegungen des Tieres bei dem durch den Filmtransport verursachten Geräusch und um einer Überempfindlichkeitsreaktion gegenüber dem Jodkontrastmittel vorzubeugen, ist es zweckmäßig, die Untersuchung am narkotisierten Patienten vorzunehmen.

Der Patient ist in linker Seitenlage auf dem Seriographen so zu lagern und zu fixieren, daß sich die Medianebene des Brustkorbs parallel zur Ober-

fläche des Geräts befindet (s. Abb. 205). Der Fokus-Film-Abstand sollte 2 m betragen, um einer unnötigen Vergrößerung des Herzschattens vorzubeugen.

Angiocardiography

General considerations: For diagnostic radiography of the heart and blood vessels, X-ray pictures taken in rapid succession are necessary. This requires special equipment.

The passage of the contrast medium can be depicted with a machine equipped with image intensifier and movie camera or tape recorder and with a rapid cassette changer. Such a machine can produce multiple exposures, each less than 1/2 second apart.

The radiographs in fig. 214 and fig. 216 are part of a series of picture taken with an angio-cardio-seriograph after BUCHTALA. This machine features two speeds for film transportation (0.4 and 0.8 seconds per picture, including the film transportation). In 6 or 12 seconds, respectively, 16 radiographs, size 24 × 30 cm, can be taken. Since this apparatus also has an exact setting for slow-down of the film-passage, late filling or delayed passage of the contrast medium can be depicted.

There is special equipment on the market allowing rapid injection of the contrast medium under high pressure.

Should an angio-seriograph be employed the syringe (with contrast medium) should be connected to the machine. In this way, the first radiograph can be taken automatically as soon as a given amount of the contrast medium has been injected.

Contrast media: Tri-iodized organic compounds, 3 – 15 ml depending on the size of the dog, intravenous injection.

Technique: The patient should be anesthetized and placed in the left lateral position upon the seriograph. The median plane of the thorax should be parallel to the top of the table (fig. 205). The focus film distance should be 2 m in order to eliminate unnecessary magnification of the cardiac silhouette.

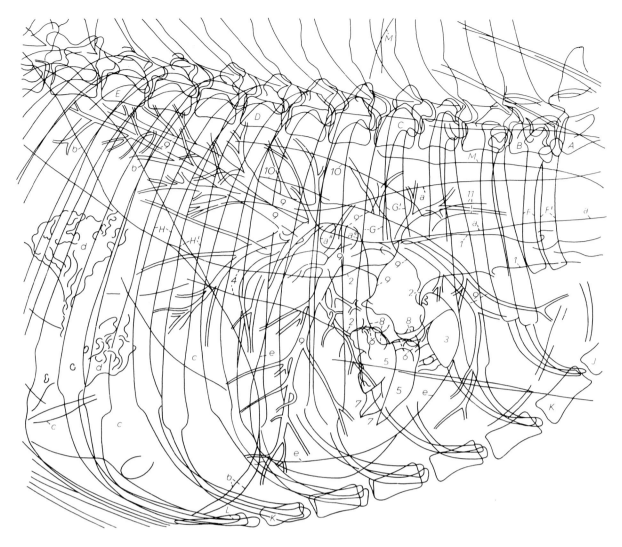

Abb. 215*   Röntgenskizze zu Abb. 214   Fig. 215*   X-ray sketch to fig. 214

A	7. Vertebra cervicalis;		c	Hepar;
B	1. Vertebra thoracica;		d	Ventriculus;
C	4. Vertebra thoracica;		e	Cor;
D	7. Vertebra thoracica;			
E	10. Vertebra thoracica;		1	V. cava cranialis;
F	1. Os costale;		2	Atrium dextrum;
G	4. Os costale;		3	Auricula dextra;
H	7. Os costale;		4	V. cava caudalis;
J	Manubrium sterni;		5	Ventriculus dexter;
K	Corpus sterni, Sternebrae;		6	Valva atrioventricularis dextra, kammerseitige Begrenzung – Valva atrioventricularis dextra, ventricular border;
L	Processus xiphoideus;			
M	Scapula;		7	M. papillaris;
			8	Sinus trunci pulmonalis;
a	Trachea;		9	Truncus pulmonalis et Aa. pulmonales;
a″	Bifurcatio tracheae;		10	Aorta thoracica;
b	Diaphragma;		11	A. subclavia sinistra.

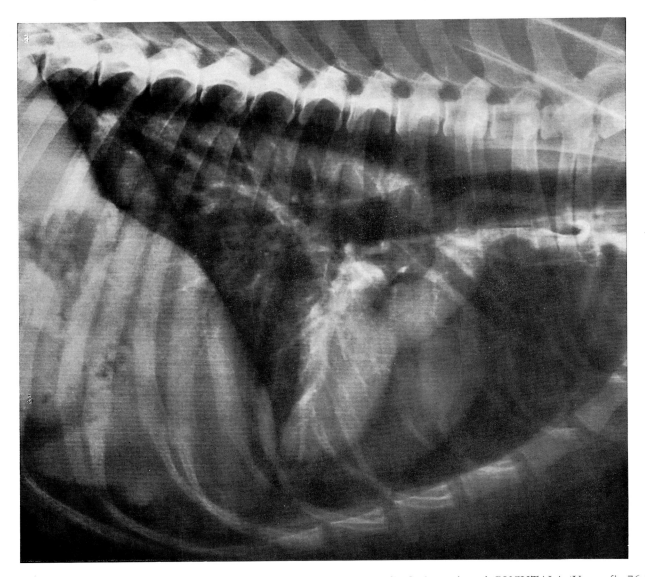

Abb. 216   Angiokardiographie. Aufnahme aus Bildserie, aufgenommen mit Angio-Kardio-Seriograph nach BUCHTALA (Urografin 76 %, Schering). Injektion in die V. jugularis externa. Arterielle Seite, beginnende Systole. Liegend. Latero-lateral. Deutscher Schäferhund, 4 Jahre.
Mittelverstärkende Folie – FFA 200 cm – 75 kV – 36 mAs
Verkleinerung von 24 × 30 cm
Lagerung Abb. 205

Fig. 216   Angiocardiography. Radiograph out of series, taken by angio-cardio-seriograph after BUCHTALA (76 % Urografin, Schering). Injection into external jugular vein. Arterial side, initial phase of systole. Recumbent. Laterolateral. German shepherd dog, 4 years old.
Standard screens – FFD 200 cm – 75 kV – 36 mAs
Diminution of 24 × 30 cm
Positioning fig. 205

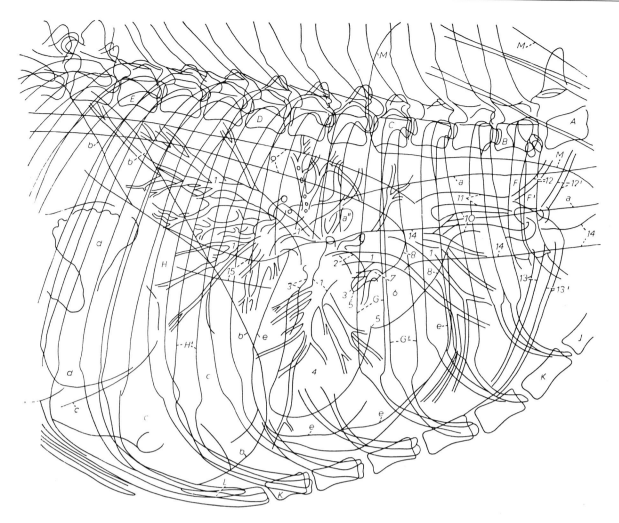

Abb. 217*   Röntgenskizze zu Abb. 216   Fig. 217*   X-ray sketch to fig. 216

A	7. Vertebra cervicalis;	1	Vv. pulmonales;
B	1. Vertebra thoracica;	2	Atrium sinistrum;
C	4. Vertebra thoracica;	3	Ostium atrioventriculare sinistrum;
D	7. Vertebra thoracica;	4	Ventriculus sinister;
E	10. Vertebra thoracica;	5	Ostium aortae;
F	1. Os costale;	6	Bulbus aortae;
G	4. Os costale;	7	A. coronaria sinistra;
H	7. Os costale;	8	Arcus aortae;
J	Manubrium sterni;	9	Aorta thoracica;
K	Corpus sterni, Sternebrae;	10	Truncus brachiocephalicus;
L	Processus xiphoideus;	11	A. subclavia sinistra;
M	Scapula;	12	Truncus costocervicalis;
		13	A. thoracica interna;
a	Trachea;	14	V. cava cranialis;
a″	Bifurcatio tracheae;	15	V. cava caudalis.
b	Diaphragma;		
c	Hepar;		
d	Ventriculus;		
e	Cor;		

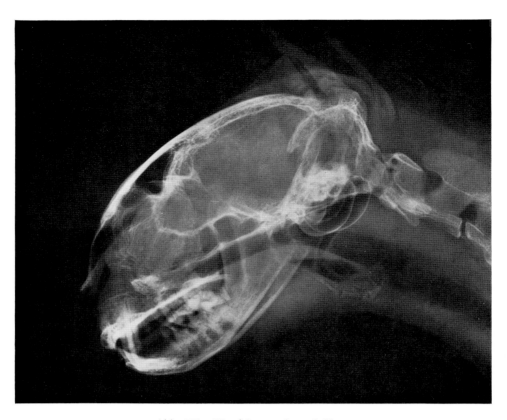

Abb. 218   Kopf. Latero-lateral. Katze.
Folienloser Film – FFA 120 cm – 60 kV – 20 mAs
Originalgröße (Ausschnitt aus 13 × 18 cm)
Lagerung Abb. 219

Fig. 218   Head. Laterolateral. Cat.
Non-screen film – FFD 120 cm – 60 kV – 20 mAs
Original size (section of 13 × 18 cm)
Positioning fig. 219

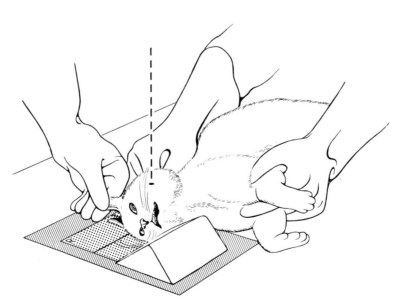

Abb. 219   Lagerung zur Aufnahme des Kopfes. Latero-lateral.

Der Kopf ist auf der Kassette (Bucky-Blende, folienloser Film) so zu lagern, daß die Medianebene parallel zum Film und damit die beiden Unterkieferhälften senkrecht übereinanderliegen. Mit Hilfe des unter den rostralen Kopfbereich gelegten Schaumgummikeils kann der Kopf korrekt gelagert werden.
Der Zentralstrahl sollte den Kopf in Höhe des Kiefergelenks treffen und im rechten Winkel auf den Film einfallen.

Fig. 219   Positioning of head. Laterolateral.

The head should be aligned with the median plane parallel to the cassette (Bucky diaphragm, non-screen film) and both halves of the mandible perpendicularly to each other. Foam rubber pads placed underneath the rostral part of the head will facilitate proper positioning and fixation.
The central beam should strike the head on a level of the temporomandibular joint and fall at the right angle on the film.

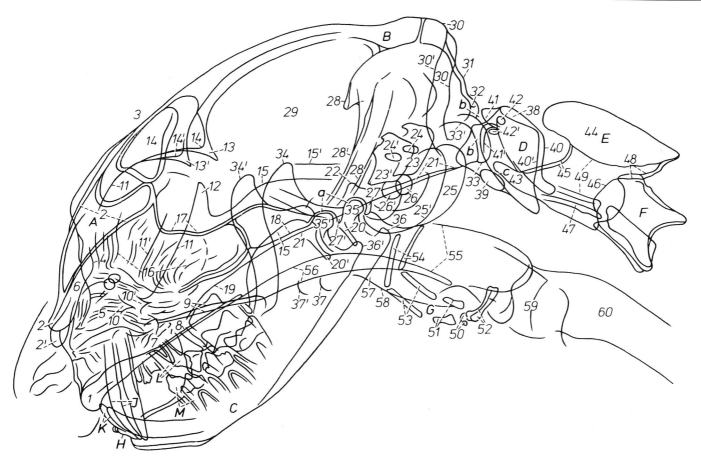

Abb. 220*   Röntgenskizze zu Abb. 218   Fig. 220*   X-ray sketch to fig. 218

A   Ossa faciei;
B   Ossa cranii;
C   Mandibula;
D   Atlas;
E   Axis;
F   3. Vertebra cervicalis;
G   Os hyoideum;
H   Mandibulare Dentes incisivi – Mandibular incisor teeth;
J   Maxillare Dentes canini – Maxillary canine teeth;
K   Mandibulare Dentes canini – Mandibular canine teeth;
L   Maxillare Dentes praemolares III – 3rd maxillary premolar teeth;
M   Mandibulare Dentes praemolares III – 3rd mandibular premolar teeth;

a   Articulatio temporomandibularis;
b   Articulatio atlantooccipitalis;
c   Articulatio atlantoaxialis;

Am Cranium – On the cranium:

1   Os incisivum;
2   Os nasale;
3   Os frontale;
4   Concha nasalis dorsalis;
5   Concha nasalis ventralis;
6   Meatus nasi dorsalis;
7   Meatus nasi ventralis;
8   Palatum durum;
9   Orbita, ventrale Begrenzung – Orbit, ventral border;
10 Foramen infraorbitale;
11 Aditus orbitae, Margines supraorbitalis et infraorbitalis;
12 Os zygomaticum, Processus frontalis;
13 Os frontale, Processus zygomaticus;
14 Sinus frontalis;
15 Arcus zygomaticus;
16 Labyrinthus ethmoidalis;
17 Fossa ethmoidalis;
18 Sinus sphenoidalis;
19 Vomer;
20 Hamulus pterygoideus;
21 Basis cranii, rostral: Os praesphenoidale; kaudal: Os occipitale, Pars basilaris – Basis cranii, rostrally: Os praesphenoidale; caudally: Os occipitale, Pars basilaris;
22 Dorsum sellae;
23 Os temporale, Pars petrosa;
24 Porus acusticus internus;
25 Os temporale, Pars tympanica;

26 Porus acusticus externus;
27 Processus retroarticularis;
28 Tentorium cerebelli osseum;
29 Cavum cranii;
30 Crista nuchae, seitlich in die Crista temporalis übergehend – Crista nuchae, laterally merging into the temporal crest;
31 Squama occipitalis;
32 Foramen magnum, dorsaler Rand – Foramen magnum, dorsal border;
33 Condylus occipitalis;

An der Mandibula – On the mandible:

34 Processus coronoideus;
35 Processus condylaris;
36 Processus angularis;
37 Foramen mandibulae;

Am Atlas – On the atlas:

38 Arcus dorsalis;
39 Arcus ventralis;
40 Ala atlantis;
41 Fovea articularis cranialis;
42 Foramen vertebrale;

Am Axis – On the axis:

43 Dens;
44 Processus spinosus;
45 Incisura vertebralis cranialis;
46 Incisura vertebralis caudalis;
47 Processus transversus;
48 Processus articularis caudalis;
49 Canalis vertebrae;

Am Os hyoideum – On the hyoid bone:

50 Basihyoideum;
51 Ceratohyoideum;
52 Thyreohyoideum;
53 Epihyoideum;
54 Stylohyoideum;

An Pharynx und Larynx – On the pharynx and larynx:

55 Pars nasalis;
56 Velum palatinum;
57 Pars oralis;
58 Radix linguae;
59 Larynx;
60 Trachea.

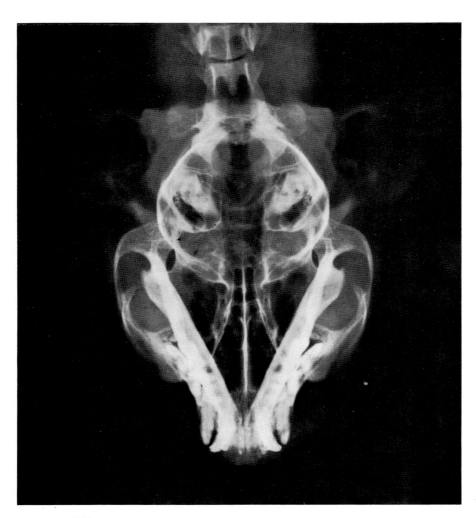

Abb. 221   Kopf. Nasenrücken anliegend. Ventro-dorsal. Katze.
Bucky-Blende – Feinzeichnende Folie – FFA 120 cm – 58 kV – 30 mAs
Originalgröße (Ausschnitt aus 13 × 18 cm)
Lagerung Abb. 222

Fig. 221   Head. Dorsum nasi resting on cassette. Ventrodorsal. Cat.
Bucky diaphragm – High definition screens – FFD 120 cm – 58 kV – 30 mAs
Original size (section of 13 × 18 cm)
Positioning fig. 222

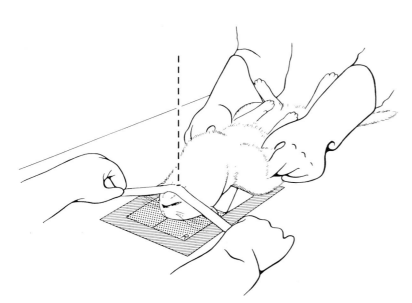

Abb. 222   Lagerung zur Aufnahme des Kopfes. Nasenrücken anliegend.
Ventro-dorsal.

Voraussetzung für eine sichere Orientierung ist eine gleichmäßige Abbildung der Kopfhälften. Da schon eine geringe Verkantung erhebliche Verzeichnungen und Überlagerungen verursacht, sollte der Patient narkotisiert werden, wenn er sich aus Angst oder wegen Schmerzen gegen die Lagerung sträubt.
Bei der Lagerung mit auf der Kassette (folienloser Film, Bucky-Blende) anliegendem Nasenrücken sind die seitlichen Buchten der Stirnhöhle übersichtlicher dargestellt. Die Lagerung des Nasenrückens auf dem Film läßt sich durch Unterlegen eines Schaumgummikeils unter die ersten Halswirbel erleichtern.
Es ist darauf zu achten, daß die Kehlränder des Unterkiefers in gleicher Höhe liegen.
Der Zentralstrahl sollte den Kopf in der Medianebene in Höhe des Angulus mandibulae treffen und im rechten Winkel auf den Film einfallen.

Fig. 222   Positioning of head with dorsum nasi resting on cassette.
Ventrodorsal.

A symmetrical reproduction of both halves of the head is essential for proper radiographic evaluation. Even a minor tilting of the head causes considerable distortions and superimposures. Because of this, restless patients (fear, pain) should be anesthetized.
In this position with the dorsum nasi on the cassette (non-screen film, Bucky diaphragm) the lateral recesses of the frontal sinuses are more clearly defined. Positioning of the dorsum nasi can be facilitated by placing a wedge-shaped foam rubber pad underneath the cranial cervical vertebrae. The ventral borders of the mandible must lie on the same level.
The central beam should strike the midline of the head on a level of the mandibular angle and fall at the right angle on the film.

A Axis;
B Atlas;
C Ossa cranii;
D Ossa faciei;
E Mandibula;
F Maxillarer Dens praemolaris IV – 4th maxillary premolar tooth;
G Mandibularer Dens molaris I – 1st mandibular molar tooth;
H Maxillarer Dens caninus – Maxillary canine tooth;
J Mandibularer Dens caninus – Mandibular canine tooth;
K Maxillare Dentes incisivi – Maxillary incisor teeth;
L Mandibulare Dentes incisivi – Mandibular incisor teeth;
M Anteile des Os hyoideum – Parts of hyoid bone;
N Meatus acusticus externus cartilagineus;

a Articulatio atlantoaxialis;
b Articulatio atlantooccipitalis;
c Articulatio temporomandibularis;

Am Axis – On the axis:

1 Dens;
2 Processus articularis cranialis;
3 Processus spinosus;
4 Incisura vertebralis cranialis;
5 Pediculus arcus vertebrae, nach medial zugleich seitliche Begrenzung des Foramen vertebrale – Pediculus arcus vertebrae, medially forming the lateral margin of the vertebral foramen;
6 Extremitas caudalis;
7 Processus transversus;

Am Atlas – On the atlas:

8 Arcus dorsalis, kaudaler Rand – Arcus dorsalis, caudal border;
9, 10 Arcus ventralis:
9 Kranialer Rand – Cranial border,
10 Kaudaler Rand – Caudal border;
11 Fovea articularis caudalis;
12 Ala atlantis;
13 Foramen transversarium;
14 Incisura alaris;

Am Cranium – On the cranium:

15 Crista nuchae, seitlich in die Crista temporalis übergehend – Crista nuchae, laterally merging into the temporal crest;
16 Cavum cranii, Wand – Cavum cranii, wall;
17 Foramen magnum;
18 Condylus occipitalis;
19 Fossa condylaris ventralis;
20 Processus retrotympanicus;
21 Foramen jugulare;
22 Tentorium cerebelli osseum;
23, 24 Os temporale:
23 Pars petrosa,
24 Porus acusticus internus;
25 Ossicula auditus;
26 Porus acusticus externus;
27 Ostium tympanicum tubae auditivae;
28 Dorsum sellae;
29 Foramen ovale;
30 Foramen rotundum;
31 Fissura orbitalis;
32 Canalis opticus;
33 Crista orbitosphenoidea;
34 Sinus sphenoideus;
35 Sinus frontalis;
36 Choanenrand mit Spina nasalis caudalis – Choanal border with the caudal nasal spine of the palatine bone;
37 Hamulus pterygoideus;
38 Fossa ethmoidalis;
39 Sutura palatina mediana et Vomer;
40 Crista vomeris;
41 Foramen sphenopalatinum;
42 Fissura palatina;

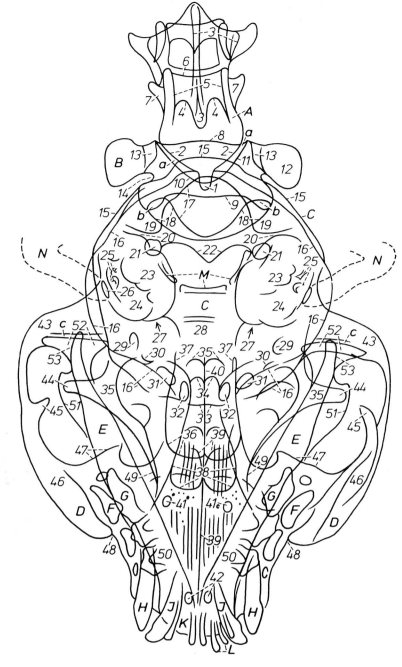

Abb. 223 Röntgenskizze zu Abb. 221
Fig. 223 X-ray sketch to fig. 221

43 Os temporale, Processus zygomaticus;
44 Os frontale, Processus zygomaticus;
45 Processus frontalis;
46 Processus temporalis;
47 Os maxillare, kaudaler Rand – Os maxillare, caudal border;
48 Foramen infraorbitale;
49 Cavum nasi, seitliche Begrenzung, zugleich mediale Wand der Orbita – Cavum nasi, lateral border forming the medial wall of the orbit;

An der Mandibula – On the mandible:

50 Corpus mandibulae;
51 Processus coronoideus;
52 Processus condylaris;
53 Processus angularis.

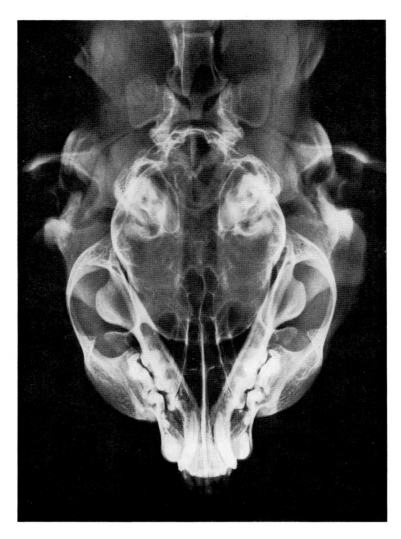

Abb. 224   Kopf. Dorso-ventral. Katze.
Folienloser Film – FFA 120 cm – 65 kV – 30 mAs
Originalgröße (Ausschnitt aus 13 × 18)
Lagerung Abb. 225

Fig. 224   Head. Dorsoventral. Cat.
Non-screen film – FFD 120 – 65 kV – 30 mAs
Original size (section of 13 × 18 cm)
Positioning fig. 225

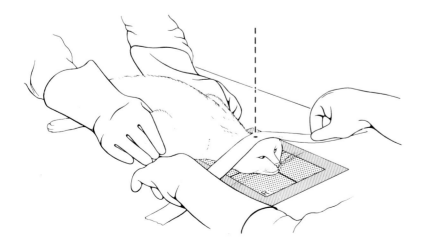

Abb. 225   Lagerung zur Aufnahme des Kopfes. Dorso-ventral.

Die Lagerung des Kopfes ist für den Patienten in Brust-Bauchlage bequemer als in Rückenlage. Daher kann bei dieser Lagerung meistens auf eine Sedierung der Katze verzichtet werden.
Die Kehlränder sollten der Kassette (folienloser Film, Bucky-Blende) anliegen, so daß die Medianebene senkrecht zur Kassette steht.
Der Zentralstrahl sollte den Kopf in der Mittellinie in Höhe des temporalen Augenwinkels treffen und im rechten Winkel auf den Film einfallen.

Fig. 225   Positioning of head. Dorsoventral.

Abdominal recumbency is more comfortable for the patient than dorsal recumbency. With few exceptions, this positioning does not require sedation of the patient.
The ventral borders of the mandible must rest on the cassette (non-screen film, Bucky diaphragm) in order to align the median plane perpendicularly to the cassette.
The central beam should strike the midline of the head on a level of the lateral commissure of the eyelid and fall at the right angle on the film.

A   Axis;
B   Atlas;
C   Ossa cranii;
D   Ossa faciei;
E   Mandibula;
F   Maxillarer Dens praemolaris IV – 4th maxillary premolar tooth;
G   Mandibularer Dens molaris I – 1st mandibular molar tooth;
H   Maxillarer Dens caninus – Maxillary canine tooth;
J   Mandibularer Dens caninus – Mandibular canine tooth;
K   Maxillare Dentes incisivi – Maxillary incisor teeth;
L   Mandibulare Dentes incisivi – Mandibular incisor teeth;
M   Anteile des Os hyoideum – Parts of hyoid bone;
N   Meatus acusticus externus cartilagineus;

a   Articulatio atlantoaxialis;
b   Articulatio atlantooccipitalis;
c   Articulatio temporomandibularis;

Am Axis – On the axis:

1   Dens;
2   Processus articularis cranialis;
3   Processus spinosus;
4   Incisura vertebralis cranialis;
5   Foramen vertebrale, seitliche Begrenzung – Foramen vertebrale, lateral margin;
6   Extremitas caudalis;
7   Processus transversus;

Am Atlas – On the atlas:

8, 9   Arcus dorsalis:
8   Kranialer Rand – Cranial border,
9   Kaudaler Rand – Caudal border;
10, 11   Arcus ventralis:
10   Kranialer Rand – Cranial border,
11   Kaudaler Rand – Caudal border;
12   Fovea articularis cranialis;
13   Fovea articularis caudalis;
14   Ala atlantis;
15   Foramen transversarium;
16   Incisura alaris;
17   Foramen vertebrale, seitliche Begrenzung – Foramen vertebrale, lateral margin;

Am Cranium – On the cranium:

18   Crista nuchae;
19   Foramen magnum;
20   Condylus occipitalis;
21   Fossa condylaris ventralis;
22   Tentorium cerebelli osseum;
23 – 29   Os temporale:
23   Pars petrosa,
24   Porus acusticus internus,
25   Pars tympanica,
26   Porus acusticus externus,
27   Ostium tympanicum tubae auditivae,
28   Foramen stylomastoideum,
29   Processus retrotympanicus;
30   Foramen jugulare;
31   Dorsum sellae;
32   Foramen ovale;
33   Foramen rotundum;
34   Fissura orbitalis;
35   Canalis opticus;
36   Crista orbitosphenoidea;
37   Sinus sphenoideus;
38   Choanae;
39   Choanenrand mit Spina nasalis caudalis – Choanal border with caudal nasal spine of the palatine bone;
40   Cavum cranii, Wand – Cavum cranii, wall;
41   Fossa ethmoidalis;
42   Processus retroarticularis;
43   Os temporale, Processus zygomaticus;
44   Sinus frontalis;

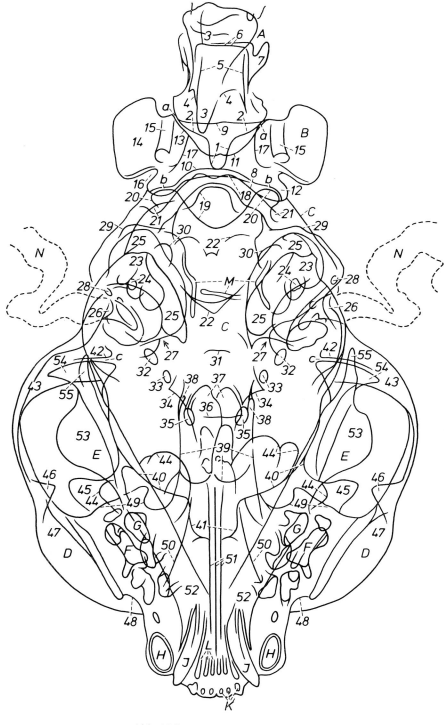

Abb. 226   Röntgenskizze zu Abb. 224
Fig. 226   X-ray sketch to fig. 224

45   Os frontale, Processus zygomaticus;
46, 47   Os zygomaticum:
46   Processus frontalis,
47   Processus temporalis;
48   Foramen infraorbitale, Rand – Foramen infraorbitale, margin;
49   Os maxillare, Kaudalrand – Os maxillare, caudal border;
50   Cavum nasi, laterale Begrenzung, zugleich mediale Wand der Orbita – Cavum nasi, lateral border forming the medial wall of the orbit;
51   Sutura palatina mediana et Vomer;

An der Mandibula – On the mandible:

52   Corpus mandibulae;
53   Processus coronoideus;
54   Processus condylaris;
55   Processus angularis.

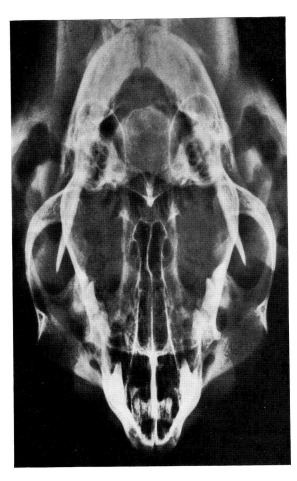

Abb. 227   Oberkiefer bei geöffneter Mundspalte.
Ventro-dorsal. Schrägprojektion. Katze.
Folienloser Film – FFA 115 cm – 60 kV – 20 mAs
Originalgröße (Ausschnitt aus 13 × 18 cm)
Lagerung Abb. 228

Fig. 227 Upper jaw with open mouth.
Obliquely ventrodorsal. Cat.
Non-screen film – FFD 115 cm – 60 kV – 20 mAs
Original size (section of 13 × 18 cm)
Positioning fig. 228

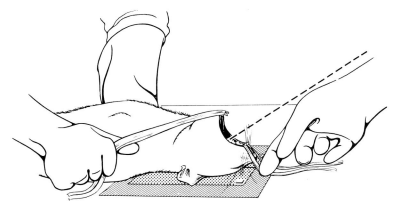

Abb. 228   Lagerung zur Aufnahme des Oberkiefers bei geöffneter
Mundspalte. Ventro-dorsal. Schrägprojektion.

Die Lagerung ist nur bei duldsamen Tieren ohne Sedierung bzw. Narkose
möglich. Person 1 fixiert die auf dem Rücken liegende Katze und das den
Unterkiefer abziehende Band. Person 2 lagert den Kopf mit der Stirn auf
dem Film und hält ihn mit dem über die Schneidezähne des Oberkiefers gelegten Band. Es ist darauf zu achten, daß die Medianebene des Kopfes und
die des Rumpfes senkrecht zum Film stehen.
Die Aufnahme ist mit einem folienlosen Film (Bleiunterlage nicht vergessen!) anzufertigen.
Der Zentralstrahl sollte in einem Winkel von 45° in die weit geöffnete
Mundhöhle einfallen und den harten Gaumen in der Mittellinie in Höhe
des 3. Prämolaren treffen.

Fig. 228   Positioning of the upper jaw with open mouth.
Obliquely ventrodorsal.

This positioning is only possible in docile animals without sedation or
narcosis. One assistant secures the cat in dorsal recumbency and holds the
strap pulling on the lower jaw. A second assistant places the head with
the frontal region on the film and secures it with a strap drawn over the
upper incisor teeth. Care should be taken to align the median plane of the
head and the trunk perpendicularly to the film.
The radiograph should be taken with a non-screen film (lead blocker not
to be forgotten!).
The central beam should fall at an angle of 45° into the widely opened
mouth and strike the hard palate in the midline on a level of the 3rd premolar.

160

A   Ossa cranii;
B   Mandibula;
C   Atlas;
D   Axis;
Ca   Dens caninus;
P₃   Dens praemolaris III;
P₄   Dens praemolaris IV;

a   Articulatio temporomandibularis;
b   Articulatio atlantoaxialis;

Am Cranium – On the cranium:

1   Foramen magnum;
2   Os occipitale, Pars basilaris;
3   Cavum cranii, seitliche Begrenzung – Cavum cranii, lateral limit;
4   Foramen jugulare;
5 – 9   Os temporale:
5   Pars tympanica,
6   Pars petrosa,
7   Porus acusticus externus,
8   Semicanalis tubae auditivae,
9   Processus zygomaticus;
10   Sella turcica;
11   Foramen ovale;
12   Foramen rotundum;
13   Fissura orbitalis;
14   Foramen opticum;
15   Sinus sphenoidalis;
16   Hamulus pterygoideus;
17   Choanae, lateraler Rand – Lateral border of the choanae;
18   Os palatinum, Lamina horizontalis;
19, 20   Os zygomaticum:
19   Processus temporalis,
20   Processus frontalis;
21   Sutura zygomaticomaxillaris;
22   Os maxillare, Processus alveolaris;
23   Foramen infraorbitale, zugleich laterale Begrenzung des Canalis infraorbitalis – Foramen infraorbitale forming the lateral border of the infraorbital canal;
24   Recessus maxillaris;
25   Os ethmoidale, Endoturbinalia;
26   Cavum cranii, rostrale Begrenzung – Cavum cranii, rostral limit;
27   Septum nasi;
28   Cavum nasi et Os conchae nasalis ventralis;
29   Fissura palatina;
30   Os incisivum, rostraler Rand des Corpus – Os incisivum, rostral border of the body;
31   Os nasale;

An der Mandibula – On the mandible:

32   Processus coronoideus;
33   Processus condylaris;
34   Corpus mandibulae;

Am Atlas – On the atlas:

35   Ala atlantis;
36   Foramen transversarium;
37   Fovea articularis caudalis;
38   Arcus ventralis, kaudaler Rand – Arcus ventralis, caudal border;

Am Axis – On the axis:

39   Processus articularis cranialis;
40   Dens;
41   Processus spinosus;
42   Lingua;
43   Os hyoideum.

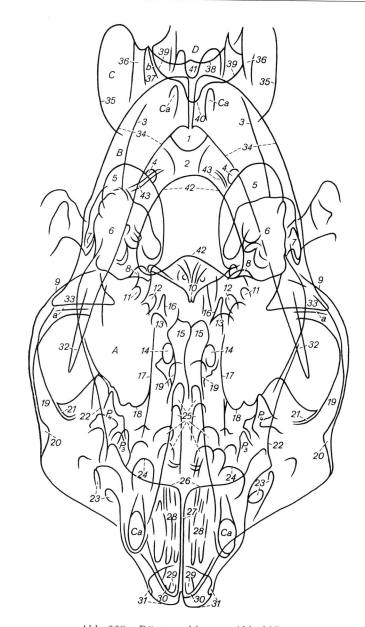

Abb. 229   Röntgenskizze zu Abb. 227
Fig. 229   X-ray sketch to fig. 227

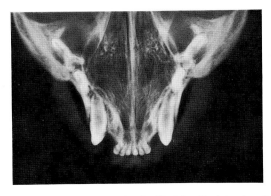

Abb. 230   Oberkiefer (Film in die Mundhöhle eingelegt).
Dorso-ventral. Katze.
Folienloser Film – FFA 115 cm – 57 kV – 15 mAs
Originalgröße
Lagerung Abb. 232

Fig. 230   Upper jaw (film placed in oral cavity).
Dorsoventral. Cat.
Non-screen film – FFD 115 cm – 57 kV – 15 mAs
Original size
Positioning fig. 232

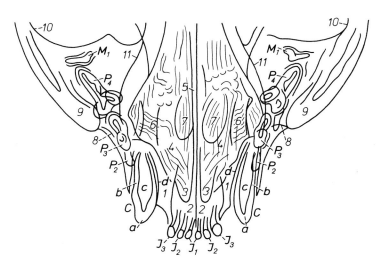

Abb. 231   Röntgenskizze zu Abb. 230

Fig. 231   X-ray sketch to fig. 230

J₁ – J₃	Dentes incisivi I – III;
C	Dens caninus;
P₂ – P₄	Dentes praemolares II – IV;
M₁	Dens molaris I;
a	Corona dentis;
b	Radix dentis;
c	Cavum dentis, an C beschriftet – Cavum dentis, marked on C only;
d	Alveolus dentalis;

1   Apertura nasi, seitliche Begrenzung – Apertura nasi, lateral border;
2   Os incisivum;
3   Fissura palatina;
4   Cavum nasi et Conchae nasales;
5   Septum nasi;
6   Recessus maxillaris;
7   Sinus frontalis;
8   Foramen infraorbitale;
9   Os maxillare et Os zygomaticum;
10   Arcus zygomaticus;
11   Mediale Orbitawand – Media wall of orbit.

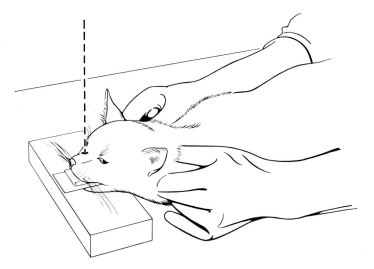

Abb. 232   Lagerung zur Aufnahme des Oberkiefers
(Film in die Mundhöhle eingelegt). Dorso-ventral.

Für die Lagerung ist, von seltenen Ausnahmen abgesehen, eine Sedation oder Narkose notwendig. Der Helfer fixiert die Katze in Brust-

Bauchlage mit nach kaudal gestreckten Schultergliedmaßen. Das Einlegen des Filmes und Bleilatzes in die Mundhöhle bis hinter den letzten Backenzahn ist nicht schwierig, wenn die Lippen im Bereich des Mundwinkels etwas nach außen gezogen werden. Die korrekte Lagerung der beiden Unterkieferknochen ist auf einem unter den Kopf gelegten Block (Holz, Schaumgummi) vorteilhafter.
Der Zentralstrahl sollte den Nasenrücken in der Mittellinie in Höhe des 3. Prämolaren treffen und im rechten Winkel auf den in der Mundhöhle liegenden Film einfallen.

Fig. 232   Positioning of upper jaw (film placed in oral cavity).
Dorsoventral.

With few exceptions, this positioning must be done with sedation or narcosis. The assistant secures the cat in abdominal recumbency with the front limbs pulled caudally. It is not difficult to place the non-screen film and lead blocker in the oral cavity to a level behind the last molar tooth if the lips are pulled sideways at the oral angle. The correct positioning of the two lower jaw bones is facilitated by placing a wooden of foam rubber block underneath the head.
The central beam should strike the dorsum nasi in the midline on a level of the 3rd premolar and fall at the right angle on the film lying in the mouth.

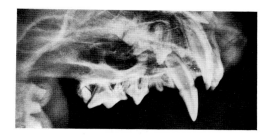

Abb. 233 Oberkiefer bei geöffneter Mundspalte.
Schräglagerung. Medio-lateral. Katze.
Folienloser Film – FFA 120 cm – 60 kV – 20 mAs
Originalgröße (Ausschnitt aus 9 × 12 cm)
Lagerung Abb. 235

Fig. 233 Upper jaw with open mouth.
Oblique positioning. Mediolateral. Cat.
Non-screen film – FFD 120 cm – 60 kV – 20 mAs
Original size (section of 9 × 12 cm)
Positioning fig. 235

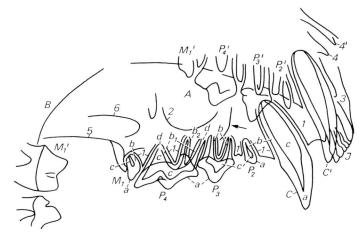

Abb. 234* Röntgenskizze zu Abb. 233
Fig. 234* X-ray sketch to fig. 233

A   Ossa faciei;
B   Mandibula;
J   Dentes incisivi;
C   Dens caninus;
P₂ Dens praemolaris II;
P₃ Dens praemolaris III;
P₄ Dens praemolaris IV;
M₁ Dens molaris I;

a   Corona dentis;
b   Radix dentis;
b₁ Radices buccales;
b₂ Radix lingualis;
c   Cavum dentis;
d   Foramen apicis dentis;

1   Alveolus dentalis;
2   Recessus maxillaris;
3   Corpus ossis incisivi;
4   Os nasale;
5   Arcus zygomaticus, Ventralrand – Arcus zygomaticus, ventral border;
6   Choanenrand – Choanal border.

Pfeil kennzeichnet rostralen Rand des Foramen infraorbitale – Arrow points to the rostral margin of the foramen infraorbitale.

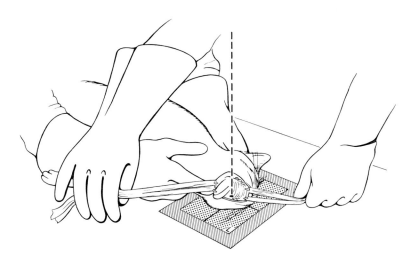

Abb. 235 Schräglagerung zur Aufnahme des Oberkiefers bei geöffneter Mundspalte. Medio-lateral.

Für die Lagerung ist, von seltenen Ausnahmen abgesehen, eine Sedation oder Narkose notwendig. Person 1 lagert und fixiert die Katze in Seitenlage oder halber Rückenlage. Person 2 öffnet mit je einem auf die Schneidezähne des Ober- und Unterkiefers gelegten Band die Mundhöhle und lagert den Kopf so, daß sich der harte Gaumen in einem Winkel von 45° zum Film befindet.
Die Aufnahme ist mit einem folienlosen Film (Bleiunterlage nicht vergessen!) anzufertigen.
Der Zentralstrahl sollte den Alveolarrand in Höhe des 3. Prämolaren treffen und im rechten Winkel auf den Film einfallen.

Fig. 235 Oblique positioning of upper jaw with open mouth.
Mediolateral.

With few exceptions, this positioning must be done with sedation or narcosis. One assistant secures the cat in the lateral or dorsolateral position. A second assistant opens the mouth with straps drawn over the upper and lower incisor teeth and secures the head in such a manner that the hard palate forms an angle of 45° with the film.
The radiograph should be taken with a non-screen film (lead-blocker not to be forgotten!).
The central beam should strike the alveolar border on a level of the 3rd premolar and fall at the right angle on the film.

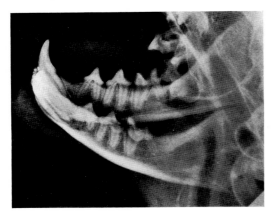

Abb. 236   Unterkiefer bei geöffneter Mundspalte.
Schräglagerung. Medio-lateral. Katze.
Folienloser Film – FFA 115 cm – 55 kV – 12 mAs
Originalgröße (Ausschnitt aus 9 × 12 cm)
Lagerung Abb. 238

Fig. 236   Lower jaw with open mouth.
Oblique positioning. Mediolateral. Cat.
Non-screen film – FFD 115 cm – 55 kV – 12 mAs
Original size (section of 9 × 12 cm)
Positioning fig. 238

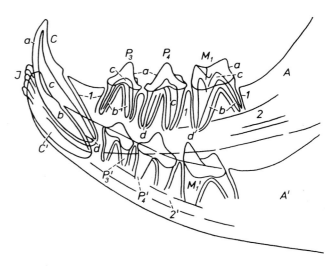

Abb. 237*   Röntgenskizze zu Abb. 236
Fig. 237*   X-ray sketch to fig. 236

A	Mandibula;	a	Corona dentis;
J	Dentes incisivi;	b	Radix dentis;
C	Dens caninus;	c	Cavum dentis;
P_3	Dens praemolaris III;	d	Foramen apicis dentis;
P_4	Dens praemolaris IV;		
M_1	Dens molaris I;	1	Alveolus dentalis;
		2	Canalis mandibulae.

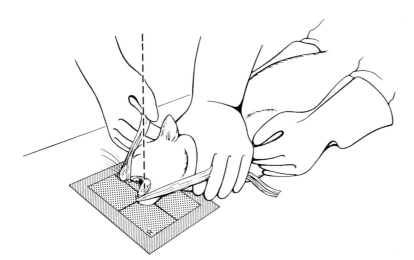

Abb. 238 Schräglagerung zur Aufnahme des Unterkiefers
bei geöffneter Mundspalte. Medio-lateral.

Nur bei duldsamen Tieren ist die Lagerung ohne Sedation bzw. Narkose möglich. Person 1 fixiert die Katze in Brust-Bauchlage. Person 2

öffnet die Mundspalte des Tieres mit je einem über die Schneidezähne des Ober- bzw. Unterkiefers gelegten Band und lagert den Unterkiefer im Winkel von 45° auf dem Film.
Die Aufnahme ist mit einem folienlosen Film (Bleiunterlage nicht
vergessen!) anzufertigen.
Der Zentralstrahl sollte den Alveolarrand in Höhe des 4. Prämolaren
treffen und im rechten Winkel auf den Film einfallen.

Fig. 238   Oblique positioning of lower jaw with open mouth.
Mediolateral.

This positioning is only possible in docile animals without sedation
or narcosis. One assistant secures the cat in abdominal recumbency. A
second assistant opens the mouth with straps placed over the upper
and lower incisor teeth respectively and secures the lower jaw at an
angle of 45° on the film.
The radiograph should be taken with a non-screen film (lead blocker
not to be forgotten!).
The central beam should strike the alveolar border on a level of the
4th premolar and fall at the right angle on the film.

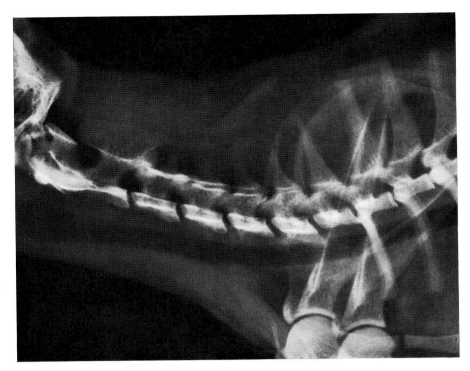

Abb. 239 Halswirbelsäule. Latero-lateral. Katze.
Bucky-Blende – Feinzeichnende Folie – FFA 115 cm – 58 kV – 30 mAs
Originalgröße (Ausschnitt aus 13 × 18 cm)
Lagerung Abb. 240

Fig. 239 Cervical vertebral column. Laterolateral. Cat.
Bucky diaphragm – High definition screens – FFD 115 – 58 kV – 30 mAs
Original size (section of 13 × 18 cm)
Positioning fig. 240

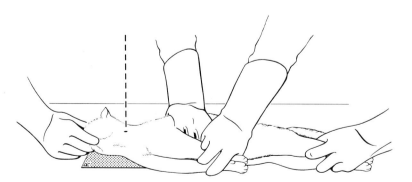

Abb. 240 Lagerung zur Aufnahme der Halswirbelsäule.
Latero-lateral.

Person 1 lagert und fixiert mit einer Hand den Kopf und mit der anderen Hand den Thorax der Katze in Seitenlage. Eine 2. am Rücken oder hinter der Katze stehende Person zieht mit einer Hand die Schultergliedmaßen gering nach hinten; dabei kann die Katze zusätzlich mit dem Unterarm auf die Unterlage gedrückt werden. Mit der anderen Hand fixiert sie die

Beckengliedmaßen und den Schwanz. Wichtig ist, daß die Medianebene des Kopfes und die des Rumpfes parallel zur Kassette gelagert sind.
Zur Reduzierung der Streustrahlung ist eine Bucky-Blende oder eine Kassette mit stehendem Raster erforderlich.
Der Zentralstrahl sollte den Hals in Höhe des 3. Halswirbels treffen und im rechten Winkel auf die Kassette einfallen.

Fig. 240 Positioning of cervical vertebral column. Laterolateral.

One assistant secures the head and the chest of the cat. The second assistant, standing behind the animal pulls the thoracic limbs slightly caudally with one hand and fixes the pelvic limbs and tail with the other hand. At the same time the forearm gently presses the cat against the table. The median plane of the head and trunk should be aligned parallel to the cassette.
To reduce scattered radiation, a Bucky diaphragm or stationary grid should be used.
The central beam should strike the neck on a level of the 3rd cervical vertebra and fall at the right angle on the cassette.

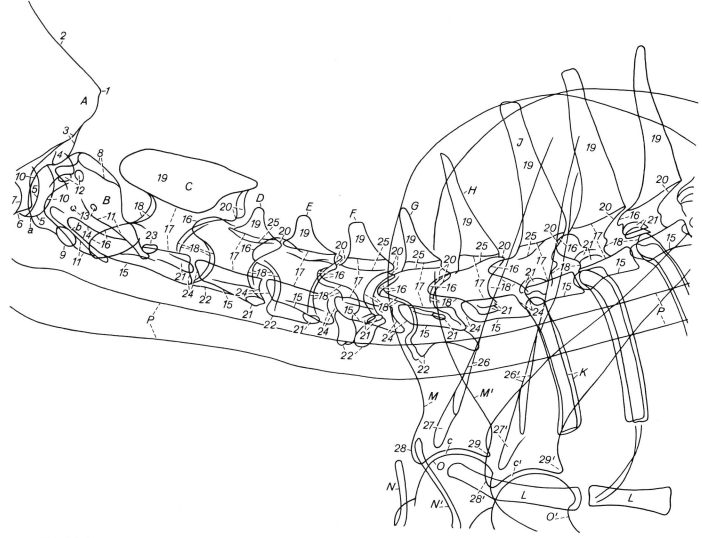

Abb. 241*   Röntgenskizze zu Abb. 239

Fig. 241*   X-ray sketch to fig. 239

A   Os occipitale;
B   Atlas;
C   Axis;
D   3. Vertebra cervicalis;
E   4. Vertebra cervicalis;
F   5. Vertebra cervicalis;
G   6. Vertebra cervicalis;
H   7. Vertebra cervicalis;
J   1. Vertebra thoracica;
K   1. Os costale;
L   Sternum;
M   Scapula;
N   Clavicula;
O   Humerus;
P   Trachea;

a   Articulatio atlantooccipitalis;
b   Articulatio atlantoaxialis;
c   Articulatio humeri;

Am Cranium – On the cranium:

1   Crista nuchae;
2   Crista sagittalis externa;
3   Squama occipitalis;
4   Foramen magnum, dorsaler Rand – Foramen magnum, dorsal margin;
5   Condylus occipitalis;
6   Os temporale, Pars basilaris;
7   Bulla tympanica;
8   Arcus dorsalis;

9   Arcus ventralis;
10   Fovea articularis cranialis;
11   Ala atlantis;
12   Foramen vertebrale laterale;
13   Gefäßloch – Vascular foramen;

Am Axis und an den folgenden Vertebrae cervicales – On the axis and the succeeding cervical vertebrae:

14   Dens;
15   Corpus vertebrae;
16   Processus articularis cranialis;
17   Foramen vertebrale, ventrale bzw. dorsale Begrenzung – Foramen vertebrale, ventral and dorsal margins respectively;
18   Incisurae vertebrales craniales et caudales, zugleich Begrenzung der Foramina intervertebralia – Incisurae vertebrales craniales et caudales forming the margins of the intervertebral foramina;
19   Processus spinosus;
20   Processus articularis caudalis;
21   Processus transversus;
22   Tuberculum ventrale;
23   Foramen transversarium, nur am Axis dargestellt – Foramen transversarium only shown on the axis;
24   Facies terminalis cranialis resp. caudalis;
25   Verschattung, die sich aus der Knochenleiste zum kranialen und kaudalen Gelenkfortsatz ergibt – Shadow formed by the connecting ridge between the cranial and caudal articular processes;
26   Spina scapulae;
27   Acromion;
28   Tuberculum supraglenoidale;
29   Tuberculum infraglenoidale.

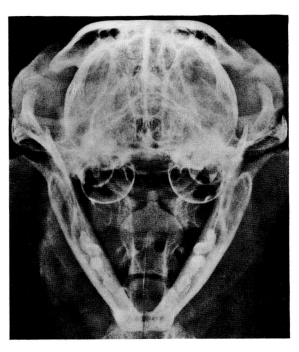

Abb. 242   1. und 2. Halswirbel durch die geöffnete Mundhöhle.
Ventro-dorsal. Katze.
Folienloser Film – FFA 115 cm – 65 kV – 30 mAs
Originalgröße (Ausschnitt aus 13 × 18 cm)
Lagerung Abb. 243

Fig. 242   1st and 2nd cervical vertebrae through the open mouth.
Ventrodorsal. Cat.
Non-screen film – FFD 115 cm – 65 kV – 30 mAs
Original size (part of 13 × 18 cm)
Positioning fig. 243

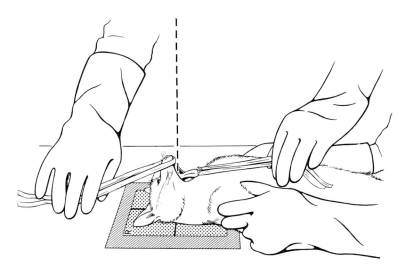

Abb. 243   Lagerung zur Aufnahme des 1. und 2. Halswirbels
durch die geöffnete Mundhöhle. Ventro-dorsal.

Die Lagerung ist nur bei einem sehr duldsamen Tier ohne Sedation bzw. Narkose möglich. Zur Erleichterung der Atmung ist die Zunge ggf. etwas vorzuziehen. Person 1 fixiert die auf dem Rücken liegende Katze mit nach kaudal gestreckten Schultergliedmaßen. Person 2 öffnet die Mundhöhle mit je einem im Bereich der Schneidezähne um den Ober- und den Unterkiefer gelegten Band und fixiert den Kopf mit Hilfe der Bänder so, daß der harte Gaumen im Winkel von etwa 80° zum Film steht. Wichtig ist, daß die Medianebene des Kopfes und die des Rumpfes senkrecht zur Kassette stehen.

Die Aufnahme ist mit einem folienlosen Film (Bleiunterlage nicht vergessen!) anzufertigen.

Der Zentralstrahl sollte die geöffnete Mundhöhle in der Mittellinie in Höhe des Zungenwulstes treffen und im rechten Winkel auf den Film einfallen.

Fig. 243   Positioning of 1st and 2nd cervical vertebrae through
the open mouth. Ventrodorsal.

This positioning is only possible in a docile animal without sedation or narcosis. To facilitate breathing the tongue is pulled out slightly. One assistant secures the cat in dorsal recumbency with the front legs pulled caudally. The second assistant opens the mouth with straps placed over the upper and lower incisor teeth and secures the head in such a manner that the hard palate forms an angle of approximately 80° with the film. The median plane of the head and trunk should be aligned perpendicularly to the film.

The radiograph should be taken with a non-screen film (lead blocker not to be forgotten!).

The central beam should strike the open mouth in the midline on a level of the root of the tongue and fall at the right angle on the film.

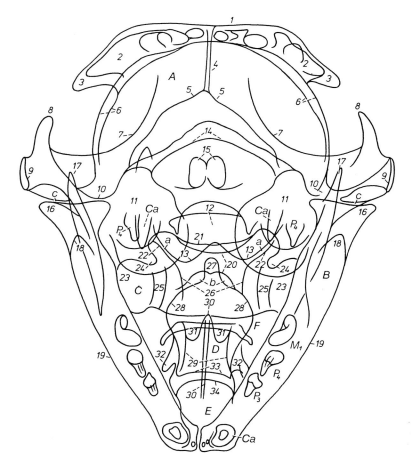

Abb. 244   Röntgenskizze zu Abb. 242   Fig. 244   X-ray sketch to fig. 242

A   Ossa cranii et faciei;
B   Mandibula;
C   Atlas;
D   Axis;
E   3. Vertebra cervicalis;
F   Os hyoideum;
Ca   Dens caninus;
P₃   Dens praemolaris III;
P₄   Dens praemolaris IV;
M₁   Dens molaris I;

a   Articulatio atlantooccipitalis;
b   Articulatio atlantoaxialis;
c   Articulatio temporomandibularis;

Am Cranium – On the cranium;

1   Os frontale, Facies externa;
2   Sinus frontalis;
3   Os frontale, Processus zygomaticus;
4   Crista sagittalis externa;
5   Crista nuchae;
6   Seitenwand der Schädelhöhle – Lateral wall of the cranial cavity;
7   Orbita, mediale Wand – Orbita, medial border;
8   Os zygomaticum, Processus frontalis;
9   Arcus zygomaticus, orthograph getroffen – Arcus zygomaticus, orthographically struck;
10   Tuberculum articulare, zugleich Basis des Processus zygomaticus der Pars squamosa des Os temporale – Tuberculum articulare forming the base of the zygomatic process of the squamous temporal bone;
11   Os temporale, Partes petrosa et tympanica;
12   Foramen magnum;
13   Condylus occipitalis;

14   Tentorium cerebelli osseum;
15   Os ethmoidale, Lamina cribrosa;

An der Mandibula – On the mandible:

16   Processus condylaris;
17   Processus coronoideus;
18   Verschattung, die sich aus dem Angulus mandibulae dorsal vom Processus angularis ergibt – Shadow formed by the angle of the mandible dorsally to the angular process;
19   Corpus mandibulae;

Am Atlas – On the atlas:

20   Arcus ventralis;
21   Arcus dorsalis, kranialer Rand – Arcus dorsalis, cranial border;
22   Fovea articularis cranialis;
23   Ala atlantis;
24   Incisura alaris;
25   Canalis alaris;
26   Foramen vertebrale;

Am Axis – On the axis:

27   Dens;
28   Processus articularis cranialis;
29   Foramen vertebrale, zugleich Pediculus arcus vertebrae – Foramen vertebrale and Pediculus arcus vertebrae;
30   Processus spinosus;
31   Incisura vertebralis cranialis;
32   Processus transversus;
33   Facies terminalis caudalis;

An der 3. Vertebra cervicalis – On the 3rd cervical vertebra:

34   Facies terminalis cranialis.

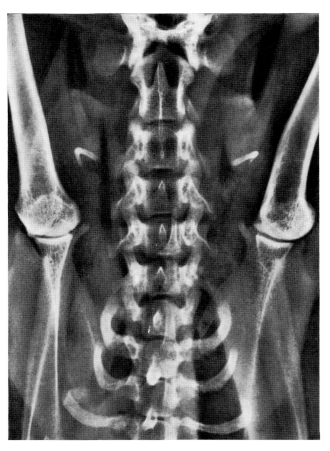

Abb. 245   Halswirbelsäule. Ventro-dorsal. Katze.
Bucky-Blende – Feinzeichnende Folie – FFA 115 cm – 60 kV – 30 mAs
Originalgröße (Ausschnitt aus 18 × 24 cm)
Lagerung Abb. 246

Fig. 245   Cervical vertebral column. Ventrodorsal. Cat.
Bucky diaphragm – High definition screens – FFD 115 cm – 60 kV – 30 mAs
Original size (section of 18 × 24 cm)
Positioning fig. 246

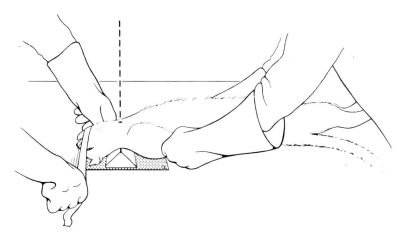

Abb. 246   Lagerung zur Aufnahme der Halswirbelsäule.
Ventro-dorsal.

Person 1 hält die Katze in Rückenlage und streckt die Schultergliedmaßen nach hinten. Person 2 zieht den Kopf nach vorn, wobei die Kehlränder des Unterkiefers in gleicher Höhe liegen sollen. Der Kopf wird in dieser Lage mit einem quer über den Unterkiefer gelegten und durch die Hände straff gespannten Band fixiert. Die korrekte horizontale Lagerung der Halswirbelsäule wird durch einen Schaumgummikeil unter den ersten Halswirbel erleichtert. Es ist darauf zu achten, daß die Medianebene des Kopfes und die des Rumpfes senkrecht zur Kassette stehen.
Zur Reduzierung der Streustrahlung ist eine Bucky-Blende oder eine Kassette mit stehendem Raster zweckmäßig.
Der Zentralstrahl sollte den Hals median in Höhe des 3. Halswirbels treffen und im rechten Winkel auf die Kassette einfallen.

Fig. 246   Positioning of cervical vertebral column. Ventrodorsal.

One assistant holds the cat in dorsal recumbency, stretching the thoracic limbs caudally. The second assistant pulls the head forward, making sure that the ventral borders of the mandible lie on the same level. The head is secured in this position by means of a band stretched tightly across the mandible with the hands. Placing a foam rubber wedge under the cranial cervical vertebrae ensures correct horizontal positioning of the cervical vertebral column. The median plane of the head and trunk should be aligned perpendicularly to the cassette.
A Bucky diaphragm or a cassette with a stationary grid is used to reduce scattered radiation.
The central beam should strike the midline of the neck on a level of the 3rd cervical vertebra and fall at the right angle on the cassette.

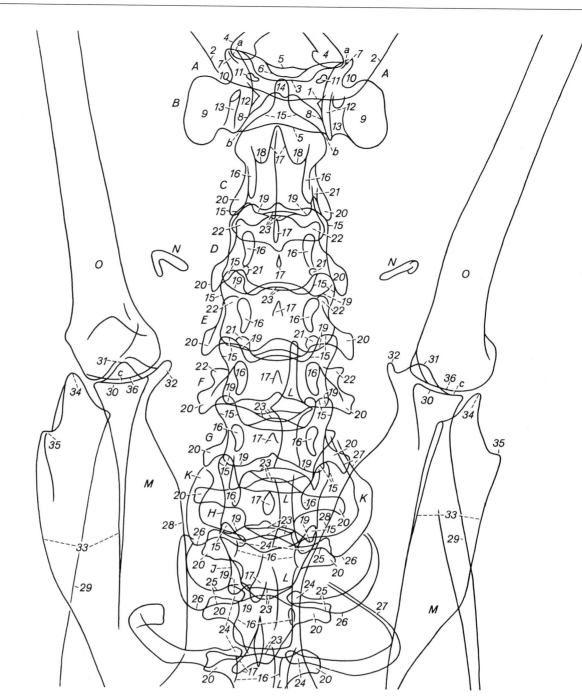

Abb. 247   Röntgenskizze zu Abb. 245   Fig. 247   X-ray sketch to fig. 245

A   Os occipitale;
B   Atlas;
C   Axis;
D   3. Vertebra cervicalis;
E   4. Vertebra cervicalis;
F   5. Vertebra cervicalis;
G   6. Vertebra cervicalis;
H   7. Vertebra cervicalis;
J   1. Vertebra thoracica;
K   1. Os costale;
L   Sternum;
M   Scapula;
N   Clavicula;
O   Humerus;

a   Articulatio atlantooccipitalis;
b   Articulatio atlantoaxialis;
c   Articulatio humeri;

Am Cranium – On the cranium:

1   Crista nuchae;
2   Crista temporalis;
3   Squama occipitalis;
4   Condylus occipitalis;

Am Atlas – On the atlas:

5   Arcus dorsalis;
6   Arcus ventralis;
7   Fovea articularis cranialis;
8   Fovea articularis caudalis;
9   Ala atlantis;
10   Incisura alaris;
11   Foramen vertebrale laterale;
12   Foramen vertebrale, seitliche Begrenzung, zugleich orthograph getroffene Abschnitte von 5 und 6 – Foramen vertebrale, lateral margin forming the orthographically struck parts of 5 and 6;
13   Foramen transversarium;

Am Axis und an den folgenden Vertebrae cervicales – On the axis and the succeeding cervical vertebrae:

14   Dens;
15   Processus articularis cranialis;
16   Pediculus arcus vertebrae, zugleich seitliche Begrenzung des Foramen vertebrale – Pediculus arcus vertebrae forming the lateral margin of the vertebral foramen;
17   Processus spinosus;
18   Incisura vertebralis cranialis;

19   Processus articularis caudalis;
20   Processus transversus;
21   Tuberculum dorsale;
22   Tuberculum ventrale;
23   Facies terminalis cranialis bzw. caudalis zweier aufeinanderfolgender Wirbel – Facies terminales cranialis and caudalis of two succeeding vertebrae respectively;

An den Rippen – On the ribs:

24   Caput costae;
25   Collum costae;
26   Tuberculum costae;
27   Cartilago costae;

An der Scapula – On the scapula:

28   Margo cranialis;
29   Margo caudalis;
30   Cavitas glenoidalis;
31   Tuberculum supraglenoidale;
32   Processus coracoideus;
33   Spina scapulae;
34   Acromion, Processus hamatus;
35   Processus suprahamatus;
36   Caput humeri.

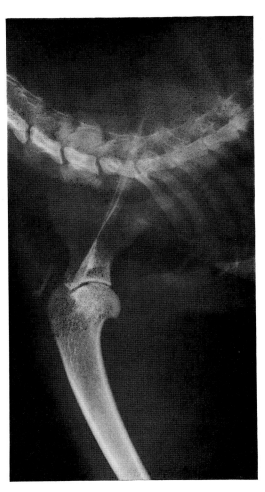

Abb. 248   Rechtes Schultergelenk. Medio-lateral. Katze.
Folienloser Film – FFA 115 cm – 70 kV – 40 mAs
Originalgröße (Ausschnitt aus 13 × 18 cm)
Lagerung Abb. 249

Fig. 248   Right shoulder joint. Mediolateral. Cat.
Non-screen film – FFD 115 cm – 70 kV – 40 mAs
Original size (section of 13 × 18 cm)
Positioning fig. 249

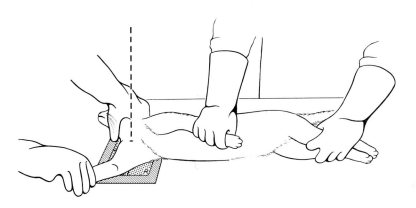

Abb. 249   Lagerung zur Aufnahme des Schultergelenks und des
Oberarms. Medio-lateral.

Die Katze befindet sich in Seitenlage. Die obenliegende Schultergliedmaße ist so weit nach hinten und die zu untersuchende, untenliegende Schultergliedmaße so weit nach vorn zu ziehen, daß deren Schultergelenk ohne Über-

lagerung dargestellt wird. Dabei sind Kopf und Hals zum Rücken hin gestreckt zu fixieren.
Die Aufnahme kann mit einem folienlosen Film (Bleiunterlage nicht vergessen!) angefertigt werden. Nur bei einem sehr korpulenten Tier ist eine Bucky-Blende oder eine Kassette mit stehendem Raster zur Verringerung der Streustrahlung notwendig.
Schultergelenk: Der Zentralstrahl sollte das Gelenk treffen und im rechten Winkel auf den Film einfallen.
Humerus: Der Zentralstrahl sollte den Humerus in Schaftmitte treffen und im rechten Winkel auf den Film einfallen.

Fig. 249   Positioning of shoulder joint and arm. Mediolateral.

The cat is placed in lateral recumbency. The upper thoracic limb is pulled far enough caudally, the lower one cranially to avoid superimposition of the shoulder joints. The head and neck are pulled backward.
A non-screen film may be used (lead blocker not to be forgotten!). In obese cats, a Bucky diaphragm or a cassette with stationary grid should be used to reduce scattered radiation.
Shoulder joint: The central beam should strike the joint and fall at the right angle on the film.
Humerus: The central beam should strike the mid-shaft of the humerus and fall at the right angle on the film.

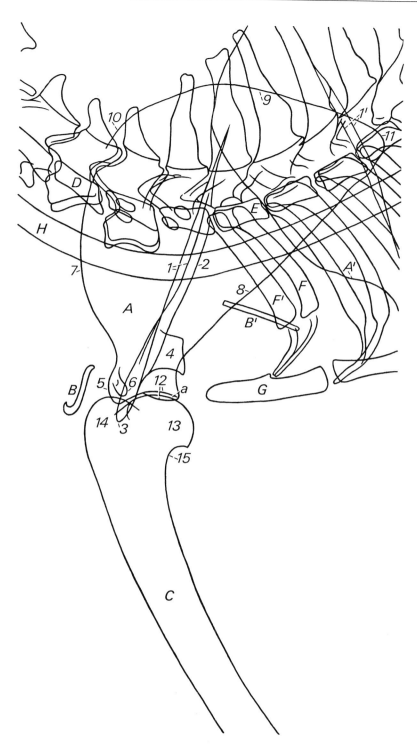

Abb. 250* Röntgenskizze zu Abb. 248
Fig. 250* X-ray sketch to fig. 248

A   Scapula;
B   Clavicula;
C   Humerus;
D   5. Vertebra cervicalis;
E   1. Vertebra thoracica;
F   1. Os costale;
G   Manubrium sterni;
H   Trachea;

a   Articulatio humeri;

1, 2   Spina scapulae:
1   Basis – Base,
2   Rand – Border;

3   Acromion;
4   Processus suprahamatus;
5   Tuberculum supraglenoidale;
6   Processus coracoideus;
7   Margo cranialis;
8   Margo caudalis;
9   Margo dorsalis;
10   Angulus cranialis;
11   Angulus caudalis;
12   Cavitas glenoidalis;
13   Caput humeri;
14   Tuberculum majus;
15   Collum humeri.

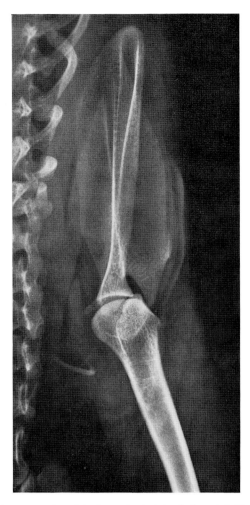

Abb. 251 Linkes Schultergelenk. Kaudo-kranial. Katze.
Folienloser Film – FFA 115 cm – 65 kV – 30 mAs
Originalgröße (Ausschnitt aus 13 × 18 cm)
Lagerung Abb. 253

Fig. 251 Left shoulder joint. Caudocranial. Cat.
Non-screen film – FFD 115 cm – 65 kV – 30 mAs
Original size (section of 13 × 18 cm)
Positioning fig. 253

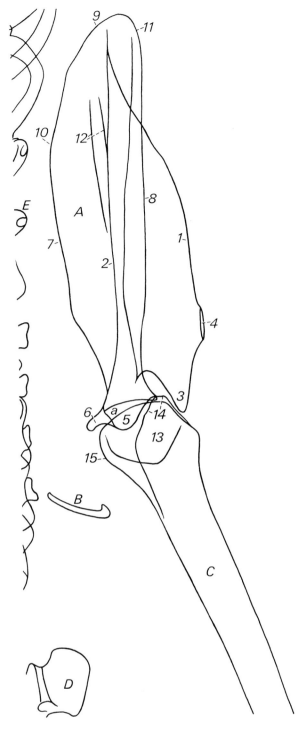

Abb. 252   Röntgenskizze zu Abb. 251
Fig. 252   X-ray sketch to fig. 251

A   Scapula;
B   Clavicula;
C   Humerus;
D   Atlas;
E   2. Os costale;

a   Articulatio humeri;

1   Spina scapulae;
2   Verschattung, die sich aus der Konkavität der Fossa subscapularis er-
     gibt – Shadow formed by the concavity of the subscapular fossa;
3   Acromion;

4   Processus suprahamatus;
5   Tuberculum supraglenoidale;
6   Processus coracoideus;
7   Margo cranialis;
8   Margo caudalis;
9   Margo dorsalis;
10   Angulus cranialis;
11   Angulus caudalis;
12   Muskelleisten – Muscular lines;
13   Caput humeri;
14   Tuberculum majus;
15   Tuberculum minus.

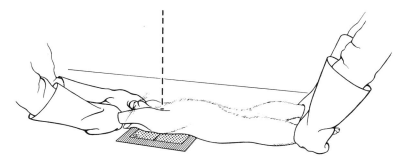

Abb. 253   Lagerung zur Aufnahme des Schultergelenks.
Kaudo-kranial.

Die Schultergliedmaßen des auf dem Rücken liegenden Tieres werden von einer Person gestreckt nach vorn gezogen. Die haltenden Hände stützen sich auf der Tischplatte. Die Gliedmaßen des Tieres werden gegen den Kopf gedrückt, um diesen mitzufixieren.
Beim nicht narkotisierten Tier faßt der am Kopf der Katze stehende Helfer mit einer Hand die zu untersuchende Gliedmaße, mit der anderen Hand sollte er die zweite Schultergliedmaße sowie den Hals mit Gabelgriff unmittelbar hinter dem Kopf fixieren. Zu beachten ist, daß der Ellbogenhöcker der zu untersuchenden Gliedmaße nach oben zeigt und die Medianebene des Rumpfes senkrecht zur Kassette steht.
Die Aufnahme kann mit einem folienlosen Film (Bleiunterlage nicht vergessen!) angefertigt werden. Bei fettleibigen Tieren ist eine Bucky-Blende oder eine Kassette mit stehendem Raster zur Verringerung der Streustrahlung nützlich.
Der Zentralstrahl sollte kaudal auf die Mitte der Gliedmaße in die Beuge des Schultergelenks treffen und im rechten Winkel auf den Film einfallen.

Fig. 253   Positioning of shoulder joint. Caudocranial.

The animal is placed on its back. An assistant pulls the extended thoracic limbs forward supporting his hands on the table. The head of the animal is secured by pressing the thoracic limbs firmly against it. In non-sedated animals the assistant standing at the head of the cat secures the thoracic limb to be radiographed with one hand while the other hand holds the second thoracic limb and the neck firmly in a fork grip immediately behind the head. The olecranon of the leg to be examined should point upward. The median plane of the trunk must be perpendicular to the cassette.
A non-screen film may be used (lead blocker not to be forgotten!).
In obese animals a Bucky diaphragm or a cassette with stationary grid is used to reduce scattered radiation.
The central beam should strike the middle of the flexion surface of the joint and fall at the right angle on the film.

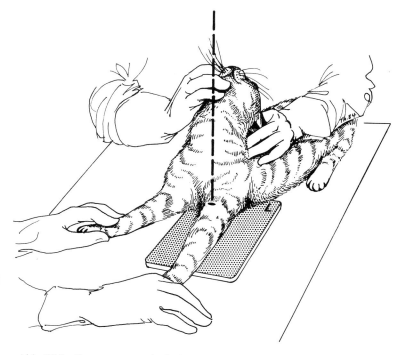

Abb. 254   Lagerung zur Aufnahme des Oberarms bei Frakturverdacht.
Kranio-kaudal.

Person 1 fixiert mit einer Hand den Kopf mit möglichst weit zum Rücken hin abgebogenem Hals. Mit der anderen Hand wird die Katze am Rücken fixiert und dabei mit dem Unterarm auf den Tisch gedrückt. Die 2. Person fixiert die Schultergliedmaßen an den Pfoten.
Die Aufnahme kann mit einem folienlosen Film (Bleiunterlage nicht vergessen!) angefertigt werden. Bei fettleibigen Tieren ist eine Bucky-Blende oder eine Kassette mit stehendem Raster zur Verringerung der Streustrahlung nützlich.
Der Zentralstrahl sollte den Humerus in Schaftmitte treffen und im rechten Winkel auf den Film einfallen.

Fig. 254   Positioning of arm in case of suspected fracture of the humerus.
Craniocaudal.

One assistant holds the head with one hand, pulling it backward as far as possible. The other hand secures the cat on the back, at the same time pressing it with the forearm against the table. The second assistant secures both of the thoracic limbs at the paws.
A non-screen film should be used (lead blocker not to be forgotten!). In obese animals a Bucky diaphragm or a cassette with stationary grid is necessary to reduce scattered radiation.
The central beam should strike the mid-shaft of the humerus and fall at the right angle on the film.

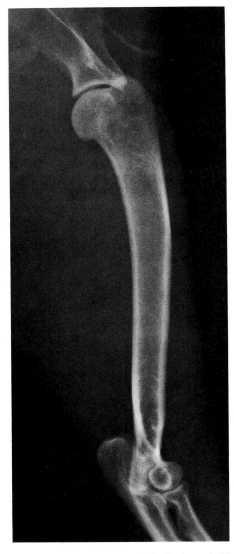

Abb. 255 Linker Oberarm. Medio-lateral. Katze.
Bucky-Blende – Feinzeichnende Folie – FFA 120 cm – 60 kV – 10 mAs
Originalgröße (Ausschnitt aus 13 × 18 cm)
Lagerung Abb. 249

Fig. 255 Left arm. Mediolateral. Cat.
Bucky-diaphragm – High definition screens – FFD 120 cm – 60 kV – 10 mAs
Original size (section of 13 × 18 cm)
Positioning fig. 249

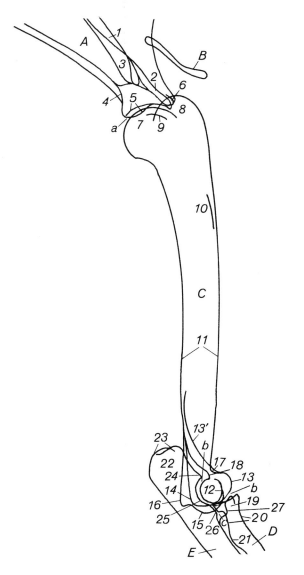

Abb. 256 Röntgenskizze zu Abb. 255
Fig. 256 X-ray sketch to fig. 255

A   Scapula;
B   Clavicula;
C   Humerus;
D   Radius;
E   Ulna;

a   Articulatio humeri;
b   Articulatio cubiti;
c   Articulatio radioulnaris proximalis;

An der Scapula – On the scapula:

1   Spina scapulae;
2   Acromion;
3   Processus suprahamatus;
4   Collum scapulae;
5   Cavitas glenoidalis;
6   Tuberculum supraglenoidale mit Processus coracoideus – Tuberculum supraglenoidale with the coracoid process;

Am Humerus – On the humerus:

7   Caput humeri;
8   Tuberculum majus;
9   Tuberculum minus;
10 Tuberositas deltoidea;
11 Corpus humeri;

12 Condylus humeri, orthograph getroffene Gelenkfläche – Condylus humeri, articular surface is struck orthographically;
13 Lateraler Rand des Condylus humeri – Lateral border of the condyle of the humerus;
13' Crista supracondylaris lateralis;
14 Epicondylus lateralis;
15 Medialer Rand des Condylus humeri – Medial border of the condyle of the humerus;
16 Epicondylus medialis;
17 Fossa radialis;
18 Fossa olecrani;

Am Radius – On the radius:

19 Caput radii;
20 Collum radii;
21 Tuberositas radii;

An der Ulna – On the ulna:

22 Olecranon;
23 Tuber olecrani;
24 Processus anconaeus;
25 Incisura trochlearis;
26 Processus coracoideus lateralis;
27 Processus coracoideus medialis.

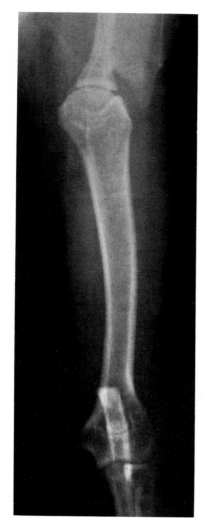

Abb. 257 Linker Oberarm. Kranio-kaudal. Katze.
Bucky-Blende – Feinzeichnende Folie – FFA 120 cm – 60 kV – 20 mAs
Originalgröße (Ausschnitt aus 18 × 24 cm)
Lagerung Abb. 254

Fig. 257 Left arm. Craniocaudal. Cat.
Bucky diaphragm – High definition screens – FFD 120 cm – 60 kV – 20 mAs
Original size (section of 18 × 24 cm)
Positioning fig. 254

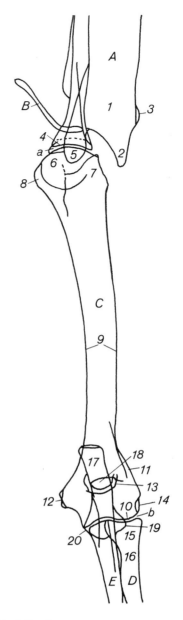

Abb. 258 Röntgenskizze zu Abb. 257
Fig. 258 X-ray sketch to fig. 257

A   Scapula;
B   Clavicula;
C   Humerus;
D   Radius;
E   Ulna;

a   Articulatio humeri;
b   Articulatio cubiti;

An der Scapula – On the scapula:

1   Spina scapulae;
2   Acromion;
3   Processus suprahamatus;
4   Cavitas glenoidalis;
5   Tuberculum supraglenoidale;

Am Humerus – On the humerus:

6   Caput humeri;
7   Tuberculum majus;
8   Tuberculum minus;
9   Corpus humeri;

10   Condylus humeri;
11   Epicondylus lateralis;
12   Epicondylus medialis;
13   Facies articularis, kaudaler Rand, zugleich Begrenzung der Fossa ole-crani – Facies articularis, caudal border forming the border of the ole-cranon fossa;
14   Laterale Bandgrube – Lateral depression for ligamentous attachment;

Am Radius – On the radius:

15   Caput radii;
16   Collum radii;

An der Ulna – On the ulna:

17   Olecranon;
18   Processus anconaeus;
19   Processus coronoideus lateralis;
20   Processus coronoideus medialis.

177

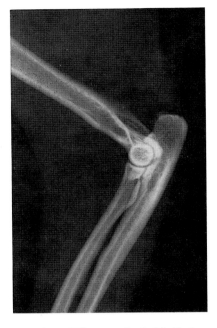

Abb. 259   Rechtes Ellbogengelenk. Medio-lateral. Katze.
Folienloser Film – FFA 115 cm – 65 kV – 25 mAs
Originalgröße (Ausschnitt aus 13 × 18 cm)
Lagerung Abb. 267

Fig. 259   Right elbow joint. Mediolateral. Cat.
Non-screen film – FFD 115 cm – 65 kV – 25 mAs
Original size (section of 13 × 18 cm)
Positioning fig. 267

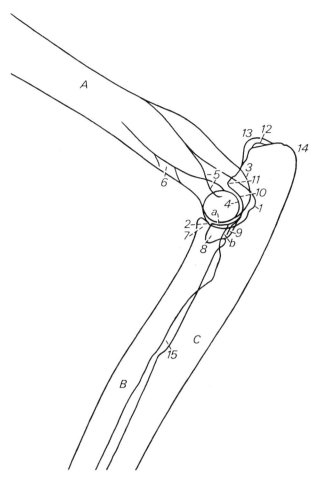

Abb. 260   Röntgenskizze zu Abb. 259
Fig. 260   X-ray sketch to fig. 259

A	Humerus;
B	Radius;
C	Ulna;
a	Articulatio cubiti;
b	Articulatio radioulnaris proximalis;
1	Epicondylus medialis;
2	Trochlea humeri;
3	Epicondylus lateralis;
4	Kompaktaschatten, der sich aus der Konkavität der Führungsrinne am Condylus humeri ergibt – Shadow of compacta formed by the concavity of the groove of the humeral condyle;

5	Fossa olecrani;
6	Foramen supracondylare;
7	Tuberositas radii;
8	Processus coronoideus medialis;
9	Processus coronoideus lateralis;
10	Incisura trochlearis;
11	Processus anconaeus;
12, 13, 14 Tuber olecrani:	
12	Kraniolateraler Höcker – Craniolateral prominence,
13	Kraniomedialer Höcker – Craniomedial prominence,
14	Kaudaler Höcker – Caudal prominence;
15	Spatium interosseum antebrachii.

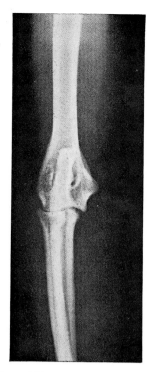

Abb.   261 Rechtes Ellbogengelenk. Kranio-kaudal. Katze.
Folienloser Film – FFA 115 cm – 65 kV – 30 mAs
Originalgröße (Ausschnitt aus 13 × 18 cm)
Lagerung Abb. 268

Fig. 261   Right elbow joint. Craniocaudal. Cat.
Non-screen film – FFD 115 cm – 65 kV – 30 mAs
Original size (section of 13 × 18 cm)
Positioning fig. 268

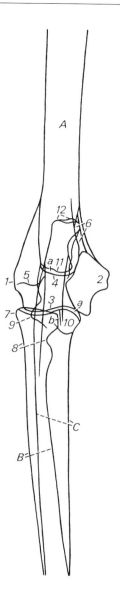

Abb. 262   Röntgenskizze zu Abb. 261
Fig. 262   X-ray sketch to fig. 261

A   Humerus;
B   Radius;
C   Ulna;

a   Articulatio cubiti;
b   Articulatio radioulnaris proximalis;

1   Epicondylus lateralis;
2   Epicondylus medialis;
3   Condylus humeri, Rand – Condylus humeri, border;
4   Facies articularis, kaudaler Rand, zugleich Begrenzung der Fossa ole-
     crani – Facies articularis, caudal border forming the border of the ole-
     cranon fossa;

5   Kraniale Begrenzung der Gelenkfläche des Condylus humeri – Cra-
     nial border of the articular surface of the condyle of the humerus;
6   Foramen supracondylare;
7   Caput radii;
8   Collum radii;
9   Processus coronoideus lateralis;
10   Processus coronoideus medialis;
11   Processus coracoideus;
12   Dreigeteiltes Tuber olecrani – Trifid olecranon tuberosity.

179

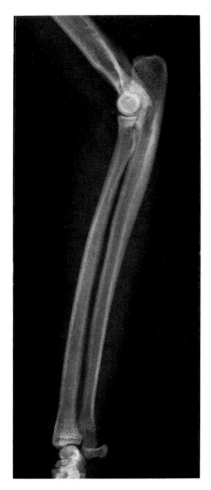

Abb. 263 Rechter Unterarm. Medio-lateral. Katze.
Folienloser Film – FFA 120 cm – 60 kV – 16 mAs
Originalgröße (Ausschnitt aus 13 × 18 cm)
Lagerung Abb. 267

Fig. 263 Right forearm. Mediolateral. Cat.
Non-screen film – FFD 120 cm – 60 kV – 16 mAs
Original size (section of 13 × 18 cm)
Positioning fig. 267

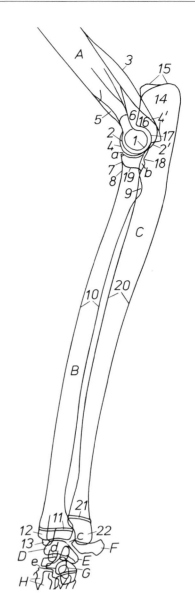

Abb. 264 Röntgenskizze zu Abb. 263
Fig. 264 X-ray sketch to fig. 263

A   Humerus;
B   Radius;
C   Ulna;
D   Os carpi intermedioradiale;
E   Os carpi ulnare;
F   Os carpi accessorium;
G   Ossa carpalia I – IV mit Überlagerungen – Ossa carpalia I – IV with
    overshadows;
H   Ossa metacarpalia;

a   Articulatio cubiti;
b   Articulatio radioulnaris proximalis;
c   Articulatio antebrachiocarpea;
d   Articulatio mediocarpea;
e   Articulationes carpometacarpeae;

Am Humerus – On the humerus:

1   Kompaktaschatten, der sich aus der Konkavität der Führungsrinne
    am Condylus humeri ergibt – Shadow of compacta formed by the
    groove of the condyle of the humerus;
2   Lateraler Rand des Condylus humeri – Lateral border of the condyle
    of the humerus;
2' Epicondylus lateralis;
3   Crista supracondylaris lateralis;
4   Medialer Rand des Condylus humeri – Medial border of the condyle
    of the humerus;
4' Epicondylus medialis;
5   Foramen supracondylare;
6   Fossa olecrani;

Am Radius – On the radius:

7   Caput radii;
8   Collum radii;
9   Tuberositas radii;
10 Corpus radii;
11 Distale Epiphysenfuge – Distal epiphyseal groove;
12 Distale Epiphyse mit Trochlea radii – Distal epiphysis with the ra-
    dial trochlea;
13 Processus styloideus radii;

An der Ulna – On the ulna:

14 Olecranon;
15 Tuber olecrani;
16 Processus anconaeus;
17 Incisura trochlearis;
18 Processus coronoideus lateralis;
19 Processus coronoideus medialis;
20 Corpus ulnae;
21 Distale Epiphysenfuge – Distal epiphyseal groove;
22 Caput ulnae mit Processus styloideus – Caput ulnae with the styloid
    process.

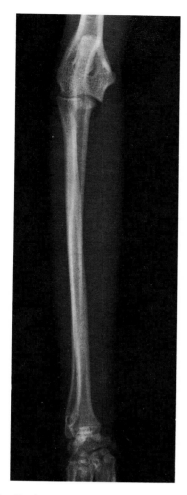

Abb. 265   Rechter Unterarm. Kranio-kaudal. Katze.
Folienloser Film – FFA 120 cm – 60 kV – 20 mAs
Originalgröße (Ausschnitt aus 13 × 18 cm)
Lagerung Abb. 268

Fig. 265   Right forearm. Craniocaudal. Cat.
Non-screen film – FFD 120 cm – 60 kV – 20 mAs
Original size (section of 13 × 18 cm)
Positioning fig. 268

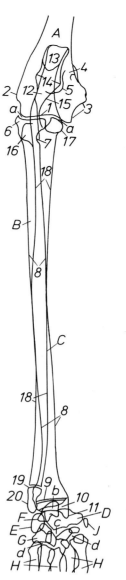

Abb. 266   Röntgenskizze zu Abb. 265
Fig. 266   X-ray sketch to fig. 265

A   Humerus;
B   Radius;
C   Ulna;
D   Os carpi intermedioradiale;
E   Os carpi ulnare;
F   Os carpi accessorium;
G   Ossa carpalia I – IV;
H   Ossa metacarpalia I – V;
J   Os sesamoideum m. abductoris pollicis longi;

a   Articulatio cubiti;
b   Articulatio antebrachiocarpea;
c   Articulatio mediocarpea;
d   Articulationes carpometacarpeae;

Am Humerus – On the humerus:

1   Condylus humeri;
2   Epicondylus lateralis;
3   Epicondylus medialis;
4   Foramen supracondylare;
5   Fossa olecrani;

Am Radius – On the radius:

6   Caput radii;
7   Collum radii;
8   Corpus radii;
9   Distale Epiphysenfuge – Distal epiphyseal groove;
10   Trochlea radii;
11   Processus styloideus radii;

An der Ulna – On the ulna:

12   Olecranon;
13   Tuber olecrani;
14   Processus anconaeus;
15   Incisura trochlearis;
16   Processus coronoideus lateralis;
17   Processus coronoideus medialis;
18   Corpus ulnae;
19   Distale Epiphysenfuge – Distal epiphyseal groove;
20   Caput ulnae mit Processus styloideus – Caput ulnae with the styloid process.

181

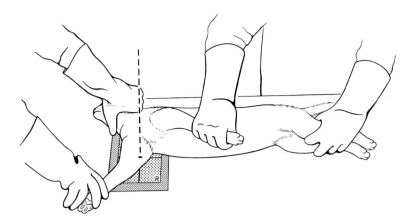

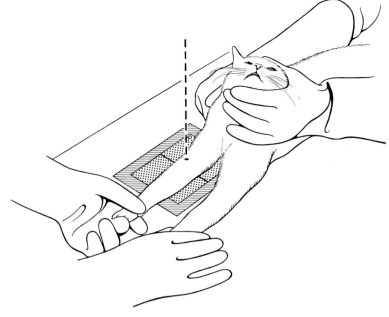

Abb. 267 Lagerung zur Aufnahme des Ellbogengelenks und des Unterarms. Medio-lateral.

Die Katze befindet sich in Seitenlage. Der am Kopf stehende Helfer fixiert mit einer Hand im Gabelgriff Kopf und Hals und mit der anderen die etwas nach vorn gezogene, untenliegende, zu untersuchende Schultergliedmaße. Der am Rücken der Katze stehende 2. Helfer fixiert mit einer Hand beide Beckengliedmaßen und mit der anderen die nach hinten gezogene, obenliegende Schultergliedmaße. Er drückt dabei mit dieser Hand die Katze auf den Tisch.
Die Aufnahme ist mit einem folienlosen Film (Bleiunterlage nicht vergessen!) anzufertigen.
Ellbogengelenk: Der Zentralstrahl sollte das Ellbogengelenk unmittelbar distal des Epicondylus medialis humeri treffen und im rechten Winkel auf den Film einfallen.
Unterarm: Der Zentralstrahl sollte den Radius in Schaftmitte treffen und im rechten Winkel auf den Film einfallen.

Fig. 267 Positioning of elbow joint and forearm. Mediolateral.

The cat lies in lateral recumbency. The assistant at the head of the cat secures the head and neck in a fork grip with one hand and with the other hand he pulls the lower thoracic limb to be radiographed sligtly cranially. The second assistant, standing at the back of the cat holds both pelvic limbs with one hand and secures the upper thoracic limb with the other hand, pulling it caudally and pressing the cat against the table. A non-screen film should be used (lead blocker not to be forgotten!).
Elbow joint: The central beam should strike the elbow joint immediately distal to the medial humeri epicondyle and fall at the right angle on the film.
Forearm: The central beam should strike the mid-shaft of the radius and fall at the right angle on the film.

Abb. 268 Lagerung zur Aufnahme des Ellbogengelenks und des Unterarms. Kranio-kaudal.

Person 1 fixiiert mit einer Hand den Kopf, indem der Hals möglichst weit zum Rücken hin abgebogen wird, und mit der anderen Hand die zu untersuchende Gliedmaße in Höhe des Schultergelenks. Zusätzlich sollte die Person zwischen ihren Unterarmen die Katze stabilisieren. Die 2. Person fixiert die Schultergliedmaßen an den Pfoten. Wichtig ist, daß das zu untersuchende Ellbogengelenk nur mit dem Kaudalrand des Ellbogengelenkfortsatzes dem Film aufliegt. Bei einer Verkantung, also bei seitlicher Auflage des Olekranons und des Epicondylus lateralis bzw. medialis, wird die Ulna von einem Epikondylus überlagert.
Die Aufnahme ist mit einem folienlosen Film (Bleiunterlage nicht vergessen!) anzufertigen.
Ellbogengelenk: Der Zentralstrahl sollte die Mitte der Gliedmaße unmittelbar distal der Epikondylen treffen und im rechten Winkel auf den Film einfallen.
Unterarm: Der Zentralstrahl sollte den Radius in der Schaftmitte treffen und im rechten Winkel auf den Film einfallen.

Fig. 268 Positioning of elbow joint and of forearm. Craniocaudal.

One assistant secures the head with one hand bending the neck backward as far as possible. The other hand supports the limb to be examined at the level of the shoulder joint. At the same time the cat is stabilized between his forearms. The second assistant secures the thoracic limbs at the paws. It is important that the elbow joint to be examined rests on the film with the caudal border of the olecranon only. Should the olecranon be tilted sideways, one of the epicondyles will be superimposed upon the ulna.
A non-screen film should be used (lead blocker not to be forgotten!).
Elbow joint: The central beam should strike the center of the limb immediately distal to the epicondyles and fall at the right angle on the film.
Forearm: The central beam should strike the forearm at the mid-shaft of the radius and fall at the right angle on the film.

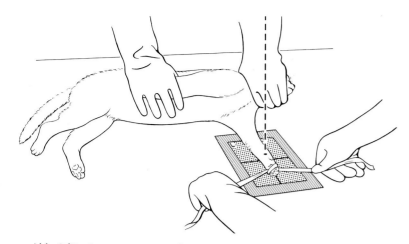

Abb. 269   Lagerung zur Aufnahme des Vorderfußes. Medio-lateral.

Die Katze befindet sich in Seitenlage. Die am Rücken der Katze stehende Person 1 fixiert mit einer Hand Kopf und Hals im Gabelgriff. Mit der anderen Hand wird zunächst die obenliegende Schultergliedmaße dem Rumpf seitlich angelegt und dann die Katze am Rumpf gefaßt und fixiert. Person 2 streckt die zu untersuchende Schultergliedmaße. Dabei wird um die Pfote eine Schlaufe gelegt und diese von beiden Händen der 2. Person so über dem Film fixiert, daß die Zehen möglichst senkrecht übereinanderliegen.
Bei einer Katze, die sich gegen diese Lagerung sehr wehrt, wird es von der Fragestellung und der Situation abhängen, ob die Beckengliedmaßen und der Schwanz von einer 3. Person zu fixieren sind oder ob das Tier besser zu sedieren ist.
Die Aufnahme ist mit einem folienlosen Film (Bleiunterlage) nicht vergessen!) anzufertigen.
Der Zentralstrahl sollte die Mitte der Gliedmaße in Höhe des Karpalgelenks treffen und im rechten Winkel auf den Film einfallen.

Fig. 269   Positioning of forepaw. Mediolateral.

The cat lies in lateral recumbency. One assistant at the back of the animal secures the head and neck with one hand in a fork grip. With the other hand the upper thoracic limb is pulled backward and pressed against the side of the trunk, securing the latter at the same time. The second assistant extends the thoracic limb to be examined over the film by means of a band tied in a loop around the paw. Care must be taken that the digits lie perpendicularly on top of each other.
In restless cats a third assistant may be required to secure the hind limbs and tail or the animal may be sedated if necessary.
A non-screen film should be used (lead blocker not to be forgotten!).
The central beam should strike the center of the limb on a level of the carpal joint and fall at the right angle on the film.

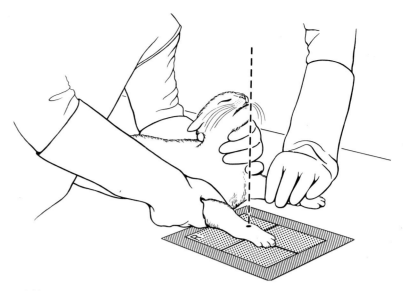

Abb. 270   Lagerung zur Aufnahme des Vorderfußes. Dorso-palmar.

Person 1 fixiert mit einer Hand Kopf und Hals, wobei der Hals zum Rücken hin abgebogen wird. Mit der anderen Hand wird die zu untersuchende Schultergliedmaße in Höhe des Ellbogengelenks umfaßt und die Pfote auf den Film gedrückt. Außerdem sollte Person 1 mit den Unterarmen die Katze stabilisieren. Person 2 fixiert mit der einen Hand die andere Schultergliedmaße und mit der anderen ggf. beide Beckengliedmaßen.
Die Aufnahme ist mit einem folienlosen Film (Bleiunterlage nicht vergessen!) anzufertigen.
Der Zentralstrahl sollte die Mitte der Gliedmaße in Höhe des Karpalgelenks treffen und im rechten Winkel auf den Film einfallen.

Fig. 270   Positioning of forepaw. Dorsopalmar.

One assistant secures the head and neck with one hand, bending the neck backward. The other hand holds the limb to be examined at the elbow, pressing the forearm firmly against the film. At the same time the cat is stabilized between his forearms. The second assistant secures the other thoracic limb with one hand and with the other hand both pelvic limbs.
A non-screen film should be used (lead blocker not to be forgotten!).
The central beam should strike the center of the limb on a level of the carpal joint and fall at the right angle on the film.

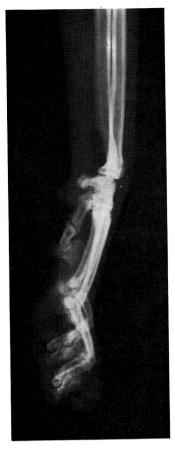

Abb. 271   Linker Vorderfuß. Medio-lateral. Katze.
Folienloser Film – FFA 120 cm – 60 kV – 15 mAs
Originalgröße (Ausschnitt aus 13 × 18 cm)
Lagerung Abb. 269

Fig. 271   Left forepaw. Mediolateral. Cat.
Non-screen film – FFD 120 cm – 60 kV – 15 mAs
Original size (section of 13 × 18 cm)
Positioning fig. 269

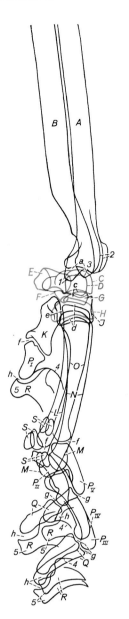

Abb. 272   Röntgenskizze zu Abb. 271
Fig. 272   X-ray sketch to fig. 271

A   Radius;
B   Ulna;
C   Os carpi intermedioradiale;
D   Os carpi ulnare;
E   Os carpi accessorium;
F   Os carpale I;
G   Os carpale II;
H   Os carpale III;
J   Os carpale IV;
K   Os metacarpale I;
L   Os metacarpale II;
M   Os metacarpale III;
N   Os metacarpale IV;
O   Os metacarpale V;
P   Phalanx proximalis (I–V kennzeichnen die Skelettelemente an der 1. bis 5. Zehe) – Phalanx proximalis (I–V show the skeletal parts of the 1st to 5th digits);
Q   Phalanx media;

R   Phalanx distalis;
S   Ossa sesamoidea proximalia;
T   Os sesamoideum m. abductoris pollicis longi;

a   Articulatio antebrachiocarpea;
b   Articulatio radioulnaris distalis;
c   Articulatio mediocarpea;
d   Articulationes carpometacarpeae;
e   Articulationes intermetacarpeae;
f   Articulatio metacarpophalangea;
g   Articulatio interphalangea proximalis manus;
h   Articulatio interphalangea distalis manus;

1   Processus styloideus radii;
2   Sehnenrinne an der Trochlea radii – Groove on the radial trochlea;
3   Processus styloideus ulnae;
4   Crista unguicularis;
5   Tuberculum flexorium.

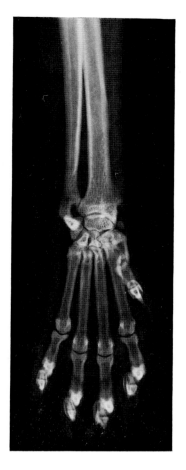

Abb. 273 Rechter Vorderfuß. Dorso-palmar. Katze.
Folienloser Film – FFA 120 cm – 60 kV – 15 mAs
Originalgröße (Ausschnitt aus 13 × 18 cm)
Lagerung Abb. 270

Fig. 273 Right forepaw. Dorsopalmar. Cat.
Non-screen film – FFD 120 cm – 60 kV – 15 mAs
Original size (section of 13 × 18 cm)
Positioning fig. 270

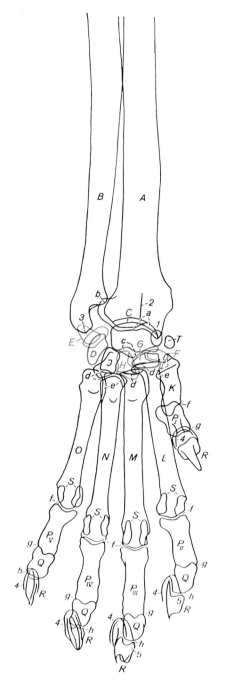

Abb. 274 Röntgenskizze zu Abb. 273
Fig. 274 X-ray sketch to fig. 273

A Radius;
B Ulna;
C Os carpi intermedioradiale;
D Os carpi ulnare;
E Os carpi accessorium;
F Os carpale I;
G Os carpale II;
H Os carpale III;
J Os carpale IV;
K Os metacarpale I;
L Os metacarpale II;
M Os metacarpale III;
N Os metacarpale IV;
O Os metacarpale V;
P Phalanx proximalis (I–V kennzeichnen die Skelettelemente an der 1. bis 5. Zehe) – Phalanx proximalis (I–V show the skeletal parts of the 1st to 5th digits);
Q Phalanx media;

R Phalanx distalis;
S Ossa sesamoidea proximalia;
T Os sesamoideum m. abductoris pollicis longi;

a Articulatio antebrachiocarpea;
b Articulatio radioulnaris distalis;
c Articulatio mediocarpea;
d Articulationes carpometacarpeae;
e Articulationes intermetacarpeae;
f Articulatio metacarpophalangea;
g Articulatio interphalangea proximalis manus;
h Articulatio interphalangea distalis manus;

1 Processus styloideus radii;
2 Begrenzung einer dorsalen Sehnenrinne – Border of a groove on the dorsal aspect;
3 Processus styloideus ulnae;
4 Crista unguicularis;
5 Tuberculum flexorium.

185

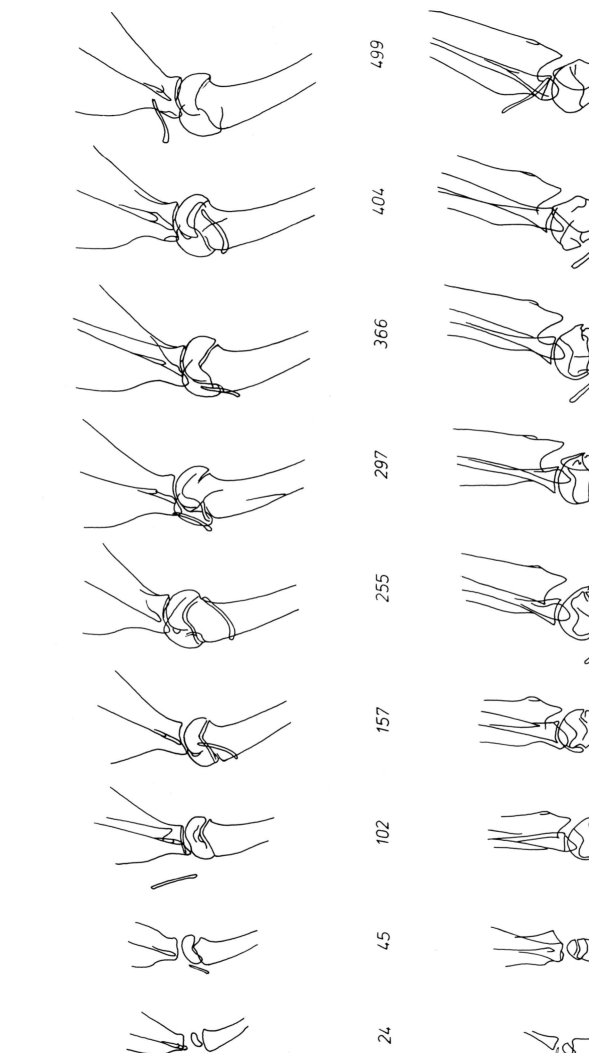

499

404

366

297

255

157

102

45

24

Fig. 275   Right shoulder joint. Mediolateral. Cat.
Postnatal development. Age in days.

Abb. 275   Rechtes Schultergelenk. Medio-lateral. Hauskatze.
Postnatale Entwicklung. Lebensalter in Tagen.

Fig. 276   Left shoulder joint. Caudocranial. Cat.
Postnatal development. Age in days.

Abb. 276   Linkes Schultergelenk. Kaudo-kranial. Hauskatze.
Postnatale Entwicklung. Lebensalter in Tagen.

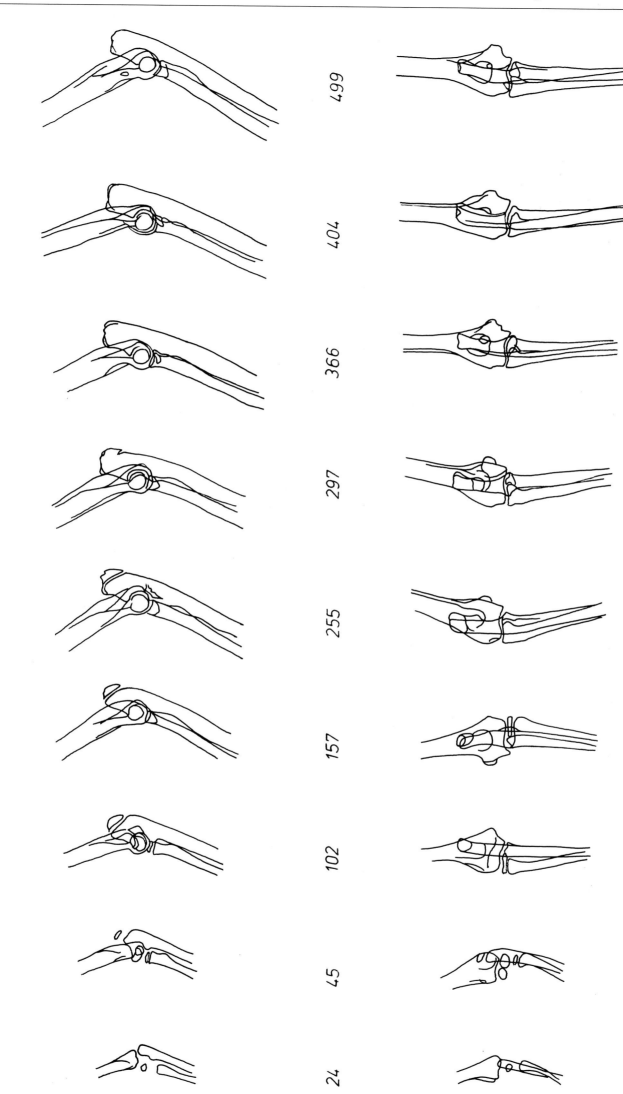

Fig. 277 Right elbow joint. Mediolateral. Cat. Postnatal development. Age in days.

Abb. 277 Rechtes Ellbogengelenk. Medio-lateral. Hauskatze. Postnatale Entwicklung. Lebensalter in Tagen.

Fig. 278 Right elbow joint. Craniocaudal. Cat. Postnatal development. Age in days.

Abb. 278 Rechtes Ellbogengelenk. Kranio-kaudal. Hauskatze. Postnatale Entwicklung. Lebensalter in Tagen.

187

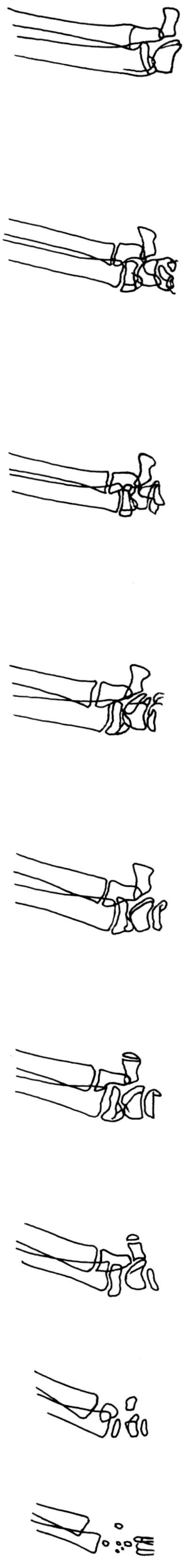

499
404
366
297
255
157
102
45
24

Fig. 279 Right carpal joint. Mediolateral. Cat. Postnatal development. Age in days.

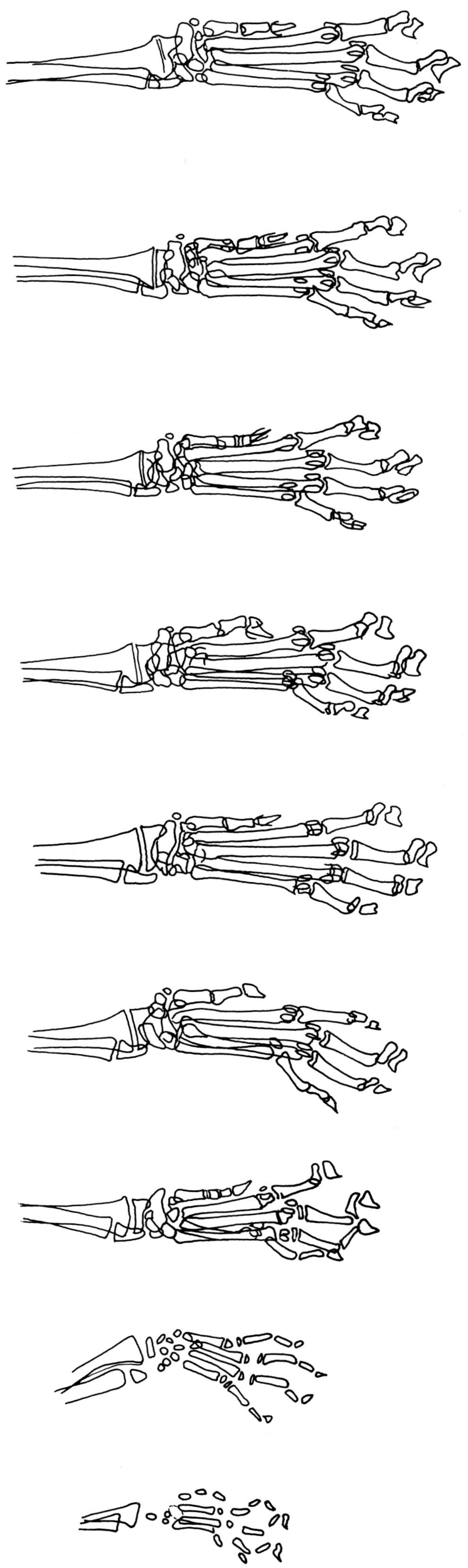

Abb. 279 Rechtes Karpalgelenk. Medio-lateral. Hauskatze. Postnatale Entwicklung. Lebensalter in Tagen.

Fig. 280 Right forepaw. Dorsopalmar. Cat. Postnatal development. Age in days.

Abb. 280 Rechter Vorderfuß. Dorso-palmar. Hauskatze. Postnatale Entwicklung. Lebensalter in Tagen.

Tabelle 3: Zeitliches Auftreten der Ossifikationspunkte sowie des Apo– und Epiphysenfugenschlusses am Skelett der Schultergliedmaße von der Hauskatze (nach I. HORVATH 1983)

Table 3: Time-table of the appearance of ossification centers and closures of apo- and epihyseal lines of the thoracic limb in the Domestic cat (after I. HORVATH 1983)

Ossifikationspunkte Apophysen und Epiphysen Ossification centers Apophyses and Epiphyses	Auftreten der Ossifikationspunkte (Angabe in Tagen) Appearance of the ossification centers (in days)	Verschmelzen der Ossifikationskerne (Angabe in Monaten) Fusion of the ossification centers (in months)	Apo- und Epiphysenfugenschluß (Angabe in Monaten) Fusion of apo- and epiphyseal lines (in months)
SCAPULA			
Tuberculum supraglenoidale	49 – 84 }	5 – 6	5 – 6
Processus coracoideus	49 – 84 }		5 – 6
HUMERUS			
Epiphysis proximalis humeri	8 – 14		22 – 24
Tuberculum minus	70 – 113		4 – 7
Condylus lateralis humeri	13 – 21 }		4 – 7
Condylus medialis humeri	28 – 36 }	4 – 5	5 – 7
Epicondylus medialis humeri	49 – 64		4 – 6
Epicondylus lateralis humeri	49 – 73		5 – 7
RADIUS			
Epiphysis proximalis radii	20 – 28		5 – 7
Epiphysis distalis radii	20 – 24		19 – 25
ULNA			
Apophysis (proximalis) ulnae	34 – 38		12 – 14
Epiphysis distalis ulnae	21 – 31		19 – 22
OSSA CARPI			
Os carpi radiale	29 – 43 }}	3 – 5	
Os carpi intermedium	22 – 36 }}		
Os carpi centrale	20 – 24 }	4 – 7	
Os carpi ulnare	29 – 42		
Os carpi accessorium	20 – 28		
Apophysis ossis carpi accessorii	43 – 59		6 – 7
Os carpale I	22 – 35		
Os carpale II et III	20 – 28		
Os carpale IV	20 – 29		
OSSA METACARPALIA			
Epiphysis proximalis ossis metacarpalis I	29 – 42		9 – 11
Epiphyses distales ossium metacarpalium II – V	20 – 29		10 – 12
OSSA DIGITORUM MANUS			
Epiphysis proximalis			
phalangis proximalis I	29 – 42		
phalangium proximalium II – V	20 – 28		} 7 – 9
phalangis proximalis V	27 – 36		
phalangium mediarum II et V	22 – 35		
phalangium mediarum III et IV	20 – 35		} 6 – 9
OSSA SESAMOIDEA			
Os sesamoideum proximale digiti I	102 – 156		
Ossa sesamoidea proximalia			
digitorum II et V	63 – 113		
digitorum III et IV	63 – 100		
Os sesamoideum m. abductoris pollicis longi	99 – 183		

Tabelle 4: Zeitliches Auftreten der Ossifikationspunkte sowie des Apo- und Epiphysenfugenschlusses am Skelett der Beckengliedmaße von der Hauskatze (nach A. HORVATH 1983)

Table 4: Time-table of the appearance of ossification centers and closures of apo- and epiphyseal lines of the pelvic limb in the Domestic cat (after A. HORVATH 1983)

Ossifikationspunkte Apophysen und Epiphysen Ossification centers Apophyses and Epiphyses	Auftreten der Ossifikationspunkte (Angabe in Tagen) Appearance of the ossification centers (in days)	Apo- und Epiphysenfugenschluß (Angabe in Monaten) Fusion of apo- and epiphyseal lines (in months)
OS COXAE		
Os ilium		8 – 9
Os ischii		
Os pubis	62 – 69	
Os acetabuli		3 – 6
Ramus caudalis ossis pubis		
Ramus ossis ischii	183 – 231	über 26
Crista iliaca	58 – 73	über 26
Tuberculum ischiadicum		
OS FEMORIS		
Epiphysis proximalis ossis femoris	13 – 17	11 – 12
Trochanter major	20 – 24	11 – 14
Trochanter minor	13 – 17	12 – 15
Epiphysis distalis ossis femoris	13 – 17	17 – 20
TIBIA		
Epiphysis proximalis tibiae	14 – 21	17 – 21
Tuberositas tibia	48 – 59	17 – 22
Epiphysis distalis tibiae	15 – 24	12 – 14
FIBULA		
Epiphysis proximalis fibulae	36 – 58	17 – 21
Epiphysis distalis fibulae	22 – 31	12 – 14
OSSA TARSI		
Tuber calcanei	34 – 45	11 – 15
Os tarsi centrale	28 – 42	
Os tarsale I	29 – 43	
Os tarsale II	35 – 43	
Os tarsale III	29 – 42	
Os tarsale IV	28 – 42	
OSSA METATARSALIA		
Os metatarsale I	58 – 113	
Epiphysis distalis		
ossium metatarsalium II et V	28 – 31	10 – 12
ossium metatarsalium III et IV	27 – 31	
OSSA DIGITORUM PEDIS		
Epiphysis proximalis		
phalangium proximalium II et V	28 – 36	7 – 9
phalangium proximalium III et IV	22 – 31	
phalangium mediarum II et V	28 – 35	6 – 8
phalangium mediarum III et IV	24 – 29	
OSSA SESAMOIDEA		
Patella	56 – 100	
Os sesamoideum laterale m. gastrocnemii	84 – 142	
Os sesamoideum mediale m. gastrocnemii	154 – 239	
Os sesamoideum m. poplitei	141 – 183	
Ossa sesamoidea proximalia		
digitorum II et V	84 – 127	
digitorum III et IV	84 – 113	

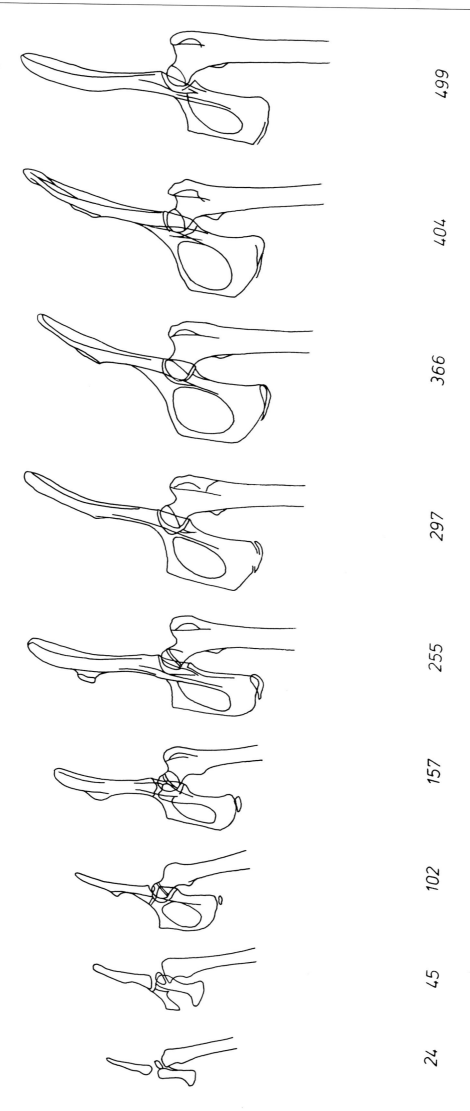

499

404

366

297

255

157

102

45

24

Abb. 281 Linke Beckenhälfte und linkes Hüftgelenk. Ventro-dorsal. Hauskatze. Postnatale Entwicklung. Lebensalter in Tagen.

Fig. 281 Left part of the pelvis and left hip joint. Ventrodorsal. Cat. Postnatal development. Age in days.

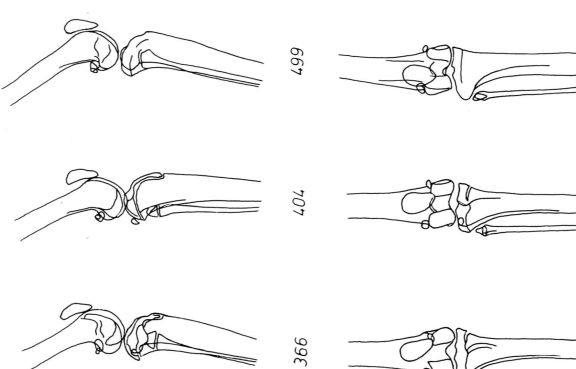

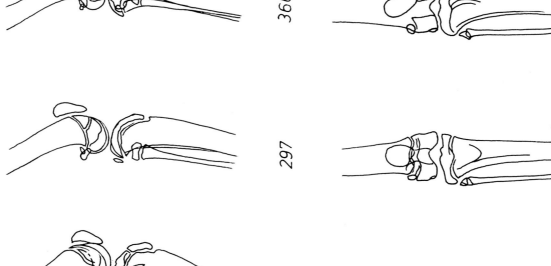

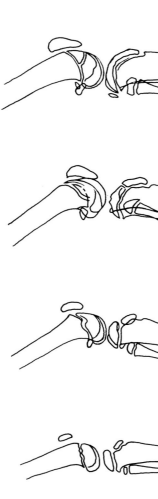

499

404

366

297

255

157

102

45

24

Fig. 282 Left stifle joint. Mediolateral. Cat.
Postnatal development. Age in days.

b. 282 Linkes Kniegelenk. Medio-lateral. Hauskatze.
Postnatale Entwicklung. Lebensalter in Tagen.

Fig. 283 Right stifle joint. Craniocaudal. Cat.
Postnatal development. Age in days.

Abb. 283 Rechtes Kniegelenk. Kranio-kaudal. Hauskatze.
Postnatale Entwicklung. Lebensalter in Tagen.

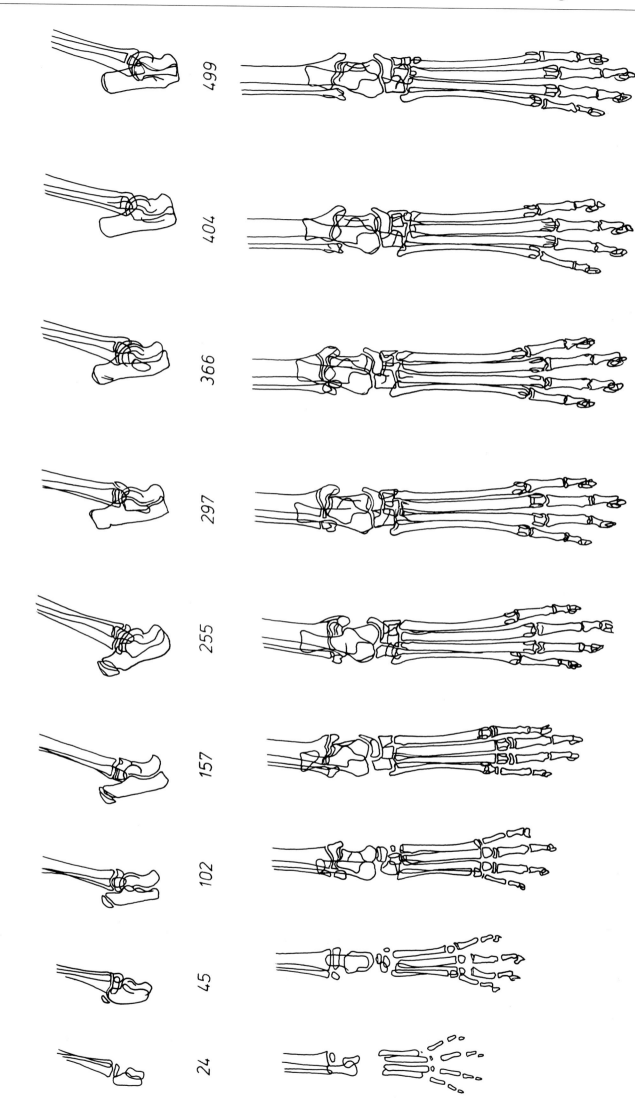

Fig. 284 Left hock joint. Mediolateral. Cat. Postnatal development. Age in days.

Abb. 284 Linkes Tarsalgelenk. Medio-lateral. Hauskatze. Postnatale Entwicklung. Lebensalter in Tagen.

Fig. 285 Right hindpaw. Dorsoplantar. Cat. Postnatal development. Age in days.

Abb. 285 Rechter Hinterfuß. Dorso-plantar. Hauskatze. Postnatale Entwicklung. Lebensalter in Tagen.

499

404

366

297

255

157

102

45

24

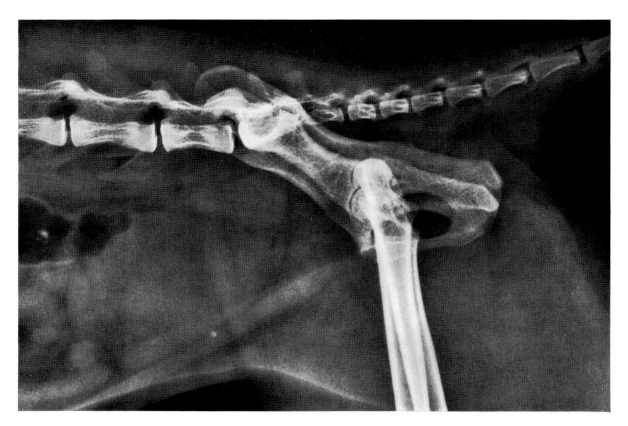

Abb. 286   Becken. Latero-lateral. Katze.
Bucky-Blende – Feinzeichnende Folie – FFA 115 cm – 55 kV – 30 mAs
Originalgröße (Ausschnitt aus 18 × 24 cm)
Lagerung Abb. 287

Fig. 286   Pelvis. Laterolateral. Cat.
Bucky diaphragm – High definition screens – FFD 115 cm – 55 kV – 30 mAs
Original size (section of 18 × 24 cm)
Positioning fig. 287

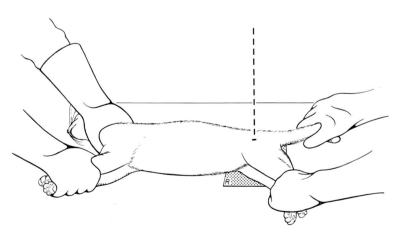

Abb. 287 Lagerung zur Aufnahme des Beckens. Latero-lateral.

Um eine Abbildung beider Hüftbeine in Überlagerung zu erreichen, muß die Medianebene des Rumpfes parallel zur Kassette liegen. Bei fettleibigen Tieren kann die korrekte Lagerung durch Schaumgummikeile unter den aufliegenden Gliedmaßen in Höhe der haltenden Hände, ggf. auch un-

ter dem Brustbein, erreicht werden. Die korrekte Lagerung läßt sich durch Zug am Schwanz unterstützen. Die Beckengliedmaßen sind nur mäßig nach hinten gestreckt zu fixieren.

Die Aufnahme kann mit einem folienlosen Film (Bleiunterlage nicht vergessen!) angefertigt werden. Bei fettleibigen Tieren ist eine Bucky-Blende oder eine Kassette mit stehendem Raster zur Verringerung der Streustrahlung nützlich.

Der Zentralstrahl sollte den Trochanter major der obenliegenden Beckengliedmaße treffen und im rechten Winkel auf die Kassette einfallen.

Fig. 287   Positioning of pelvis. Laterolateral.

To obtain a radiograph of both hip bones properly superimposed, the median plane of the trunk must be parallel to the cassette. In obese animals, wedge-shaped foam rubber pads should be placed underneath the lower limbs and the sternum if necessary. Slight pull on the tail facilitates correct positioning. The hind limbs are secured by pulling them slightly backward.

A non-screen film may be used (lead blocker not to be forgotten!). In obese animals the use of a Bucky diaphragm or a cassette with stationary grid is recommended to reduce scattered radiation.

The central beam should strike the trochanter major of the upper limb and fall at the right angle on the cassette.

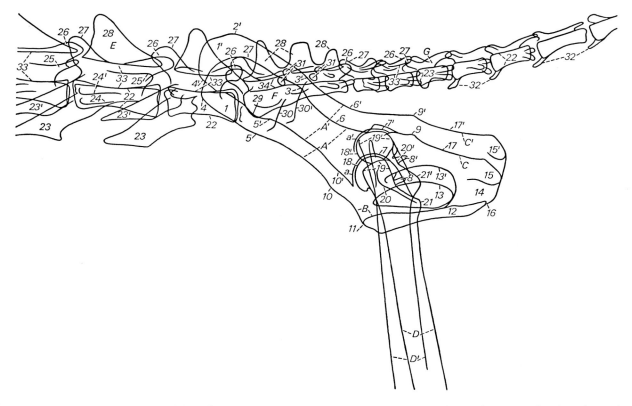

Abb. 288* Röntgenskizze zu Abb. 286   Fig. 288* X-ray sketch to fig. 286

A   Os ilium;
B   Os pubis;
C   Os ischii;
D   Os femoris;
E   6. Vertebra lumbalis;
F   Os sacrum;
G   3. Vertebra caudalis;

a   Articulatio coxae;

Am Becken – On the pelvis:

1   Ala ossis ilii;
2, 3   Tuber sacrale:
2   Spina iliaca dorsalis cranialis,
3   Spina iliaca dorsalis caudalis;
4   Tuber coxae: Spina iliaca ventralis cranialis;
5   Spina alaris;
6   Incisura ischiadica major;
7   Acetabulum;
8   Incisura acetabuli;
9   Spina ischiadica;
10   Eminentia iliopubica;
11   Pecten ossis pubis;
12   Symphysis pelvina;
13   Foramen obturatum;

14   Tabula ossis ischii;
15   Tuber ischiadicum;
16   Arcus ischiadicus;
17   Incisura ischiadica minor;

Am Os femoris – On the femur:

18   Caput ossis femoris;
19   Trochanter major;
20   Fossa trochanterica;
21   Trochanter minor;

An der Wirbelsäule – On the vertebral column:

22   Corpus vertebrae;
23   Processus transversus;
24   Basis processus transversi;
25   Processus accessorius;
26   Processus articularis caudalis;
27   Processus articularis cranialis;
28   Processus spinosus;
29   Promontorium;
30   Ala ossis sacri, Kaudalrand – Ala ossis sacri, caudal border;
31   Foramina sacralia dorsalia;
32   Arcus haemalis;
33   Canalis vertebralis;
34   Canalis sacralis.

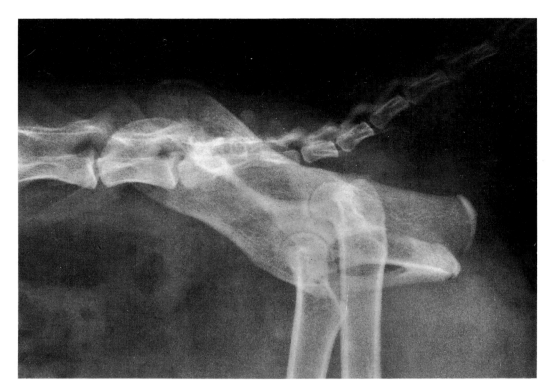

Abb. 289   Becken. Latero-lateral. Schrägprojektion (20°). Katze.
Feinzeichnende Folie – FFA 120 cm – 60 kV – 35 mAs
Originalgröße (Ausschnitt aus 18 × 24 cm)
Lagerung Abb. 290

Fig. 289   Pelvis. Obliquely laterolateral (20°). Cat.
High definition screens – FFA 120 cm – 60 kV – 35 mAs
Original size (section of 18 × 24 cm)
Positioning fig. 290

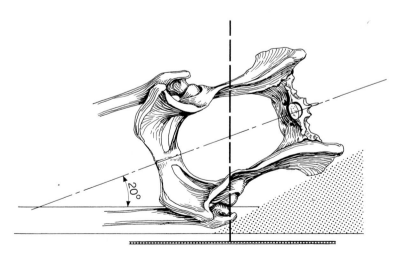

Abb. 290   Lagerung zur Aufnahme des Beckens. Latero-lateral, Schrägprojektion.

Zur Beurteilung der operativen Rekonstruktion nach Fraktur des Darmbeins, des Azetabulums und/oder Sitzbeins ist eine Schrägprojektion erforderlich. Die zu beurteilende Beckenhälfte ist plattennah zu lagern. Durch Unterlegen eines Schaumstoffkeils unter dem Rumpf im Bereich des Beckens wird die Medianebene um etwa 20° zur Unterlage hin gekantet. Die Lagerung wird durch Zug am Schwanz erleichtert. Die Beckengliedmaßen sind nur mäßig nach hinten gestreckt zu fixieren.

Zur Reduzierung der Streustrahlung ist eine Bucky-Blende oder eine Kassette mit stehendem Raster notwendig.

Der Zentralstrahl sollte fingerbreit dorsal des Trochanter major die obenliegende Gliedmaße treffen und im rechten Winkel auf die Kassette einfallen.

Fig. 290   Positioning of pelvis. Obliquely laterolateral.

An oblique projection is necessary in order to evaluate a surgical reconstruction of a fracture of ilium, of acetabulum and/or of os ischii. The median plane of the pelvis must be tilted approximately 20° toward the table with the fracture side close to the film. Pelvic limbs are moderately stretched backward. A slight pull on the tail will facilitate the positioning.

A Bucky diaphragm or a cassette with stationary grid is necessary in order to reduce scattered radiation.

The central beam should strike the upper limb one finger's width dorsally to the trochanter major and fall at the right angle on the cassette.

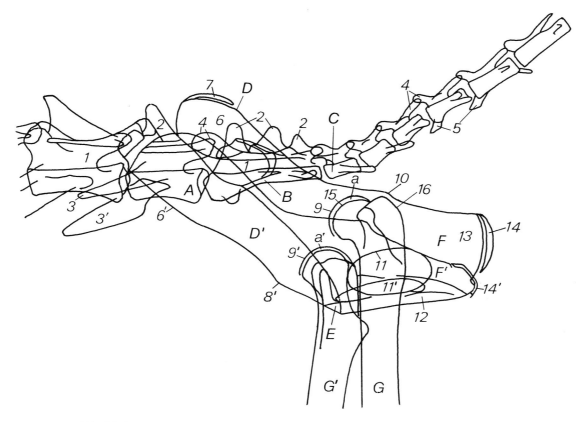

Abb. 291*   Röntgenskizze zu Abb. 289   Fig. 291*   X-ray sketch to fig. 289

A   7. Vertebra lumbalis;
B   Os sacrum;
C   1. Vertebra caudalis;
D   Os ilium;
E   Os pubis;
F   Os ischii;
G   Os femoris;

a   Articulatio coxae;

An der Wirbelsäule – On the vertebral column:

1   Canalis vertebralis;
2   Processus spinosus;
3   Processus transversus;
4   Processus articulares caudales et craniales;
5   Processus haemalis;

Am Becken – On the pelvis:

6   Ala ossis ilii;
7   Crista iliaca, Spina iliaca dorsalis cranialis, Apophysenkern – Apophyseal nucleus of the dorsal cranial iliac spine of the iliac crest;
8   Eminentia iliopubica;
9   Acetabulum;
10   Spina ischiadica;
11   Foramen obturatum;
12   Symphysis pelvina;
13   Tabula ossis ischii;
14   Tuber ischiadicum, Apophysenkern – Tuber ischiadicum, apophyseal nucleus;

Am Os femoris – On the femur:

15   Caput ossis femoris;
16   Trochanter major.

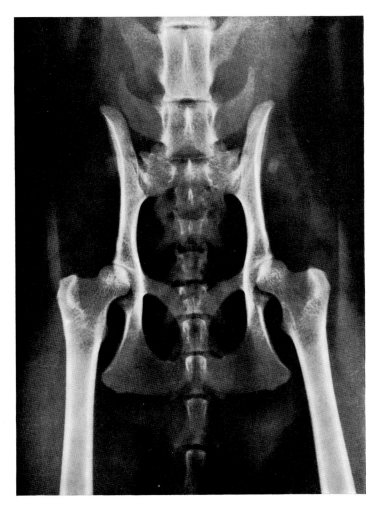

Abb. 292   Becken. Ventro-dorsal. Katze.
Bucky-Blende – Feinzeichnende Folie – FFA 120 cm – 64 kV – 30 mAs
Originalgröße (Ausschnitt aus 18 × 24 cm)
Lagerung Abb. 293

Fig. 292   Pelvis. Ventrodorsal. Cat.
Bucky diaphragm – High definition screens – FFD 120 cm – 64 kV – 30 mAs
Original size (section of 18 × 24 cm)
Positioning fig. 293

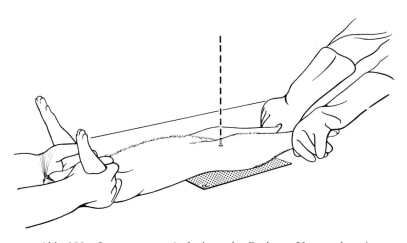

Abb. 293   Lagerung zur Aufnahme des Beckens. Ventro-dorsal.

Um eine symmetrische Abbildung beider Beckenhälften zu erhalten, muß die Medianebene des Rumpfes senkrecht zur Kassette stehen. Die Becken-

gliedmaßen sind so weit nach hinten zu strecken, daß die Fersenbeine bzw. die die Gliedmaßen haltenden Finger der Unterlage aufliegen. Die Hintergliedmaßen sollten in leichter Abduktion und geringer Außenrotation von der Hilfsperson fixiert werden.

Bei kleinen Tieren kann die Aufnahme mit einem folienlosen Film (Blei-unterlage nicht vergessen!) angefertigt werden. Bei größeren Katzen ist eine Bucky-Blende oder eine Kassette mit stehendem Raster zur Reduzierung der Streustrahlung zweckmäßig.

Der Zentralstrahl sollte die Mittellinie in Höhe der Trochanteren treffen und im rechten Winkel auf den Film einfallen.

Fig. 293   Positioning of pelvis. Ventrodorsal.

To obtain a symmetrical radiograph of both pelvic bones the median plane of the trunk must be perpendicular to the cassette. The hind limbs are stretched thus far caudally that the calcanei and the fingers holding the limbs rest on the table. The limbs should be secured in slight abduction and moderate outward rotation.

In small cats the radiograph can be taken with a non-screen film (lead blocker not to be forgotten!). In larger cats a Bucky diaphragm or stationary grid should be used to reduce scattered radiation.

The central beam should strike the midline on a level of the trochanters and fall at the right angle on the film.

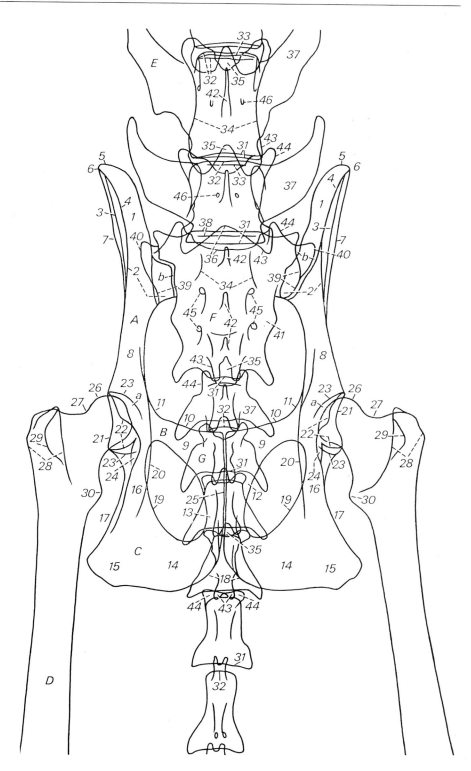

Abb. 294 Röntgenskizze zu Abb. 292
Fig. 294 X-ray sketch to fig. 292

A Os ilium;
B Os pubis;
C Os ischii;
D Os femoris;
E 6. Vertebra lumbalis;
F Os sacrum;
G 2. Vertebra caudalis;

a Articulatio coxae;
b Articulatio sacroiliaca;

Am Becken – On the pelvis:

1 Ala ossis ilii;
2 Tuber sacrale, Spina iliaca dorsalis caudalis;
3 Linea glutaea;
4 Kompaktaschatten, der sich aus der Konkavität der Facies glutaea ergibt – Shadow of compacta formed by the concavity of the gluteal surface;
5 Crista iliaca;
6 Tuber coxae, Spina iliaca ventralis cranialis;
7 Spina alaris;
8 Corpus ossis ilii;
9 Ramus cranialis ossis pubis;
10 Pecten ossis pubis;
11 Eminentia iliopubica;
12 Ramus caudalis ossis pubis;
13 Ramus ossis ischii;
14 Tabula ossis ischii;
15 Tuber ischiadicum;
16 Corpus ossis ischii;
17 Incisura ischiadica minor;
18 Arcus ischiadicus;
19 Foramen obturatum;
20 Spina ischiadica;
21 – 24 Acetabulum:
21 Dorsaler Rand – Dorsal border,
22 Ventraler Rand – Ventral border,
23 Verschattung, die sich aus der Konkavität der Facies lunata ergibt – Shadow formed by the concavity of the lunate surface,
24 Incisura acetabuli;
25 Symphysis pelvina;

Am Os femoris – On the femur:

26 Caput ossis femoris;
27 Collum ossis femoris;
28 Trochanter major;
29 Fossa trochanterica;
30 Trochanter minor;

An der Wirbelsäule – On the vertebral column:

31 Extremitas caudalis;

32 Extremitas cranialis,
33 Verschattung, die sich aus deren Konkavität ergibt – Shadow formed by its concavity;
34 Canalis vertebralis, seitliche Begrenzung – Canalis vertebralis, lateral border;
35 Spatium interarcuale;
36 Spatium interarcuale lumbosacrale;
37 Processus transversus;
38 Basis ossis sacri;
39, 40 Ala ossis sacri:
39 Dorsaler Anteil – Dorsal part,
40 Ventraler Anteil – Ventral part;
41 Os sacrum, Pars lateralis;
42 Processus spinosus;
43 Processus articularis caudalis;
44 Processus articularis cranialis;
45 Foramina sacralia dorsalia et pelvina, ineinander projiziert – Foramina sacralia dorsalia et pelvina projected into one another;
46 Gefäßloch – Vascular foramen.

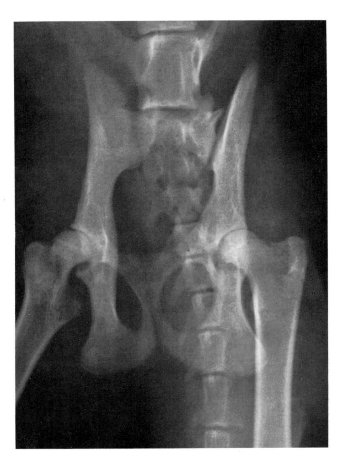

Abb. 295   Becken. Ventro-dorsal, Schräglagerung. Katze.
Feinzeichnende Folie – FFA 120 cm – 60 kV – 20 mAs
Originalgröße (Ausschnitt aus 13 × 18 cm)
Lagerung Abb. 296

Fig. 295   Pelvis. Ventrodorsal, oblique positioning. Cat.
High definition screens – FFD 120 cm – 60 kV – 20 mAs
Original size (section of 13 × 18 cm)
Positioning fig. 296

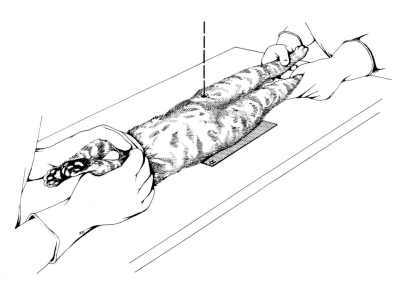

Abb. 296   Schräglagerung zur Aufnahme des Beckens bei Frakturverdacht
des Darmbeinflügels. Ventro-dorsal.

Zum Nachweis einer Fraktur am dorsokranialen Rand des Darmbeinflügels ist die Aufnahme bei schräg gelagertem Becken zweckmäßig. Das zu
beurteilende Iliosakralgelenk wird vom Zentralstrahl orthograph getroffen,
wenn die Medianebene des Beckens um 30° zur gesunden Seite hin ge-
dreht ist. Die Beckengliedmaße auf der gesunden Seite ist in leichter Abduktion und geringer Außenrotation nach hinten gestreckt zu fixieren.
Die Gliedmaße auf der zu untersuchenden Seite sollte in leichter Adduk-
tion und geringer Innenrotation auf der Unterlage fixiert werden.
Bei einer kleinen Katze kann die Aufnahme mit einem folienlosen Film
(Bleiunterlage nicht vergessen!) angefertigt werden. Bei einer größeren
Katze ist eine Bucky-Blende oder eine Kassette mit stehendem Raster zur
Reduzierung der Streustrahlung zweckmäßig.
Der Zentralstrahl sollte fingerbreit kaudal des Hüfthöckers das Iliosakral-
gelenk treffen und im rechten Winkel auf den Film einfallen.

Fig. 296   Oblique positioning of pelvis in case of suspected fracture of
the wing of iliac bone. Ventrodorsal.

In order to depict a fracture of the dorsocranial edge of the wing of iliac
bone, a radiograph of obliquely positioned pelvis is necessary. The central
beam will strike orthographically a given sacroiliac joint of the median
plane of the pelvis is turned about 30° toward the healthy side. The pelvic
limb of the healthy side is to be secured in slight abduction and outward
rotation. The opposite leg should be held on the X-ray table in slight ad-
duction and inward rotation.
Small cats may be radiographed, using non-screen film (lead blocker not
to be forgotten!); for larger cats a Bucky diaphragm or a cassette with stationary grid is indicated in order to reduce scattered radiation.
The central beam should strike the sacroiliac joint, approximately one finger's width caudally from the iliac crest and fall at the right angle on the
film.

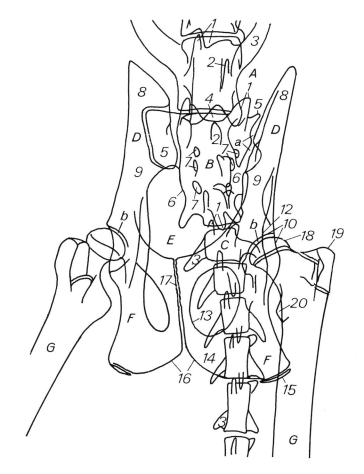

Abb. 297   Röntgenskizze zu Abb. 295   Fig. 297   X-ray sketch to fig. 295

A   7. Vertebra lumbalis;
B   Os sacrum;
C   1. Vertebra caudalis;
D   Os ilium;
E   Os pubis;
F   Os ischii;
G   Os femoris;

a   Articulatio sacroiliaca;
b   Articulatio coxae;

An der Wirbelsäule – On the vertebral column:

1   Processus articulares caudalis et cranialis:
2   Processus spinosus;
3   Processus transversus;
4   Spatium interarcuale lumbosacrale;
5   Ala sacralis;
6   Pars lateralis;
7   Foramina sacralia;

Am Becken – On the pelvis:

8   Ala ossis ilii;
9   Corpus ossis ilii;
10   Acetabulum;
11   Incisura acetabuli;
12   Spina ischiadica;
13   Foramen obturatum;
14   Tabula ossis ischii;
15   Tuber ischiadicum, Apophysenkern – Tuber ischiadicum, apophyseal nucleus;
16   Arcus ischiadicus;
17   Symphysis pelvina;

Am Os femoris – On the femur:

18   Caput ossis femoris mit Epiphysenfuge – Caput ossis femoris with the epiphyseal groove;
19   Trochanter major mit Apophysenfugen – Trochanter major with apophyseal grooves;
20   Trochanter minor.

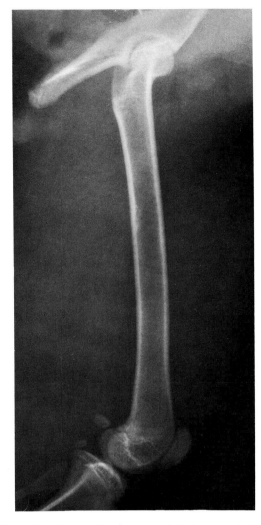

Abb. 298   Linker Oberschenkel. Medio-lateral. Katze.
Feinzeichnende Folie – FFA 120 cm – 60 kV – 12 mAs
Originalgröße (Ausschnitt aus 13 × 18 cm)
Lagerung Abb. 306

Fig. 298   Left thigh. Mediolateral. Cat.
High definition screens – FFD 120 cm – 60 kV – 12 mAs
Original size (section of 13 × 18 cm)
Positioning fig. 306

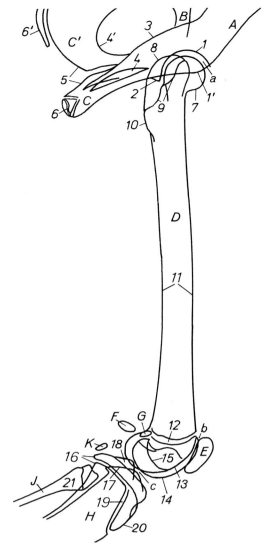

Abb. 299*   Röntgenskizze zu Abb. 298
Fig. 299*   X-ray sketch to fig. 298

A   Os ilium;
B   Os pubis;
C   Os ischii;
D   Os femoris;
E   Patella;
F, G   Ossa sesamoidea m. gastrocnemii;
H   Tibia;
J   Fibula;
K   Os sesamoideum m. poplitei;

a   Articulatio coxae;
b   Articulatio femoropatellaris;
c   Articulatio femorotibialis;

Am Becken – On the pelvis:

1, 1′   Acetabulum:
1   Facies lunata,
1′   Ventraler Rand – Ventral border;
2   Incisura acetabuli;
3   Spina ischiadica;
4   Foramen obturatum;
5   Arcus ischiadicus;
6   Tuber ischiadicum, Apophysenkern – Tuber ischiadicum, apophyseal nucleus;

Am Os femoris – On the femur:

7   Caput ossis femoris;
8   Trochanter major;
9   Fossa trochanterica;
10   Trochanter minor;
11   Corpus ossis femoris;
12   Epiphysenfuge – Epiphyseal groove;
13   Condylus lateralis;
14   Condylus medialis;
15   Verschattung am Grund der Fossa intercondylaris – Shadow formed by the bottom of the intercondylar fossa;

An der Tibia – On the tibia:

16   Condylus lateralis;
17   Condylus medialis;
18   Eminentia intercondylaris;
19   Fugenknorpel – Cartilage of the groove;
20   Tuberositas tibiae;

An der Fibula – On the fibula:

21   Caput fibulae.

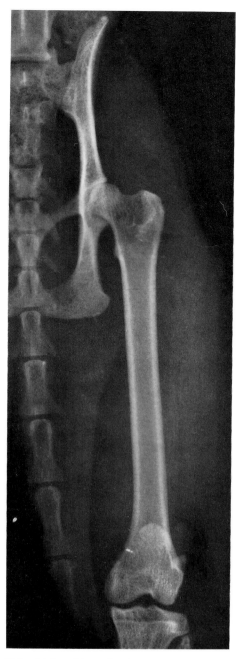

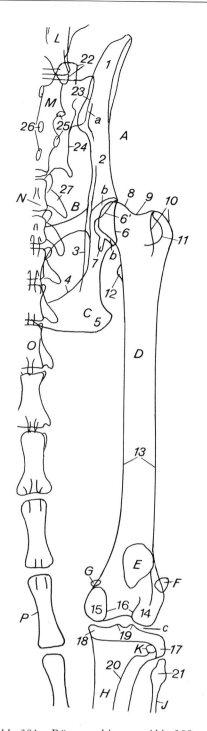

Abb. 300   Linker Oberschenkel. Kranio-kaudal. Katze.
Feinzeichnende Folie – FFA 120 cm – 63 kV – 20 mAs
Originalgröße (Ausschnitt aus 18 × 24 cm)
Lagerung Abb. 307

Fig. 300   Left thigh. Craniocaudal. Cat.
High definition screens – FFD 120 cm – 63 kV – 20 mAs
Original size (section of 18 × 24 cm)
Positioning fig. 307

Abb. 301   Röntgenskizze zu Abb. 300
Fig. 301   X-ray sketch to fig. 300

A   Os ilium;
B   Os pubis;
C   Os ischii;
D   Os femoris;
E   Patella;
F   Os sesamoideum m. gastrocnemii lateralis;
G   Os sesamoideum m. gastrocnemii medialis;
H   Tibia;
J   Fibula;
K   Os sesamoideum m. poplitei;
L   7. Vertebra lumbalis;
M   Os sacrum;
N   1. Vertebra caudalis;
O   5. Vertebra caudalis;
P   9. Vertebra caudalis;

a   Articulatio sacroiliaca;
b   Articulatio coxae;
c   Articulatio femorotibialis;
d   Articulatio tibiofibularis proximalis;

Am Becken – On the pelvis:

1   Ala ossis ilii;
2   Corpus ossis ilii;
3   Spina ischiadica;
4   Foramen obturatum;
5   Tuber ischiadicum;
6, 6'   Acetabulum:
6   Dorsaler Rand – Dorsal border,
6'   Ventraler Rand – Ventral border;
7   Incisura acetabuli;

Am Os femoris – On the femur:

8   Caput ossis femoris;
9   Collum ossis femoris;
10   Trochanter major;
11   Fossa trochanterica;
12   Trochanter minor;
13   Corpus ossis femoris;
14   Condylus lateralis;
15   Condylus medialis;
16   Fossa intercondylaris;

An der Tibia – On the tibia:

17   Condylus lateralis;
18   Condylus medialis;
19   Eminentia intercondylaris;
20   Margo cranialis;

An der Fibula – On the fibula:

21   Caput fibulae;

An der Wirbelsäule – On the vertebral column:

22   Processus articularis;
23   Ala sacralis;
24   Pars lateralis;
25   Foramina sacralia;
26   Processus spinosus;
27   Processus transversus.

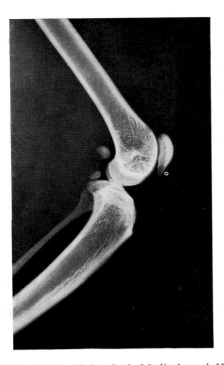

Abb. 302 Linkes Kniegelenk. Medio-lateral. Katze.
Folienloser Film – FFA 120 cm – 62 kV – 25 mAs
Originalgröße (Ausschnitt aus 13 × 18 cm)
Lagerung Abb. 306

Fig. 302 Left stifle joint. Mediolateral. Cat.
Non-screen film – FFD 120 cm – 62 kV – 25 mAs
Original size (section of 13 × 18 cm)
Positioning fig. 306

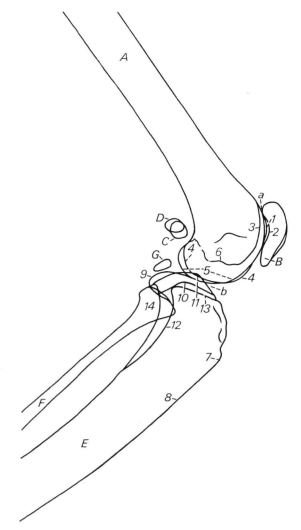

Abb. 303 Röntgenskizze zu Abb. 302
Fig. 303 X-ray sketch to fig. 302

A   Os femoris;
B   Patella;
C   Laterales Os sesamoideum m. gastrocnemii – Lateral sesamoid bone
     in the gastrocnemius muscle;
D   Mediales Os sesamoideum m. gastrocnemii – Medial sesamoid bone
     in the gastrocnemius muscle;
E   Tibia;
F   Fibula;
G   Os sesamoideum m. poplitei;

a   Articulatio femoropatellaris;
b   Articulatio femorotibialis;

Am Os femoris – On the femur:

1, 2, 3 Trochlea ossis femoris:
1   Lateraler Rollkamm – Lateral ridge,
2   Medialer Rollkamm – Medial ridge,

3   Rollfurche – Groove;
4   Condylus lateralis;
5   Condylus medialis;
6   Fossa intercondylaris;

An der Tibia – On the tibia:

7   Tuberositas tibiae;
8   Margo cranialis;
9   Condylus lateralis;
10   Condylus medialis;
11   Eminentia intercondylaris;
12   Verschattung, die sich aus der Konkavität der Incisura poplitea ergibt
     – Shadow formed by the concavity of the popliteal notch;
13   Verschattung, die sich aus der Konkavität der Facies articularis am
     Condylus medialis ergibt – Shadow formed by the concavity of the ar-
     ticular surface of the medial condyle;
14   Caput fibulae.

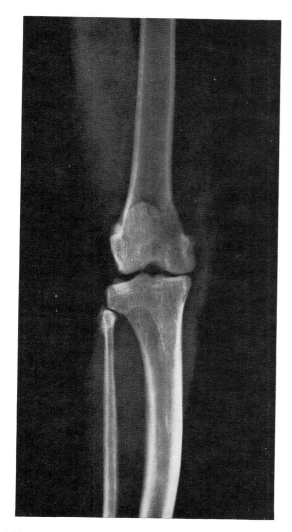

Abb. 304   Rechtes Kniegelenk. Kranio-kaudal. Katze.
Folienloser Film – FFA 120 cm – 65 kV – 30 mAs
Originalgröße (Ausschnitt aus 13 × 18 cm)
Lagerung Abb. 308

Fig. 304   Right stifle joint. Craniocaudal. Cat.
Non-screen film – FFD 120 cm – 65 kV – 30 mAs
Original size (section of 13 × 18 cm)
Positioning fig. 308

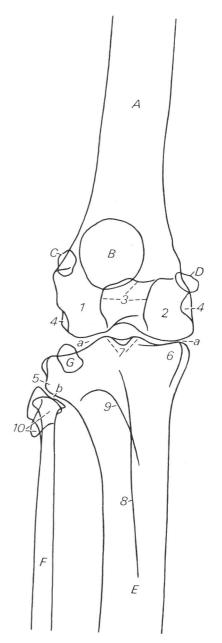

Abb. 305   Röntgenskizze zu Abb. 304
Fig. 305   X-ray sketch to fig. 304

A   Os femoris;
B   Patella;
C   Laterales Os sesamoideum m. gastrocnemii – Lateral sesamoid bone
     in the gastrocnemius muscle;
D   Mediales Os sesamoideum m. gastrocnemii – Medial sesamoid bone
     in the gastrocnemius muscle;
E   Tibia;
F   Fibula;
G   Os sesamoideum m. poplitei;

a   Articulatio femorotibialis;
b   Articulatio tibiofibularis proximalis;

Am Os femoris – On the femur:

1   Condylus lateralis;

2   Condylus medialis;
3   Fossa intercondylaris;
4   Bandgrube – Depression for ligamentous attachment;

An der Tibia – On the tibia:

5   Condylus lateralis;
6   Condylus medialis;
7   Eminentia intercondylaris et Tubercula intercondylaria lateralis et
     medialis;
8   Margo cranialis;
9   Incisura poplitea;
10   Caput fibulae.

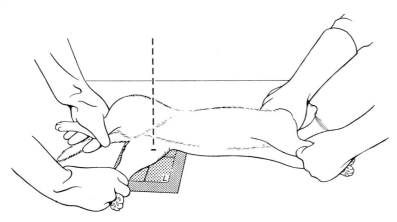

Abb. 306 Lagerung zur Aufnahme des Oberschenkels, des Kniegelenks sowie des Unterschenkels. Medio-lateral.

Die obenliegende Beckengliedmaße ist so weit abzuspreizen, daß das zu untersuchende Kniegelenk ohne Überlagerung dargestellt wird. Die Hand, die die abgespreizte Gliedmaße hält, stützt sich auf den Tisch und fixiert zusätzlich den Schwanz.
Os femoris: Der Zentralstrahl sollte das Os femoris in Schaftmitte medial treffen und im rechten Winkel auf die Kassette einfallen.
Kniegelenk: Die Aufnahme ist mit einem folienlosen Film (Bleiunterlage nicht vergessen!) anzufertigen.
Der Zentralstrahl sollte den Gelenkspalt des Kniekehlgelenks treffen und im rechten Winkel auf den Film einfallen.
Tibia: Die Aufnahme ist mit einem folienlosen Film (Bleiunterlage nicht vergessen!) anzufertigen.
Der Zentralstrahl sollte die Tibia in Schaftmitte kranial treffen und im rechten Winkel auf den Film einfallen.

Fig. 306 Positioning of thigh, of stifle joint and of lower leg. Mediolateral.

The upper hind limb is to be stretched and abducted far enough to avoid any overlaying. The hand holding the upper limb should rest on the table and secure the tail at the same time.
Femur: The central beam should strike the mid-shaft of the femur medially and fall at the right angle on the film.
Stifle joint: A non-screen film should be used (lead blocker not to be forgotten!).
The central beam should strike the articular surface of the femorotibial joint and fall at the right angle on the film.
Tibia: A non-screen film should be used (lead blocker not to be forgotten!).
The central beam should strike the mid-shaft of the tibia cranially and fall at the right angle on the film.

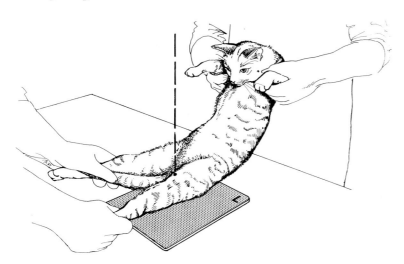

Abb. 307 Lagerung zur Aufnahme des Oberschenkels bei Frakturverdacht. Sitzend. Kranio-kaudal.

Die Lagerung in annähernd sitzender Position ist bei einem schmerzhaften Prozeß zweckmäßig, sie verursacht weniger Schmerzen und verringert die Gefahr zusätzlicher Verletzungen.
Person 1 faßt beide Oberarme vom Rücken her, fixiert den Kopf zwischen beiden Händen und hält die Katze in annähernd sitzender Position. Person 2 fixiert mit einer Hand die zu untersuchende Beckengliedmaße mit exakt kranial auf der Trochlea ossis femoris liegender Patella und umfaßt mit der anderen Hand die zweite Beckengliedmaße und den Schwanz.
Die Aufnahme ist mit einem folienlosen Film (Bleiunterlage nicht vergessen!) anzufertigen.

Bei fettleibigen Tieren ist eine Bucky-Blende oder eine Kassette mit stehendem Raster zur Verringerung der Streustrahlung nützlich.
Der Zentralstrahl sollte das Os femoris in Schaftmitte treffen und im rechten Winkel auf die Kassette einfallen.

Fig. 307 Positioning of thigh in case of suspected fracture of the femur. Sitting position. Craniocaudal.

The positioning in near sitting position is indicated in painful conditions, it is less painful and also reduces the danger of additional trauma.
One assistant grasps both upper arms from behind, secures the head between both hands and holds the cat in a near sitting position. A second assistant secures the hind limb to be examined with one hand, ensuring that the patella rests exactly on the cranial aspect of the femoral trochlea. One hand holds the other limb and the tail.
A non-screen film should be used (lead blocker not to be forgotten!).
In obese animals a Bucky diaphragm or a cassette with stationary grid is necessary to reduce scattered radiation.
The central beam should strike the mid-shaft of the femur and fall at the right angle on the cassette.

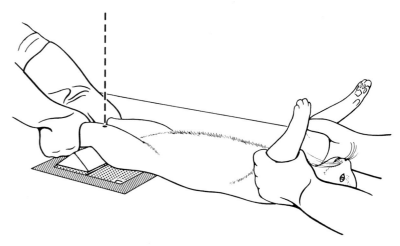

Abb. 308 Lagerung zur Aufnahme des Kniegelenks. Kranio-kaudal.

Die Beckengliedmaßen der auf dem Rücken liegenden Katze sind von einer Person so weit nach hinten zu strecken, daß sich die haltenden Hände auf den Tisch stützen können. Mit einem Schaumgummikeil unter dem Kniegelenk läßt sich die Lagerung erleichtern. Während die Patella des zu untersuchenden Gelenks durch mäßige Innenrotation der Gliedmaße exakt auf der Trochlea ossis femoris liegen muß, kann die andere Beckengliedmaße in leichter Abduktion und Außenrotation fixiert werden. Eine 2. Person umgreift beide Oberarme des Tieres.
Die Aufnahme ist mit einem folienlosen Film (Bleiunterlage nicht vergessen!) anzufertigen.
Der Zentralstrahl sollte kranial auf die Mitte der Gliedmaße dicht proximal der Tuberositas tibiae treffen und im rechten Winkel auf den Film einfallen.

Fig. 308 Positioning of stifle joint. Craniocaudal.

The cat lies in dorsal recumbency. An assistant should extent the pelvic limbs caudally far enough to allow the hands to rest on the table. The positioning is facilitated by placing foam rubber pads under the stifle joint. The patella of the joint to be examined must rest exactly on the femoral trochlea. This can be accomplished by slight inward rotation of the limb. The other limb can be secured in slight abduction and outward rotation. A second assistant holds both front limbs at the upper arms.
A non-screen film should be used (lead blocker not to be forgotten!).
The central beam should strike the center of the limb immediately proximal to the tibial tuberosity and fall at the right angle on the film.

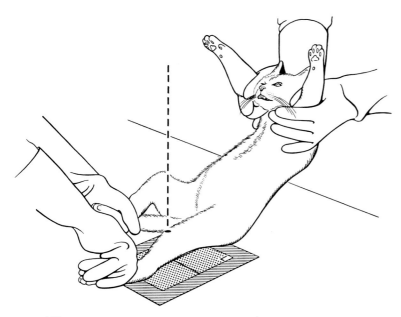

Abb. 309 Lagerung zur Aufnahme des Kniegelenks und des Unterschenkels. Sitzend. Kranio-kaudal.

Die Lagerung in annähernd sitzender Position ist bei einem sehr schmerzhaften Prozeß im Bereich des Kniegelenks, beispielsweise bei einer Gelenk- oder gelenknahen Fraktur, zweckmäßig. Diese etwas schwierige Lagerung verursacht weniger Schmerzen und verringert die Gefahr zusätzlicher Verletzungen.
Person 1 faßt beide Oberarme vom Rücken her, fixiert den Kopf zwischen beiden Händen und hält die Katze in annähernd sitzender Position. Person 2 fixiert mit einer Hand die zu untersuchende Beckengliedmaße mit exakt kranial auf der Trochlea ossis femoris liegender Patella und umfaßt mit der anderen die zweite Beckengliedmaße und den Schwanz.
Die Aufnahme ist mit einem folienlosen Film (Bleiunterlage nicht vergessen!) anzufertigen.
Kniegelenk: Der Zentralstrahl sollte kranial auf die Mitte der Gliedmaße proximal der Tuberositas tibiae treffen und im rechten Winkel auf den Film einfallen.
Tibia: Der Zentralstrahl sollte die Tibia in Schaftmitte treffen und im rechten Winkel auf den Film einfallen.

Fig. 309 Positioning of stifle joint and lower leg. Sitting position. Craniocaudal.

The positioning in near sitting position is indicated in painful conditions in the vicinity of the stifle joint, e.g. fractures. This somewhat difficult positioning is less painful and also reduces the danger of additional trauma.
One assistant grasps both arms from behind, secures the head between both hands and holds the cat in a near sitting position. A second assistant secures the hind limb to be examined with one hand, ensuring that the patella rests exactly on the cranial aspect of the femoral trochlea. One hand holds the other limb and the tail.
A non-screen film should be used (lead blocker not to be forgotten!).
Stifle joint: The central beam should strike the center of the limb immediately proximal to the tibial tuberosity and fall at the right angle on the film.
Tibia: The central beam should strike the mid-shaft of the tibia and fall at the right angle on the film.

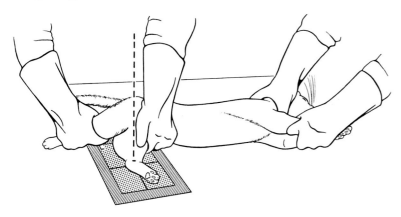

Abb. 310 Lagerung zur Aufnahme des Hinterfußes. Medio-lateral.

Die Katze liegt auf der Seite. Person 1 fixiert mit einer Hand Hals und Kopf im Gabelgriff und mit der anderen die leicht nach vorn gezogenen Schultergliedmaßen. Die am Rücken des Tieres stehende Person 2 zieht mit einer Hand die obenliegende Beckengliedmaße in leichter Abduktion

kaudal, fixiert ggf. auch den Schwanz und stützt sich mit der Hand auf den Tisch. Mit der anderen Hand faßt Person 2 die zu untersuchende, untenliegende Beckengliedmaße in Höhe des Kniegelenks und lagert das leicht gestreckte Sprunggelenk so auf dem Film, daß die Zehen möglichst senkrecht übereinanderliegen. Dabei kann die Katze zusätzlich mit dem Unterarm auf den Tisch gedrückt werden.
Die Aufnahme ist mit einem folienlosen Film (Bleiunterlage nicht vergessen!) anzufertigen.
Der Zentralstrahl sollte medial auf die Mitte der Gliedmaße in Höhe des Talus treffen und im rechten Winkel auf den Film einfallen.

Fig. 310 Positioning of hindpaw. Mediolateral.

The cat lies in lateral recumbency. One assistant secures the neck and head with one hand in a fork grip, the other hand stretches the thoracic limbs slightly cranially. The second assistant, standing at the back of the animal, pulls the upper hind limb caudally, abducting it slightly. The tail is secured simultaneously with the same hand if necessary. With the other hand he holds the lower hind limb to be radiographed at the knee joint and places the slightly extended hock joint on the film. The digits should lie perpendicularly upon each other. At the same time the forearm presses the cat against the table.
A non-screen film should be used (lead blocker not to be forgotten!).
The central beam should strike the center of the limb on a level of the tibial tarsal bone and fall at the right angle on the film.

Abb. 311 Lagerung zur Aufnahme des Hinterfußes. Dorso-plantar.

Person 1 umgreift die Schultergliedmaßen in Höhe des Ellbogengelenks vom Rücken her, fixiert den Kopf zwischen den Händen und hält die Katze aufrecht sitzend. Der neben Person 1 stehende 2. Helfer faßt die zu untersuchende Beckengliedmaße in Höhe des Kniegelenks und lagert sie mit gestrecktem Sprunggelenk so, daß der Kalkaneus dem Film aufliegt. Bei dieser Lagerung befinden sich die Malleoli lateralis und medialis in gleicher Höhe. Eine Schrägaufnahme ergibt sich, wenn der Kalkaneus seitlich und dabei entweder der Malleolus medialis oder der Malleolus lateralis dem Film anliegen.
Die Aufnahme ist mit einem folienlosen Film (Bleiunterlage nicht vergessen!) anzufertigen.
Der Zentralstrahl sollte dorsal auf die Mitte der Gliedmaße in Höhe des Talus treffen und im rechten Winkel auf den Film einfallen.

Fig. 311 Positioning of hindpaw. Dorsoplantar.

One assistant grasps the thoracic limbs from the back at the level of the elbows, secures the head between the hands and holds the cat in an upright sitting position. The second assistant holds the hind limb to be examined at the stifle joint and places the extended hock joint with the calcaneus resting on the film. In this position both lateral and medial malleoli are at the same level. If the calcaneus is tilted sideways an oblique projection will ensue.
A non-screen film should be used (lead blocker not to be forgotten!).
The central beam should strike the center of the limb on a level of the tibial tarsal bone and fall at the right angle on the film.

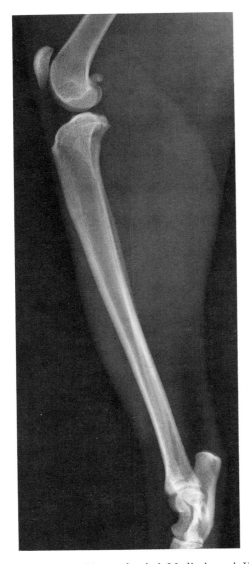

Abb. 312   Rechter Unterschenkel. Medio-lateral. Katze.
Folienloser Film – FFA 120 cm – 60 kV – 16 mAs
Originalgröße (Ausschnitt aus 18 × 24 cm)
Lagerung Abb. 306

Fig. 312   Right lower leg. Mediolateral. Cat.
Non-screen film – FFD 120 cm – 60 kV – 16 mAs
Original size (section of 18 × 24 cm)
Positioning fig. 306

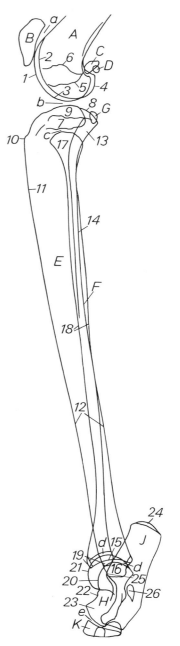

Abb. 313   Röntgenskizze zu Abb. 312
Fig. 313   X-ray sketch to fig. 312

6   Verknöcherte distale Epiphysenfuge – Ossified distal epiphyseal
groove;

An der Tibia – On the tibia:

7   Condylus lateralis;
8   Condylus medialis;
9   Eminentia intercondylaris;
10   Tuberositas tibiae;
11   Margo cranialis;
12   Corpus tibiae;
13   Margo lateralis;
14   Fossa poplitea;
15   Cochlea tibiae;
16   Malleolus medialis;

An der Fibula – On the fibula

17   Caput fibulae;
18   Corpus fibulae;
19   Malleolus lateralis;

Am Talus – On the talus:

20, 21   Trochlea tali proximalis:
20   Lateraler Rollkamm – Lateral ridge,
21   Medialer Rollkamm – Medial ridge;
22   Collum tali;
23   Caput tali;

Am Calcaneus – On the calcaneus:

24   Tuber calcanei;
25   Processus coracoideus;
26   Sustentaculum tali.

A   Os femoris;
B   Patella;
C   Laterales Os sesamoideum m. gastrocnemii – Lateral sesamoid bone
in the gastrocnemius muscle;
D   Mediales Os sesamoideum m. gastrocnemii – Medial sesamoid bone
in the gastrocnemius muscle;
E   Tibia;
F   Fibula;
G   Os sesamoideum m. poplitei;
H   Talus;
J   Calcaneus;
K   Os tarsi centrale;

a   Articulatio femoropatellaris;
b   Articulatio femorotibialis;
c   Articulatio tibiofibularis proximalis;
d   Articulatio tarsocruralis;
e   Articulationes talocalcaneocentralis et calcaneoquartalis;

Am Os femoris – On the femur:

1, 2   Trochlea ossis femoris:
1   Lateraler Rollkamm – Lateral ridge,
2   Medialer Rollkamm – medial ridge;
3   Condylus lateralis;
4   Condylus medialis;
5   Verschattung am Grund der Fossa intercondylaris – Shadow formed
by the bottom of the intercondylar fossa;

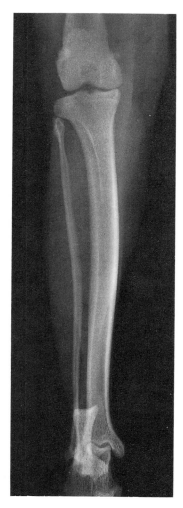

Abb. 314   Rechter Unterschenkel, Kranio-kaudal. Katze.
Folienloser Film – FFA 120 cm – 60 kV – 20 mAs
Originalgröße (Ausschnitt aus 18 × 24 cm)
Lagerung Abb. 309

Fig. 314   Right lower leg. Craniocaudal. Cat.
Non-screen film – FFD 120 cm – 60 kV – 20 mAs
Original size (section of 18 × 24 cm)
Positioning fig. 309

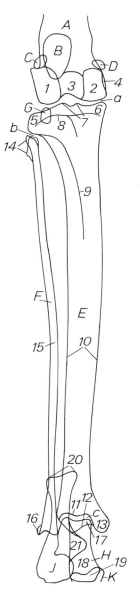

Abb. 315   Röntgenskizze zu Abb. 314
Fig. 315   X-ray sketch to fig. 314

A   Os femoris;
B   Patella;
C   Laterales Os sesamoideum m. gastrocnemii – Lateral sesamoid bone in the gastrocnemius muscle;
D   Mediales Os sesamoideum m. gastrocnemii – Medial sesamoid bone in the gastrocnemius muscle;
E   Tibia;
F   Fibula;
G   Os sesamoideum m. poplitei;
H   Talus;
J   Calcaneus;
K   Os tarsi centrale;

a   Articulatio femorotibialis;
b   Articulatio tibiofibularis proximalis;
c   Articulatio tarsocruralis;

Am Os femoris – On the femur:

1   Condylus lateralis;
2   Condylus medialis;
3   Fossa intercondylaris;
4   Bandgrube – Depression for ligamentous attachment;

An der Tibia – On the tibia:

5   Condylus lateralis;

6   Condylus medialis;
7   Eminentia intercondylaris;
8   Kaudale Begrenzung der Facies articularis – Caudal border of the articular surface;
9   Margo cranialis;
10   Corpus tibiae;
11   Cochlea tibiae;
12   Kranialer Rand – Cranial border;
13   Malleolus medialis;

An der Fibula – On the fibula:

14   Caput fibulae;
15   Corpus fibulae;
16   Malleolus lateralis;

Am Talus – On the talus:

17   Trochlea tali proximalis;
18   Collum tali;
19   Trochlea tali distalis;

Am Calcaneus – On the calcaneus:

20   Tuber calcanei;
21   Sustentaculum tali.

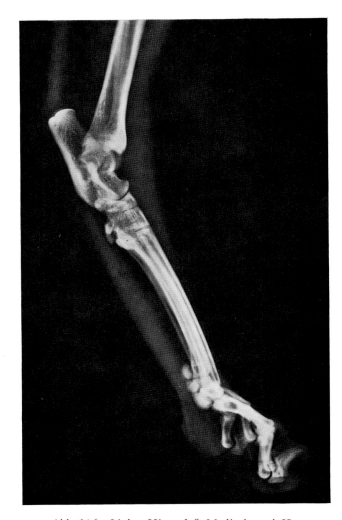

Abb. 316   Linker Hinterfuß. Medio-lateral. Katze.
Folienloser Film – FFA 120 cm – 64 kV – 15 mAs
Originalgröße (Ausschnitt aus 13 × 18 cm)
Lagerung Abb. 310

Fig. 316   Left hindpaw. Mediolateral. Cat.
Non-screen film – FFD 120 cm – 64 kV – 15 mAs
Original size (section of 13 × 18 cm)
Positioning fig. 310

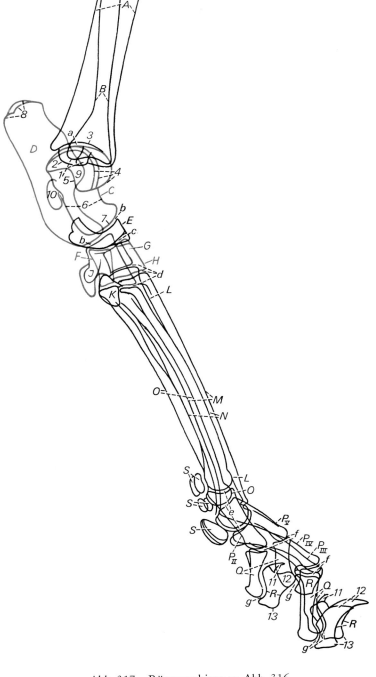

Abb. 317   Röntgenskizze zu Abb. 316
Fig. 317   X-ray sketch to fig. 316

A   Tibia;
B   Fibula;
C   Talus;
D   Calcaneus;
E   Os tarsi centrale;
F   Os tarsale I;
G   Os tarsale II;
H   Os tarsale III;
J   Os tarsale IV;
K   Os metatarsale I;
L   Os metatarsale II;
M   Os metatarsale III;
N   Os metatarsale IV;
O   Os metatarsale V;
P   Phalanx proximalis (II–V kennzeichnen die Skelettelemente an der 2. bis 5. Zehe) – Phalanx proximalis (II-V show the skeletal parts of the 2nd to 5th digits);
Q   Phalanx media;
R   Phalanx distalis;
S   Ossa sesamoidea proximalia;

a   Articulatio tarsocruralis;
b   Articulationes talocalcaneocentralis et calcaneoquartalis;
c   Articulatio centrodistalis;
d   Articulationes tarsometatarseae;
e   Articulatio metatarsophalangea;
f   Articulatio interphalangea proximalis pedis;
g   Articulatio interphalangea distalis pedis;

1   Malleolus lateralis;
2   Malleolus medialis;
3   Cochlea tibiae;
4   Trochlea tali proximalis;
5   Verschattung, die sich aus der Konkavität kaudal am Talus ergibt – Shadow formed by the concavity caudally on the talus;
6   Collum tali;
7   Trochlea tali distalis;
8   Tuber calcanei;
9   Processus coracoideus;
10   Sustentaculum tali;
11   Crista unguicularis;
12   Processus unguicularis;
13   Tuberositas flexoria.

210

Abb. 318   Rechter Hinterfuß. Dorso-plantar. Katze.
Folienloser Film – FFA 120 cm – 60 kV – 20 mAs
Originalgröße (Ausschnitt aus 13 × 18 cm)
Lagerung Abb. 311

Fig. 318   Right hindpaw. Dorsoplantar. Cat.
Non-screen film – FFD 120 cm – 60 kV – 20 mAs
Original size (section of 13 × 18 cm)
Positioning fig. 311

A   Tibia;
B   Fibula;
C   Talus;
D   Calcaneus;
E   Os tarsi centrale;
F   Os tarsale I;
G   Os tarsale II;
H   Os tarsale III;
J   Os tarsale IV;
K   Os metatarsale I;
L   Os metatarsale II;
M   Os metatarsale III;
N   Os metatarsale IV;
O   Os metatarsale V;
P   Phalanx proximalis (II-V kennzeichnen die Skelettelemente an der 2. bis 5. Zehe) – Phalanx proximalis (II-V show the skeletal parts of the 2nd to 5th digits);
Q   Phalanx media;
R   Phalanx distalis;
S   Ossa sesamoidea proximalia;

a   Articulatio tarsocruralis;
b   Articulationes talocalcaneocentralis et calcaneoquartalis;
c   Articulatio centrodistalis;
d   Articulationes tarsometatarseae;
e   Articulatio metatarsophalangea;
f   Articulatio interphalangea proximalis pedis;
g   Articulatio interphalangea distalis pedis;

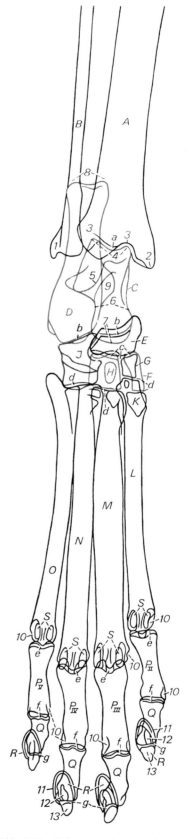

Abb. 319   Röntgenskizze zu Abb. 318

Fig. 319   X-ray sketch to fig. 318

1   Malleolus lateralis;
2   Malleolus medialis;
3   Cochlea tibiae;
4   Trochlea tali proximalis;
5   Sulcus tali;
6   Collum tali;
7   Trochlea tali distalis;
8   Tuber calcanei;
9   Sustentaculum tali;
10   Bandgruben – Depressions for ligamentous attachment;
11   Crista unguicularis;
12   Processus unguicularis;
13   Tuberositas flexoria.

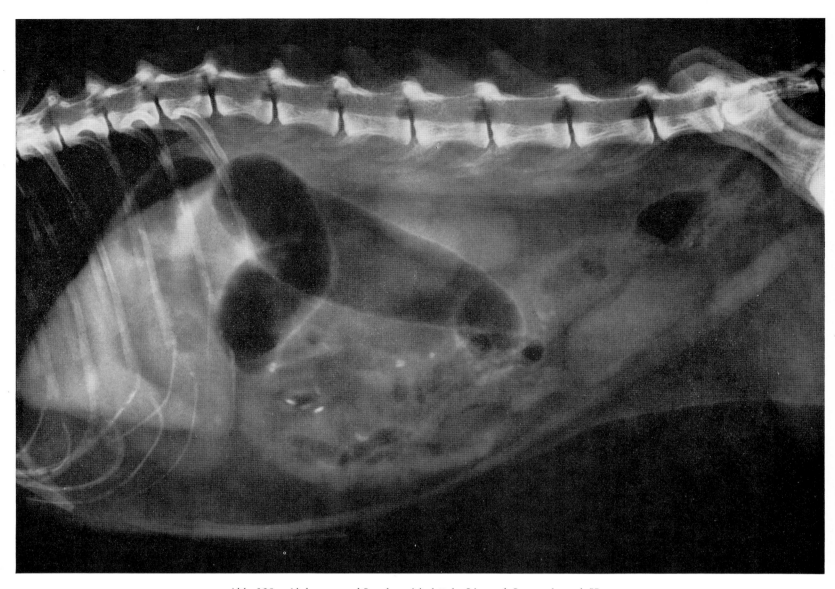

Abb. 320   Abdomen und Lendenwirbelsäule. Liegend. Latero-lateral. Katze.
Bucky-Blende – Feinzeichnende Folie – FFA 120 cm – 65 kV – 30 mAs
Verkleinerung von 24 × 30 cm
Lagerung Abb. 324

Fig. 320   Abdomen and lumbar vertebral column. Recumbent. Laterolateral.
Bucky diaphragm – High definition screens – FFD 120 cm – 65 kV – 30 mAs
Diminution of 24 × 30 cm
Positioning fig. 324

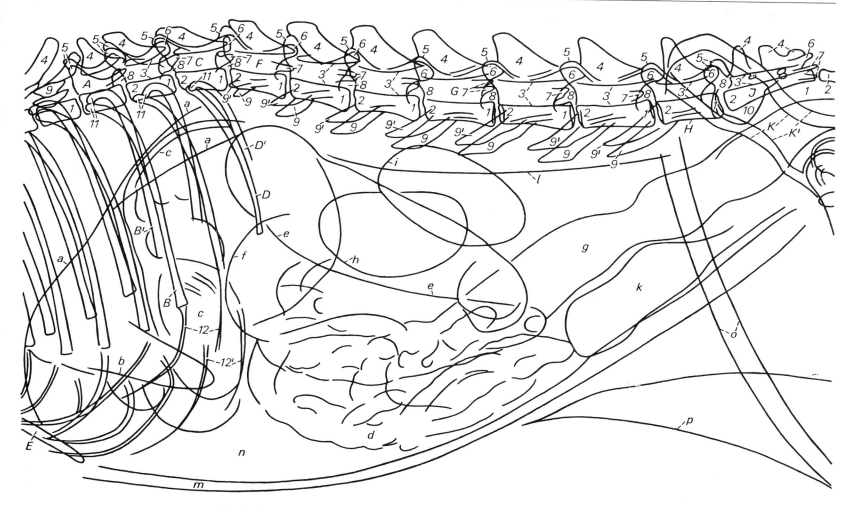

Abb. 321*　Röntgenskizze zu Abb. 320　　Fig. 321*　X-ray sketch to fig. 320

A　11. Vertebra thoracica;
B　11. Os costale;
C　13. Vertebra thoracica;
D　13. Os costale;
E　Processus xiphoideus;
F　1. Vertebra lumbalis;
G　4. Vertebra lumbalis;
H　7. Vertebra lumbalis;
J　Os sacrum;
K　Os ilium;

a　Diaphragma;
b　Hepar;
c　Ventriculus;
d　Jejunum;
e　Colon ascendens;
f　Colon transversum;
g　Colon descendens;
h　Ren dexter;
i　Ren sinister;
k　Vesica urinaria;
l　Lendenmuskulatur, ventrale Begrenzung – Ventral limit of the lumbar muscles;
m　Ventrale Bauchwand – Ventral abdominal wall;
n　Corpus adiposum;

o　Oberschenkelmuskulatur, kraniale Begrenzung – Cranial limit of the thigh musculature;
p　Kniefalte – Fold of the flank;

An den Wirbeln – On the vertebrae:

1　Facies terminalis caudalis;
2　Facies terminalis cranialis;
3　Foramen vertebrale;
4　Processus spinosus;
5　Processus mamilloarticularis;
6　Processus articularis caudalis;
7　Processus accessorius;
8　Incisurae vertebrales cranialis et caudalis, zugleich Begrenzung der Foramina intervertebralia – Incisurae vertebrales cranialis et caudalis forming the margins of the intervertebral foramina;
9　Processus transversus, plattennah – Processus transversus, next to film;
10　Ala ossis sacri;

An den Rippen – On the ribs:

11　Caput costae;
12　Cartilago costalis.

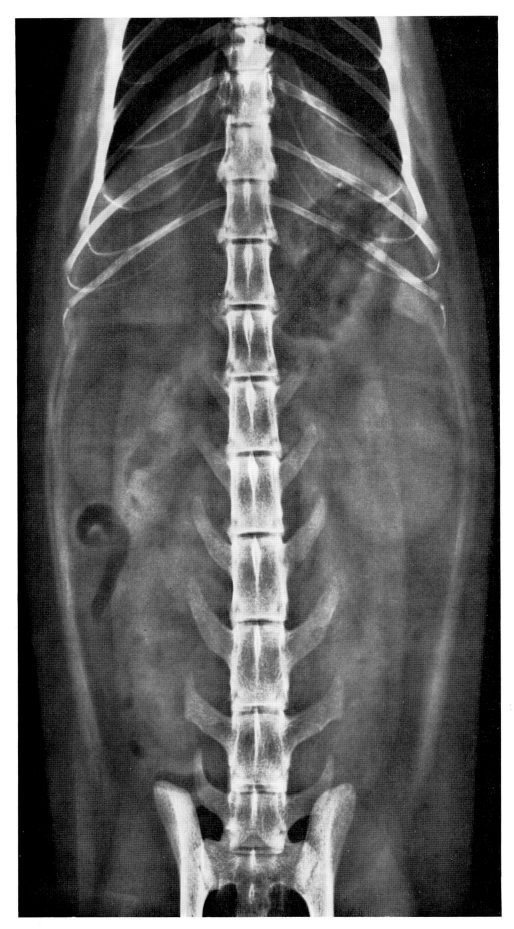

Abb. 322   Abdomen und Lendenwirbelsäule. Ventro-dorsal. Katze.
Bucky-Blende – Feinzeichnende Folie – FFA 120 cm – 65 kV – 30 mAs
Originalgröße (Ausschnitt aus 24 × 30 cm)
Lagerung Abb. 325

Fig. 322   Abdomen and lumbar vertebral column. Ventrodorsal. Cat.
Bucky diaphragm – High definition screens – FFD 120 cm – 65 kV – 30 mAs
Original size (section of 24 × 30 cm)
Positioning fig. 325

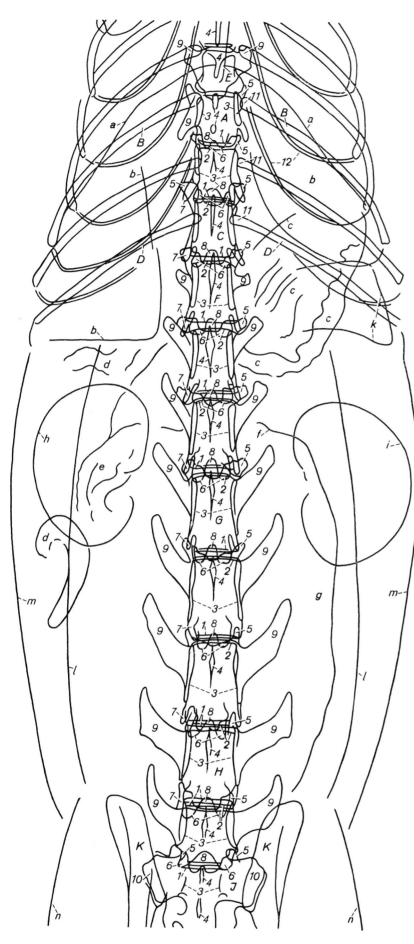

A 11. Vertebra thoracica;
B 11. Os costale;
C 13. Vertebra thoracica;
D 13. Os costale;
E Processus xiphoideus;
F 1. Vertebra lumbalis;
G 4. Vertebra lumbalis;
H 7. Vertebra lumbalis;
J Os sacrum;
K Os ilium;

a Diaphragma;
b Hepar;
c Ventriculus;
d Jejunum;
e Colon ascendens;
f Colon transversum;
g Colon descendens;
h Ren dexter;
i Ren sinister;
k Lien;
l Stammuskulatur, laterale Begrenzung – Lateral limit of the trunk musculature;
m Laterale Bauchwand – Lateral abdominal wall;
n Gesäßmuskulatur, laterale Begrenzung – Lateral limit of the gluteal muscles;

An den Wirbeln – On the vertebrae:

1 Facies terminalis caudalis;
2 Facies terminalis cranialis;
3 Pediculus arcus vertebrae, zugleich seitliche Begrenzung des Foramen vertebrale – Pediculus arcus vertebrae forming the lateral margin of the vertebral foramen;
4 Processus spinosus;
5 Processus articularis cranialis;
6 Processus articularis caudalis;
7 Processus accessorius;
8 Spatium interarcuale;
9 Processus transversus;
10 Ala ossis sacri;

An den Rippen – On the ribs:

11 Caput costae;
12 Cartilago costalis.

Beachte: Diese Katze hat 8 Lendenwirbel – Note: This cat has 8 lumbar vertebrae.

Abb. 323 Röntgenskizze zu Abb. 322

Fig. 323 X-ray sketch to fig. 322

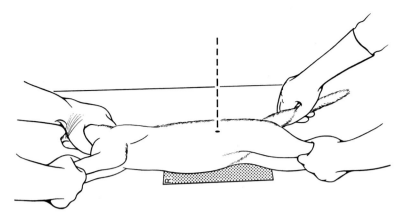

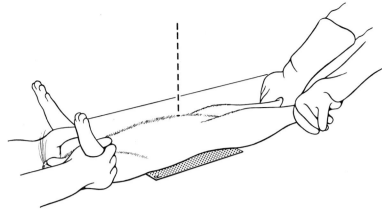

Abb. 324   Lagerung zur Aufnahme des Abdomens und der
Lendenwirbelsäule. Liegend. Latero-lateral.

Abb. 325   Lagerung zur Aufnahme des Abdomens
und der Lendenwirbelsäule. Ventro-dorsal.

Es ist zweckmäßig, die Aufnahmen immer in der gleichen Weise anzufertigen. Die Untersuchung in rechter Seitenlage hat sich bewährt.
Person 1 fixiert mit der linken Hand den etwas nach rückwärts abgebogen Kopf-Hals im Gabelgriff und mit der rechten Hand die nach vorn gezogenen Schultergliedmaßen. Person 2 streckt die mit der linken Hand in
Höhe der Sprunggelenke gefaßten Beckengliedmaßen so weit nach hinten,
daß Überlagerungen durch die Oberschenkelmuskulatur vermieden werden, und fixiert mit der rechten Hand den Schwanz.
Die Katze ist so zu lagern, daß die Medianebene parallel zur Kassette
liegt. Bei fettleibigen Tieren ist einer Verkantung durch Schaumgummikeile unter den auf der Tischplatte liegenden Gliedmaßen in Höhe der zu
haltenden Hände vorzubeugen. Leichter Zug am Schwanz unterstützt die
Lagerung und Fixation.
Zur Verminderung der Streustrahlung kann eine Bucky-Blende oder eine
Kassette mit stehendem Raster notwendig sein. Feinzeichnende Folien
sind ausreichend. Bei nicht fettleibigen Katzen kann auf einen Raster ver-
zichtet werden.
Einstellungen des Zentralstrahls:
Übersichtsaufnahme und Darm: Mitte der seitlichen Bauchwand in Höhe
des 3. Lendenwirbels.
Magen: Mitte der seitlichen Bauchwand in Höhe der 12. Rippe.
Gallenblase: Mitte der seitlichen Bauchwand in Höhe der 10. Rippe.
Nieren: Seitliche Bauchwand am Übergang vom mittleren zum dorsalen
Drittel in Höhe des 3. Lendenwirbels.
Harnblase: Mitte der seitlichen Bauchwand in Höhe des 7. Lendenwirbels.
Lendenwirbelsäule: 3. Lendenwirbel.

Fig. 324   Positioning of abdomen and lumbar vertebral
column. Recumbent. Laterolateral.

It is advisable to take radiographs always in a constant manner. Radiography in right lateral recumbency has proved to be successful.
One assistant fixes the slightly flexed head and neck with the left hand in
a fork grip and the crainally extended front limbs with the right hand.
The second assistant grasps the hind legs with the left hand at the hocks,
stretching them far enough caudally to avoid superimposure of the thigh
muscles. With the right hand he fixes the tail.
The cat must be aligned with the median plane parallel to the cassette. In
obese animals tilting can be avoided by placing foam rubber pads under
the legs where they are secured by the hands. Slight pull on the tail facilitates the positioning and fixation.
To reduce scattered radiation a Bucky diaphragm or a cassette with a stationary grid should be used. High definition screens will suffice. In lean
objects a grid is not necessary.
Directing the central beam:
Routine radiograph and intestine: Middle of the lateral abdominal wall on
a level of the 3rd lumbar vertebra.
Stomach: Middle of the lateral abdominal wall on a level of the 12th rib.
Gall-bladder: Middle of the lateral abdominal wall on a level of the 10th
rib.
Kidneys: Lateral abdominal wall at the transition of the middle and dorsal
thirds on a level of the 3rd lumbar vertebra.
Urinary bladder: Middle of the lateral abdominal wall on a level of the 7th
lumbar vertebra.
Lumbar vertebral column: 3rd lumbar vertebra.

Die Katze wird in Rückenlage gebracht. Die am Kopf des Tieres stehende
Person 1 faßt die Schultergliedmaßen in Höhe der Ellbogengelenke und
fixiert den Kopf durch Andrücken der Hände. Person 2 zieht die Beckengliedmaßen in geringer Abduktionsstellung so weit nach hinten, daß sich
die Hände, die die Gliedmaßen im Bereich der Sprunggelenke halten, auf
die Tischplatte aufstützen können. Es ist darauf zu achten, daß die Medianebene senkrecht zur Kassette steht.
Zur Verringerung der Streustrahlung ist eine Bucky-Blende oder eine Kassette mit stehendem Raster notwendig. Bei kleinen oder schlanken Tieren
kann auf den Raster verzichtet werden.
Einstellung des Zentralstrahls:
Übersichtaufnahmen, Darm, Nieren: Mittellinie in Höhe des 2. Lenden-
wirbels.
Magen: Mittellinie in Höhe des 13. Brustwirbels.
Gallenblase: Mittellinie in Höhe des 12. Brustwirbels.
Harnblase: Mittellinie in Höhe des 7. Lendenwirbels.
Lendenwirbelsäule: Mittellinie in Höhe des 3. Lendenwirbels.

Fig. 325   Positioning of abdomen and lumbar vertebral
column. Ventrodorsal.

The cat is placed on its back. One assistant standing at the head of the
animal grasps the front limbs at the elbow joints and secures the head by
pressure of the hands. The second assistant holding the hind limbs at the
hocks, pulls them caudally in slight abduction until the hands can be supported on the table. Care should be exercised that the median plane is
perpendicular to the cassette.
To reduce scattered radiation a Bucky diaphragm or a cassette with stationary grid should be used. In small or lean animals the grid can be aban-
doned.
Directing the central beam:
Routine radiograph, intestine, kidneys: Median plane on a level of the 2nd
lumbar vertebra.
Stomach: Median plane on a level of the 13th thoracic vertebra.
Gall-bladder: Median plane on a level of the 12th thoracic vertebra.
Urinary bladder: Median plane on a level of the 7th lumbar vertebra.
Lumbar vertebral column: Median plane on a level of the 3rd lumbar ver-
tebra.

Magen-Darm-Kontrastuntersuchung

Vorbereitung: Abgesehen von einer dringlich notwendigen Untersuchung (Ileusverdacht), sollte eine Magen-Darm-Passage im nüchternen Zustand nach 12stündigem Hungern und Dursten vorgenommen werden. Wird bei der Untersuchung (Durchleuchtung, Leeraufnahme) Darminhalt im Dickdarm festgestellt, der die Darstellung in einem bestimmten Abschnitt und damit die Beurteilung erschweren kann, ist ein Kontaktlaxans, ggf. ein Reinigungseinlauf zu verabreichen.

Kontrastmittel: Barium sulfuricum purissimum in einer der handelsüblichen Abfüllungen in einer wässerigen homogenen Aufschwemmung von rahmiger Konsistenz. Abhängig von der Größe der Katze sind 40–80 ml der Suspension zu applizieren.
Trijodierte Kontrastmittel in wässeriger Lösung. Es sind hochprozentige Lösungen mit Geschmackskorrigentien im Handel. Wegen der guten Verträglichkeit sind die Präparate im Falle einer Aspiration und bei perforierenden Verletzungen der Speiseröhre, des Magens oder Darmes dem Barium sulfuricum unbedingt vorzuziehen. Im Vergleich zu den Barium sulfuricum enthaltenden Präparaten passieren die trijodierten Kontrastmittel in wässeriger Lösung den Darm schneller. Außerdem sind sie weniger schattengebend und werden zum Teil resorbiert.

Applikation: Soll das Einfließen in den Magen oder die Magenfunktion untersucht werden, ist das Kontrastmittel schluckweise mit einer Schnabeltasse oder einer Spritze in die Backentasche zu verabreichen. Die Applikation mit der Magensonde ist notwendig, wenn neben dem Kontrastmittel zusätzlich Luft zur besseren Konturierung des Magens (Positiv-Negativ-Kontrast) zu insufflieren ist.

Untersuchungstechnik: Wichtig ist die Erfassung von Bewegungsabläufen. Aufschluß darüber ist nur bei der Durchleuchtung zu erhalten. Bei jeder Unregelmäßigkeit im Passageablauf ist der betreffende Abschnitt eingehend zu untersuchen. Zur Objektivierung bestimmter Zustände sind „gezielte" Aufnahmen anzufertigen. Die Zahl der notwendigen Aufnahmen und die Lagerung hängen deshalb von der jeweiligen Situation ab.

Folgendes Untersuchungsschema hat sich bewährt:
a) Durchleuchtung stehend seitlich.
b) Eingabe von 2–3 Schlucken Kontrastmittel, Untersuchung stehend seitlich, anschließend liegend seitlich und in Rückenlage (dorso-ventraler Strahlengang).
Für die Beurteilung des Reliefs der Magenschleimhaut ist die auf dem Rücken liegende Katze ggf. gering (etwa 15°) nach rechts oder links zu neigen.
c) Schluckweise Eingabe des Kontrastmittels bis zur prallen Füllung des Magens.
Bei der gesunden Katze beginnt die Entleerung des Kontrastmittels etwa 5 Minuten nach der Applikation und ist nach etwa 2 Stunden beendet.
Zweckmäßig ist die Untersuchung (Durchleuchtung, Aufnahme) in Rückenlage und in rechter Seitenlage.
d) Das Duodenum ist bei der gesunden Katze etwa 5 Minuten nach der Kontrastmittelgabe meist in ganzer Länge dargestellt. Zweckmäßig ist die Untersuchung (Durchleuchtung, Aufnahme) in Rückenlage mit einer geringen Neigung des Tieres (etwa 15°) nach der rechten Seite hin und in rechter Seitenlage.
e) Ein Barium sulfuricum enthaltendes Kontrastmittel (homogene Aufschwemmung, rahmige Konsistenz) erreicht bei der gesunden Katze nach etwa 2 1/2 Stunden das Caecum und nach etwa 4 Stunden das Colon descendens.
Ein trijodiertes Kontrastmittel in wässeriger Lösung erreicht bei der gesunden Katze nach etwa 30 Minuten das Caecum und nach etwa 45 Minuten das Colon descendens.
Zweckmäßig ist die Untersuchung (Durchleuchtung, Aufnahme) in Rückenlage und in rechter Seitenlage. Die Abstände zwischen den Untersuchungen ergeben sich aus der Fragestellung unter Berücksichtigung der Art und Menge des verabreichten Kontrastmittels.

Contrast studies of the gastro-intestinal tract

Preparation of the patient: Except in emergencies (e.g. ileus), a gastro-intestinal passage should be undertaken after withholding food and water for at least 12 hours. A contact laxative or an evacuation enema should be administered if intestinal contents can still be demonstrated in the colon by fluoroscopy or radiography prior to the administration of the contrast medium.

Contrast medium: Barium sulphate in a watery, homogeneous suspension and of creamy consistency. It is administered in doses of 40–80 ml depending on the size of the cat.
Tri-iodized contrast media in watery solution. High percentage solutions with taste-correcting components are available. Since they are well tolerated these drugs are superior to Barium sulphate if aspirated, and in perforating lesions of the esophagus, stomach and intestine. In comparison with Barium sulphate containing drugs the tri-iodized water soluble contrast media pass more rapidly through the intestine. Moreover, they produce less shadows and are partly resorbed.

Application: For examination of the inflow into the stomach or the function of the stomach the contrast medium is administered swallow by swallow by means of a feeding cup or by an injection into the cheek-pouch. Application by a stomach tube is necessary if, in addition to the contrast medium, air has to be insufflated for better definition of stomach contours (positive and negative contrast).

Technique: It is important to determine the motility-sequences. Only fluoroscopy can furnish the desired information. In case of any irregularity in the passage the particular segment should be closely examined. In order to register particular conditions, specific radiographs should be aimed at. The number of radiographs necessary and the positioning, therefore, depend on the individual case.
The following routine procedure proved to be successful:
a) Fluoroscopy – standing position, lateral.
b) Administration of 2–3 swallows of contrast medium followed by fluoroscopy in the standing, lateral recumbent and dorsal recumbent positions (dorsoventral direction of the beam). To evaluate the relief of the gastric mucosa, the cat lying in dorsal recumbency should be tilted slightly (approximately 15°) to the right or left.
c) Swallow by swallow administration of the contrast medium until the stomach is well filled. In a normal cat emptying of the stomach begins approximately 5 minutes after the medium has been swalled and is completed within 2 hours. The examination (fluoroscopy, radiograph) should be performed in the dorsal recumbent and right lateral recumbent positions.
d) The duodenum of a normal cat is outlined in its full length within approximately 5 minutes after administration of the contrast medium. The examination (fluoroscopy, radiograph) in dorsal recumbency with slight tilting to the right (15°) and in the right lateral recumbent position provides the best results.
e) Barium sulphate containing contrast medium (homogeneous suspension, creamy consistency) reaches the cecum within 2 1/2 hours and the colon descendens with 4 hours after administration.
In a normal cat a tri-iodized water soluble contrast medium reaches the cecum approximately after 30 minutes and the colon descendens after approximately 45 minutes.
The examination (fluoroscopy, radiograph) should be done in dorsal recumbent and right lateral recumbent positions. The intervals between the examination depend on the case in question and the nature and quantity of the contrast medium administered.

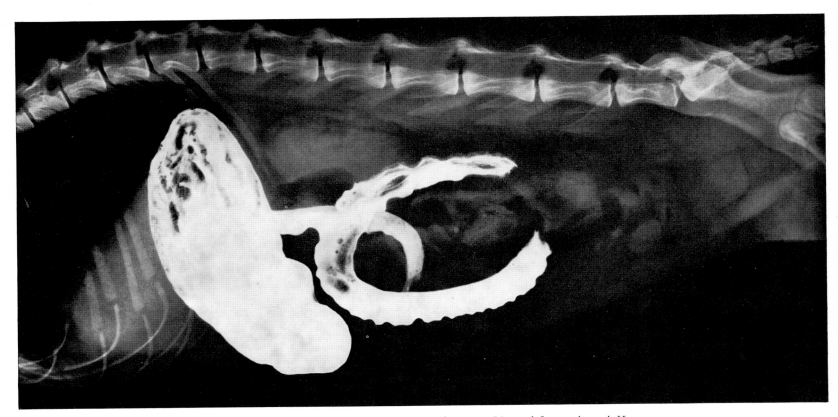

Abb. 326   Magen-Darm-Kontrast (Barium sulfuricum). Liegend. Latero-lateral. Katze.
Bucky-Blende – Feinzeichnende Folie – FFA 115 cm – 60 kV – 30 mAs
Verkleinerung von 15 × 40 cm
Lagerung Abb. 324

Fig. 326   Stomach-intestine-contrast (Barium sulphate). Recumbent. Laterolateral. Cat.
Bucky diaphragm – High definition screens – FFD 115 cm – 60 kV – 30 mAs
Diminution of 15 × 40 cm
Positioning fig. 324

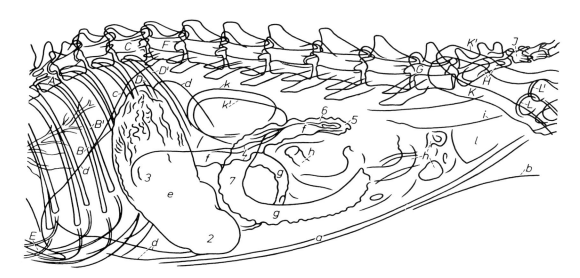

Abb. 327*   Röntgenskizze zu Abb. 326   Fig. 327*   X-ray sketch to fig. 326

A   10. Vertebra thoracica;
B   10. Os costale;
C   13. Vertebra thoracica;
D   13. Os costale;
E   Processus xiphoideus;
F   1. Vertebra lumbalis;
G   7. Vertebra lumbalis;
H   Os sacrum;
J   1. Vertebra caudalis;
K   Os ilium;
L   Os femoris;

a   Ventrale Bauchwand – Ventral abdominal wall;
b   Kniefalte – Fold of the flank;
c   Diaphragma;
d   Hepar;
e   Ventriculus;
f   Duodenum;

g   Jejunum mit Kontrastmittel – Jejunum with contrast medium;
h   Jejunumschlingen, gashaltig – Intestinal loops filled with gas;
i   Colon descendens;
k   Ren dexter;
k' Ren sinister;
l   Vesica urinaria;

Am Magen – On the stomach:

1   Fundus ventriculi;
2   Corpus ventriculi;

Am Duodenum – On the duodenum:

3   Pars cranialis;
4   Pars descendens;
5   Flexura duodeni caudalis;
6   Pars ascendens;
7   Flexura duodenojejunalis.

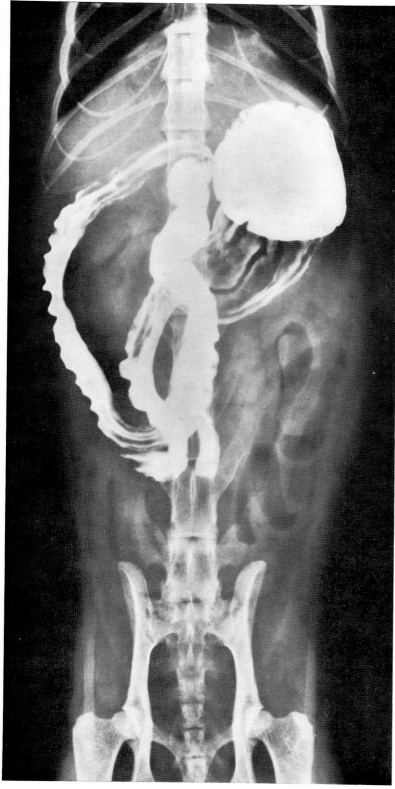

Abb. 328 Magen-Darm-Kontrast (Barium sulfuricum).
Ventro-dorsal. Katze.
Bucky-Blende – Feinzeichnende Folie – FFA 115 cm – 60 kV – 30 mAs
Verkleinerung von 15 × 40 cm
Lagerung Abb. 325

Fig. 328 Stomach-intestine-contrast (Barium sulphate).
Ventrodorsal. Cat.
Bucky diaphragm – High definition screens – FFD 115 cm –
60 kV – 30 mAs
Diminution of 15 × 40 cm
Positioning fig. 325

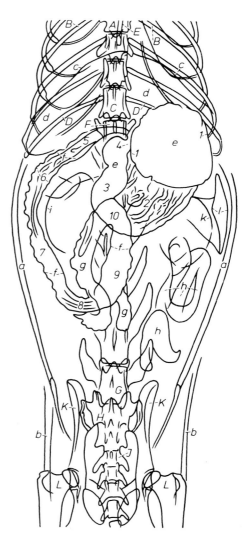

Abb. 329 Röntgenskizze zu Abb. 328
Fig. 329 X-ray sketch to fig. 328

A	10. Vertebra thoracica;
B	10. Os costale;
C	13. Vertebra thoracica;
D	13. Os costale;
E	Processus xiphoideus;
F	1. Vertebra lumbalis;
G	7. Vertebra lumbalis;
H	Os sacrum;
J	1. Vertebra caudalis;
K	Os ilium;
L	Os femoris;
a	Seitliche Bauchwand – Lateral abdominal wall;
b	Kniefalte – Fold of the flank;
c	Diaphragma;
d	Hepar;
e	Ventriculus;
f	Duodenum mit Kontrastmittel – Duodenum with contrast medium;
g	Jejunum mit Kontrastmittel – Jejunum with contrast medium;
h	Jejunumschlingen, gashaltig – Intestinal loops filled with gas;
i	Ren dexter;
k	Ren sinister;
l	Lien;

Am Magen – On the stomach:

1 Fundus ventriculi;
2 Corpus ventriculi;
3 Pars pylorica;
4 Pylorus;

Am Duodenum – On the duodenum:

5 Pars cranialis;
6 Flexura duodeni cranialis;
7 Pars descendens;
8 Flexura duodeni caudalis;
9 Pars ascendens;
10 Flexura duodenojejunalis.

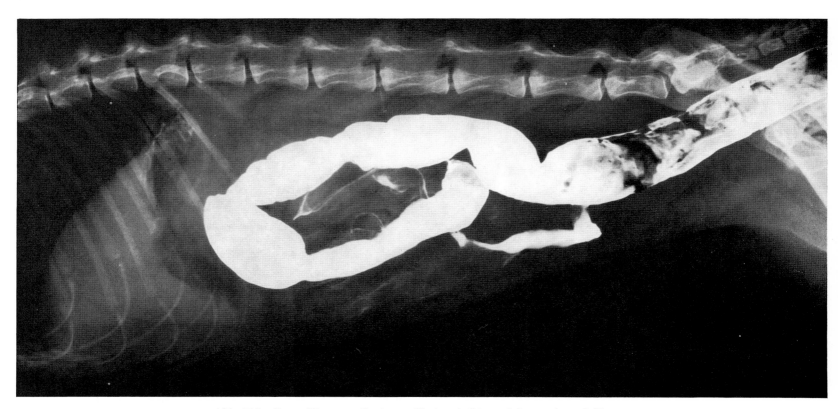

Abb. 330   Darm-Kontrast (Barium sulfuricum). Liegend. Latero-lateral. Katze.
Bucky-Blende – Feinzeichnende Folie – FFA 115 cm – 60 kV – 30 mAs
Verkleinerung von 15 × 40 cm
Lagerung Abb. 324

Fig. 330   Intestine-contrast (Barium sulphate). Recumbent. Laterolateral. Cat.
Bucky diaphragm – High definiton screens – FFD 115 cm – 60 kV – 30 mAs
Diminution of 15 × 40 cm
Positioning fig. 324

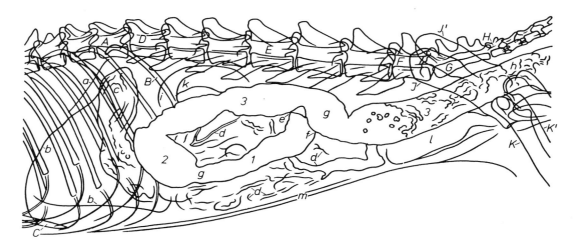

Abb. 331   Röntgenskizze zu Abb. 330   Fig. 331   X-ray sketch to fig. 330

A	13. Vertebra thoracica;	d	Jejunumschlingen mit Kontrastmittelresten – Intestinal loops with residual contrast medium;
B	13. Os costale;		
C	Processus xiphoideus;	e	Ileum;
D	1. Vertebra lumbalis;	f	Caecum;
E	4. Vertebra lumbalis;	g	Colon;
F	7. Vertebra lumbalis;	h	Rectum;
G	Os sacrum;	i	Ren dexter;
H	1. Vertebra caudalis;	k	Ren sinister;
J	Os ilium;	l	Vesica urinaria;
K	Os femoris;	m	Ventrale Bauchwand – Ventral abdominal wall;

a	Diaphragma;
b	Hepar;
c	Fundus ventriculi mit Kontrastmittelresten – Fundus ventriculi with residual contrast medium;

Am Colon – On the colon:

1   Colon ascendens;
2   Colon transversum;
3   Colon descendens.

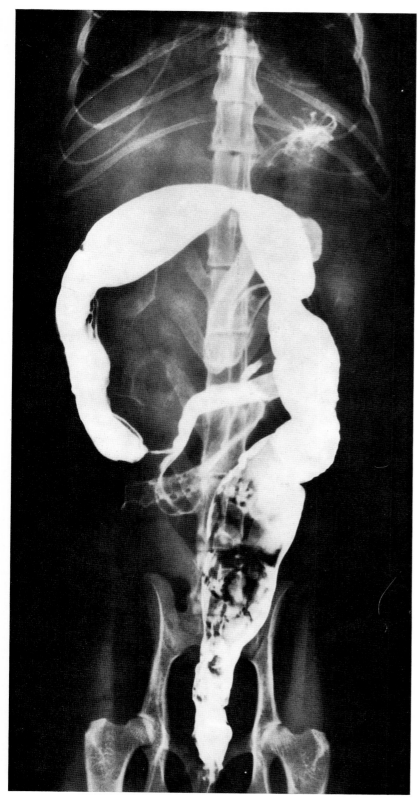

Abb. 332 Darm-Kontrast (Barium sulfuricum). Ventro-dorsal. Katze.
Bucky-Blende – Feinzeichnende Folie – FFA 115 cm – 60 kV – 30 mAs
Verkleinerung von 15 × 40 cm
Lagerung Abb. 325

Fig. 332 Intestine-contrast (Barium sulphate). Ventrodorsal. Cat.
Bucky diaphragm – High definiton screens – FFD 115 cm –
60 kV – 30 mAs
Diminution of 15 × 40 cm
Positioning fig. 325

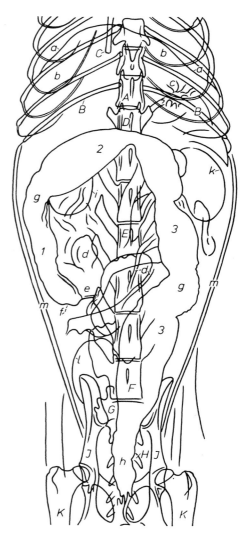

Abb. 333 Röntgenskizze zu Abb. 332
Fig. 333 X-ray sketch to fig. 332

A   13. Vertebra thoracica;
B   13. Os costale;
C   Processus xiphoideus;
D   1. Vertebra lumbalis;
E   4. Vertebra lumbalis;
F   7. Vertebra lumbalis;
G   Os sacrum;
H   1. Vertebra caudalis;
J   Os ilium;
K   Os femoris;

a   Diaphragma;
b   Hepar;
c   Fundus ventriculi mit Kontrastmittelresten – Fundus ventriculi with residual contrast medium;
d   Jejunumschlingen mit Kontrastmittelresten – Intestinal loops with residual contrast medium;
e   Ileum;
f   Caecum;
g   Colon;
h   Rectum;
i   Ren dexter;
k   Ren sinister;
l   Vesica urinaria;
m   Seitliche Bauchwand – Lateral abdominal wall;

Am Colon – On the colon:

1   Colon ascendens;
2   Colon transversum;
3   Colon descendens.

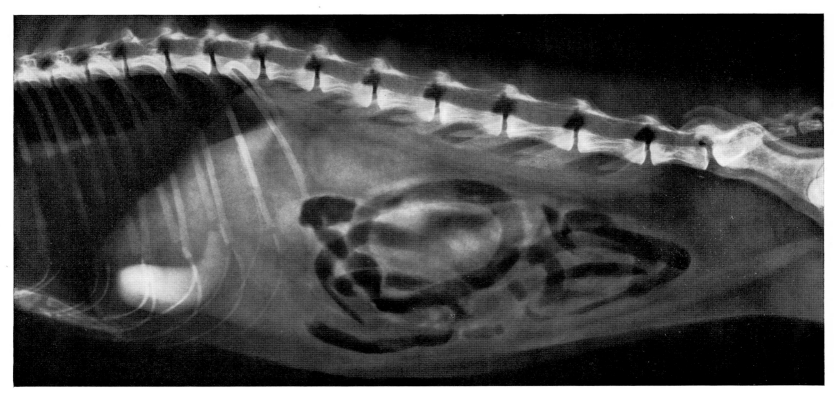

Abb. 334   Gallenblasendarstellung (Biloptin, Schering). Liegend. Latero-lateral. Katze.
Bucky-Blende – Feinzeichnende Folie – FFA 115 cm – 62 kV – 40 mAs
Verkleinerung von 15 × 40 cm
Lagerung Abb. 324

Fig. 334   Gall-bladder (Biloptin, Schering). Recumbent. Laterolateral. Cat.
Bucky diaphragm – High definition screens – FFD 115 cm – 62 kV – 40 mAs
Diminution of 15 × 40 cm
Positioning fig. 324

Orale Cholezystographie

Vorbereitung: Keine Futter- und Flüssigkeitsaufnahme 24 Stunden vor der Untersuchung. Etwa 12 Stunden vor der Untersuchung sind der Füllungszustand des Dickdarms festzustellen (Durchleuchtung, Leeraufnahme) und ggf. ein Kontaktlaxans oder ein Reinigungseinlauf zu verabreichen.

Kontrastmittel: Di- bzw. trijodierte Präparate, deren gebundenes Jod über die Gallenwege ausgeschieden wird. Perorale Applikation 12 Stunden vor der Untersuchung. Dosierung nach Vorschrift (Biloptin, Schering, 500 mg/kg KGW).

Untersuchungstechnik:
a) 12 Stunden nach der peroralen Applikation Aufnahme in 2 Ebenen. Die Gallenblase ist dargestellt, wenn das Kontrastmittel den Magen passieren und im Dünndarm ausreichend resorbiert werden konnte.
b) Reizfütterung, wenn Gallenblase dargestellt ist. Reizfutter: 2 Eigelb mit 1 Teelöffel Zucker und 1 Teelöffel Speiseöl anrühren. Geeignet ist auch eine handelsübliche Reizmahlzeit in entsprechender Dosierung. Falls das Futter nicht aufgenommen wird, Applikation mit der Magensonde.
c) 30 und 45, ggf. auch 60 Minuten nach der Reizfütterung Untersuchung (Durchleuchtung, Aufnahme) in rechter Seitenlage und in Rückenlage. Die Untersuchung gibt Aufschluß über die Entleerungsfähigkeit der Gallenblase. Da Kontrastmittel an Konkrementen länger haften können, ist es notwendig, 10–15 Minuten nach der Entleerung bzw. nach der weitgehenden Entleerung des Kontrastmittels in den Dünndarm eine Aufnahme in rechter Seitenlage und in Rückenlage anzufertigen.

Oral cholecystography

Preparation of the patient: 24 hours prior to the examination food and liquids are withhold. Approximately 12 hours prior to radiography, the degree of filling of the colon is checked (fluoroscopy, preliminary radiography) and if necessary, the cat is given a contact laxative or an evaluation enema.

Contrast media: Iodized organic compounds, excreted through the liver and concentrated in the gall-bladder. Oral administration 12 hours prior to the examination. Dosage according to prescription (Biloptin, Schering, 500 mg/kg body weight).

Technique:
a) 12 hours after oral administration of the contrast medium, radiographs are taken with two different beam directions. The gall-bladder is outlined when the contrast medium has passed the stomach and has been absorbed satisfactorily in the small intestine.
b) Challenge feeding – when the gall-bladder is outlined. The challenge food: 2 egg yolks mixed with 1 teaspoonful of sugar and 1 teaspoonful of oil. A challenge food available in the trade is also suitable in appropiate doses. If the animal refuses the food it should be administered with a stomach tube.
c) 30, 45, and if necessary 60 minutes after administration of the challenge food, examination (fluoroscopy, radiograph) in right lateral and in dorsal recumbency. This examination furnishes information on the emptying ability of the gall-bladder. Since contrast medium residues may adhere longer to the concrements, radiographs in right lateral and dorsal recumbent positions are necessary 10–15 minutes after the contrast medium has passed into the small intestine.

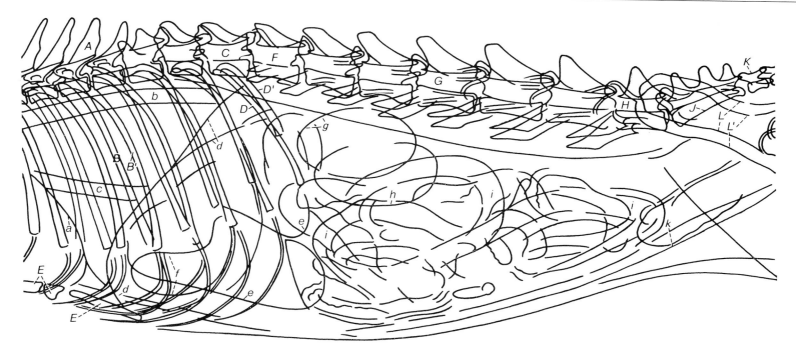

Abb. 335* Röntgenskizze zu Abb. 334   Fig. 335* X-ray sketch to fig. 334

A	9. Vertebra thoracica;	a	Cor;
B	9. Os costale;	b	Aorta thoracica;
C	13. Vertebra thoracica;	c	V. cava caudalis;
D	13. Os costale;	d	Diaphragma;
E	Sternebrae;	e	Hepar;
F	1. Vertebra lumbalis;	f	Vesica fellea;
G	4. Vertebra lumbalis;	g	Ren dexter;
H	7. Vertebra lumbalis;	h	Ren sinister;
J	Os sacrum;	i	Gashaltige Darmschlingen – Intestinal loops filled with gas;
K	1. Vertebra caudalis;	k	Vesica urinaria.
L	Os ilium;		

Intravenöse Cholezystographie

Indikation: Nach erfolgloser oraler Cholezystographie (Erbrechen, fehlende oder unzureichende Resorption im Dünndarm) und zur Bestimmung des Beginns und der Dauer der Kontrastmittelausscheidung.

Vorbereitung: 24 Stunden vor der Untersuchung keine Futter- und Flüssigkeitssaufnahme. Etwa 12 Stunden vor der Untersuchung sind der Füllungszustand des Dickdarms festzustellen (Durchleuchtung, Aufnahme) und ggf. ein Kontaktlaxans oder ein Reinigungseinlauf zu verabreichen.

Kontrastmittel: Jodreiche organische Verbindungen in wässeriger Lösung, deren Jod über die Gallenwege ausgeschieden wird. Dosierung nach Vorschrift. Die körperwarme Lösung ist langsam (2 Tropfen pro Sekunde) intravenös zu applizieren. Bei zu rascher Injektion ist mit Unverträglichkeitsreaktionen zu rechnen. Der albumingebundene Teil des Kontrastmittels wird über die Leber, der ungebundene Teil über die Nieren ausgeschieden.

Untersuchungstechnik:
a) 1. Aufnahme post injectionem, rechte Seitenlage, latero-lateraler Strahlengang.
   Die kontrastmittelhaltigen Gallengänge sind zumindest abschnittsweise, die Gallenblase ist noch nicht, nur wenig oder nur abschnittsweise dargestellt. Man achte auf Stauungen oder Aussparungen.
b) Untersuchung 2 Stunden post injectionem, in 2 Ebenen.
   Bei ungestörter Leberfunktion und unbehindertem Gallenabfluß ist die Ausscheidung abgeschlossen, also die Gallenblase deutlich dargestellt. Stellt sich die Gallenblase nicht dar, sind Untersuchungen (Durchleuchtung, Aufnahme) 3, 6 und ggf. 24 Stunden post injectionem zur Erfassung des evtl. eingedickten und dann nachweisbaren Kontrastmittels in der Gallenblase zweckmäßig.
c) Reizfütterung, wenn Gallenblase dargestellt ist. Reizfutter: 2 Eigelb mit 1 Teelöffel Zucker und 1 Teelöffel Speiseöl. Geeignet ist auch eine handelsübliche Reizmahlzeit in entsprechender Dosierung. Falls das Futter nicht aufgenommen wird, Applikation mit der Magensonde.
d) 30, 45, ggf. auch 60 Minuten nach der Reizfütterung Untersuchung (Durchleuchtung, Aufnahme) in rechter Seitenlage und in Rückenlage. Die Untersuchung gibt Aufschluß über die Entleerungsfähigkeit der Gallenblase. Da Kontrastmittelreste an Konkrementen länger haften können, ist es notwendig, 10–15 Minuten nach der Entleerung bzw. nach der weitgehenden Entleerung des Kontrastmittels in den Dünndarm eine Aufnahme in rechter Seitenlage und in Rückenlage anzufertigen.

Intravenous cholecystography

Indications: Failure of oral cholecystography (vomiting, insufficient or no resorption in the small intestine); to determine the commencement and the duration of contrast medium elimination.

Preparation of the patient: 24 hours prior to the examination no food and liquids should be administered. Approximately 12 hours prior to the examination the degree of filling of the colon is checked (fluoroscopy, preliminary radiograph), and if necessary the cat is given a contact laxative or an evacuation enema.

Contrast media: Iodized organic water soluble compounds. They are excreted through the liver (albumin-bound portion), through the biliary system (iodate) and through the kidneys (unbound portion). Dosage as prescribed. The solution must be injected intravenously at body temperature and slowly (2 drops/second). If injected too rapidly the patient may show symptoms of intolerance.

Technique:
a) Radiography follows after injection with laterolateral direction of the beam in right lateral recumbency. The biliary ducts containing contrast medium are reproduced at least partially. The gall-bladder is not yet delineated or only slightly or in part. Observe possible accumulations.
b) Examination 2 hours after injection with two different beam directions. With normal liver function and free flow of gall, the excretion of the contrast medium is completed and the gall-bladderr is clearly outlined. If the gall-bladder is not outlined, examinations (fluoroscopy, radiograph) are useful at 3, 6, and if necessary 24 hours after injection to locate the possibly thickened and then demonstrable contrast medium in the gall-bladder.
c) Challenge feeding after the gall-bladder is outlined. Challenge food: 2 egg yolks mixed with 1 teaspoonful of sugar and 1 teaspoonful of oil. A challenge food available in the trade is also suitable in appropiate doses. If the animal refuses the food it should be administered with a stomach tube.
d) 30, 45, and if necessary 60 minutes after administration of challenge food, examination (fluoroscopy, radiograph) in right lateral and dorsal recumbent positions.
   The examination furnishes information on the emptying ability of the gall-bladder. Since the residues of the contrast medium may adhere longer to the concrements, it is essential to make a radiograph in right lateral and dorsal recumbent positions 10–15 minutes after passage of the contrast medium into the small intestine.

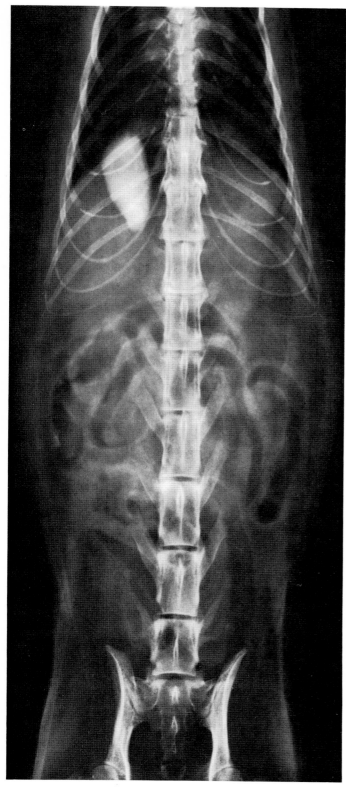

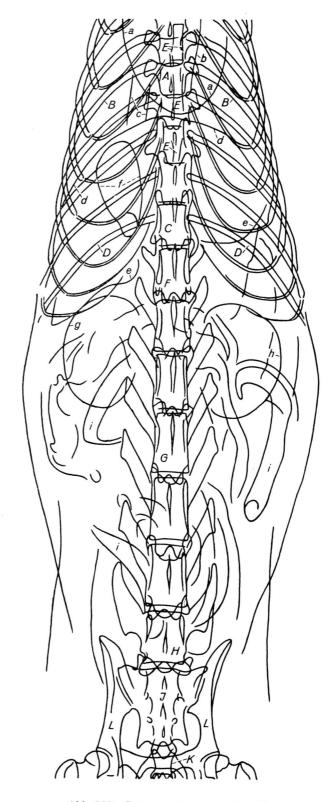

Abb. 336 Gallenblasendarstellung (Biloptin, Schering).
Ventro-dorsal. Katze.
Bucky-Blende – Feinzeichnende Folie – FFA 115 cm – 62 kV – 40 mAs
Verkleinerung von 15 × 40 cm
Lagerung Abb. 325

Fig. 336 Gall-bladder (Biloptin, Schering). Ventrodorsal. Cat.
Bucky diaphragm – High definition screens – FFD 115 cm –
62 kV – 40 mAs
Diminution of 15 × 40 cm
Positioning fig. 325

Abb. 337 Röntgenskizze zu Abb. 336
Fig. 337 X-ray sketch to fig. 336

A	9. Vertebra thoracica;	a	Cor;
B	9. Os costale;	b	Aorta thoracica;
C	13. Vertebra thoracica;	c	V. cava caudalis;
D	13. Os costale;	d	Diaphragma;
E	Sternebrae;	e	Hepar;
F	1. Vertebra lumbalis;	f	Vesica fellea;
G	4. Vertebra lumbalis;	g	Ren dexter;
H	7. Vertebra lumbalis;	h	Ren sinister;
J	Os sacrum;	i	Gashaltige Darmschlingen – Intestinal loops filled with gas.
K	1. Vertebra caudalis;		
L	Os ilium;		

225

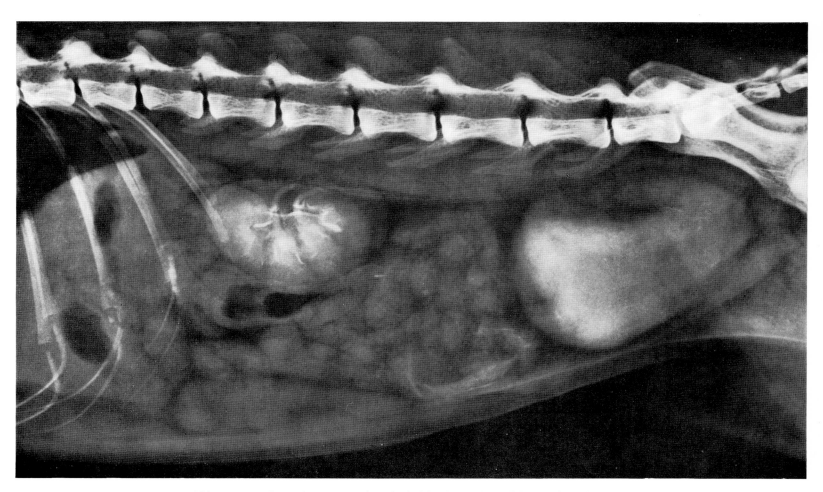

Abb. 338   Pyelographie (Urografin 76 %, Schering). Liegend. Latero-lateral. Katze.
Bucky-Blende – Feinzeichnende Folie – FFA 115 cm – 60 kV – 30 mAs
Verkleinerung von 24 × 30 cm
Lagerung Abb. 324

Fig. 338   Pyelography (Urografin 76 %, Schering). Recumbent. Laterolateral. Cat.
Bucky diaphragm – High definition screens – FFD 115 cm – 60 kV – 30 mAs
Diminution of 24 × 30 cm
Positioning fig. 324

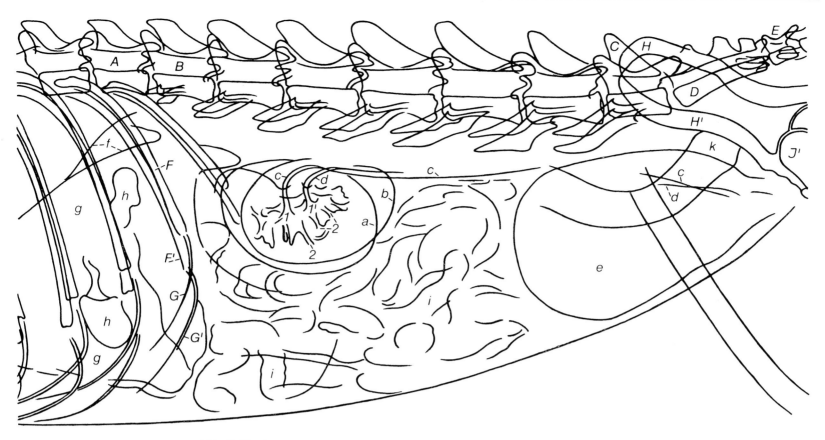

Abb. 339*   Röntgenskizze zu Abb. 338   Fig. 339*   X-ray sketch to fig. 338

A	13. Vertebra thoracica;		c	Ureter dexter;
B	1. Vertebra lumbalis;		d	Ureter sinister;
C	7. Vertebra lumbalis;		e	Vesica urinaria;
D	Os sacrum;		f	Diaphragma;
E	1. Vertebra caudalis;		g	Hepar;
F	12. Os costale;		h	Gasblase im Magen – Gas bubble in the stomach;
G	12. Cartilago costalis;		i	Intestinum tenue;
H	Os ilium;		k	Colon descendens;
J	Os femoris;			
a	Ren dexter;		1	Pelvis renalis;
b	Ren sinister;		2	Recessus pelvis.

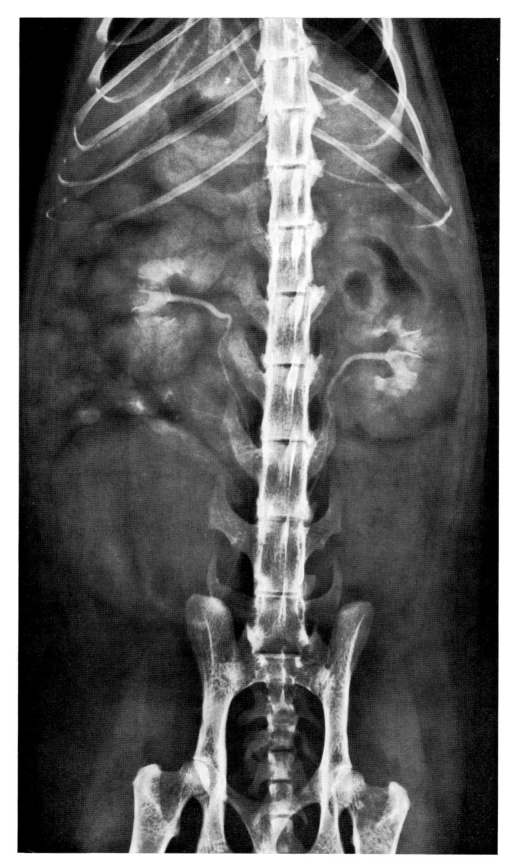

Abb. 340   Pyelographie (Urografin 76 %, Schering). Ventro-dorsal. Katze.
Bucky-Blende – Feinzeichnende Folie – FFA 115 cm – 60 kV – 30 mAs
Verkleinerung von 24 × 30 cm
Lagerung Abb. 325

Fig. 340   Pyelography (Urografin 76 %, Schering). Ventrodorsal. Cat.
Bucky diaphragm – High definition screens – FFD 115 cm – 60 kV – 30 mAs
Diminution of 24 × 30 cm
Positioning fig. 325

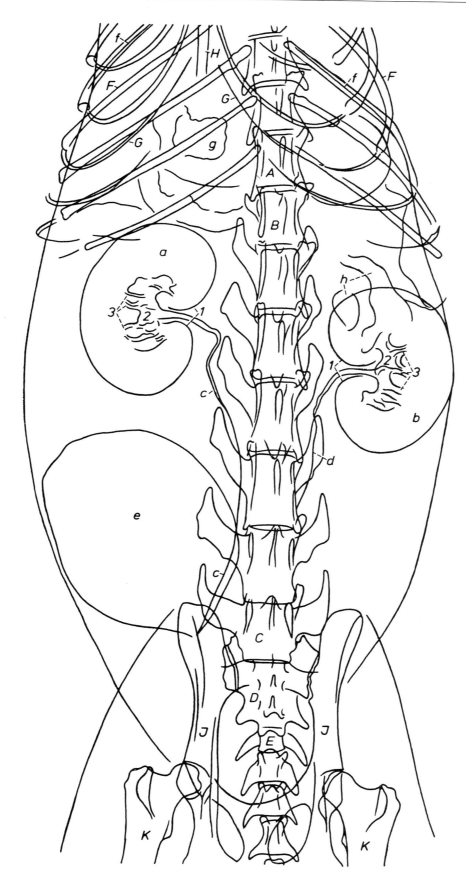

Abb. 341   Röntgenskizze zu Abb. 340   Fig. 341   X-ray sketch to fig. 340

A	13. Vertebra thoracica;		a	Ren dexter;
B	1. Vertebra lumbalis;		b	Ren sinister;
C	7. Vertebra lumbalis;		c	Ureter dexter;
D	Os sacrum;		d	Ureter sinister;
E	1. Vertebra caudalis;		e	Vesica urinaria;
F	10. Os costale;		f	Diaphragma;
G	10. Cartilago costalis;		g	Ventriculus;
H	Processus xiphoideus;		h	Gashaltige Darmschlingen – Gas-containing intestinal loops;
J	Os ilium;			
K	Os femoris;		1	Hilus renalis;
			2	Pelvis renalis;
			3	Recessus pelvis.

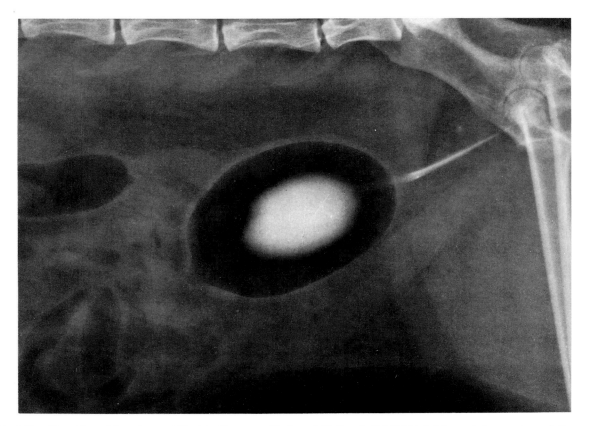

Abb. 342   Harnblase. Negativ- und Positiv-Kontrast (Luft und Kaliumjodid 10%ig). Liegend. Latero-lateral. Katze ♂.
Bucky-Blende – Feinzeichnende Folie – FFA 120 cm – 63 kV – 25 mAs
Originalgröße (Ausschnitt aus 18 × 24 cm)
Lagerung Abb. 324

Fig. 342   Urinary bladder. Negative and positive contrast (air and 10 % potassium iodide). Recumbent. Laterolateral. Cat ♂.
Bucky diaphragm – High definition screens – FFD 120 cm – 63 kV – 25 mAs
Original size (section of 18 × 24 cm)
Positioning fig. 324

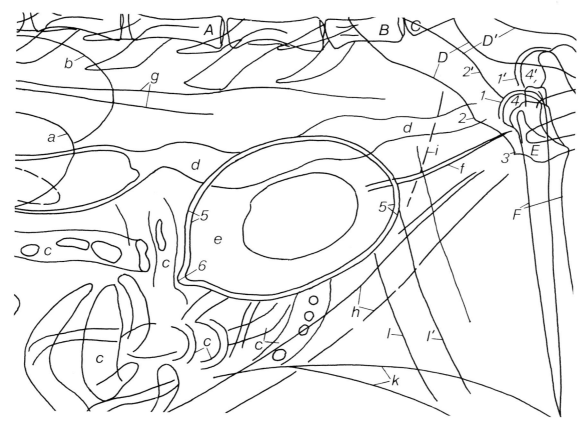

Abb. 343*   Röntgenskizze zu Abb. 342   Fig. 343*   X-ray sketch to fig. 342

A	5. Vertebra lumbalis;
B	7. Vertebra lumbalis;
C	Os sacrum;
D	Os ilium;
E	Os pubis;
F	Os femoris;

a Ren dexter;
b Ren sinister;
c Dünndarmschlingen, zum Teil mit Gas gefüllt – Intestinal loops, partly filled with gas;
d Colon descendens, zum Teil mit Gas gefüllt – Colon descendens, partly filled with gas;
e Vesica urinaria, mit Kontrastmittel und Gas gefüllt – Vesica urinaria, filled with contrast medium and gas;

f Urethra;
g Lendenmuskulatur – Lumbar musculature;
h Ventrale Bauchwand – Ventral abdominal wall;
i M. transversus abdominis;
k Kniefalte – Fold of the flank;
l Kraniale Kontur des Oberschenkels – Cranial outline of the thigh;

Am Becken – On the pelvis:

1 Acetabulum;
2 Eminentia iliopubica;
3 Pecten ossis pubis;
4 Caput ossis femoris;
5 Harnblasenwand – Wall of the urinary bladder;
6 Urachusnabel – Umbilicus of the urachus.

Ausscheidungsurographie

Vorbereitung: 12 Stunden vor der Untersuchung sollten der Füllungszustand des Darmes festgestellt (Durchleuchtung, Aufnahme), ggf. ein Kontaktlaxans oder ein Reinigungseinlauf verabreicht werden; der Patient sollte hungern und dursten. Wenn die Vorbereitung nicht möglich oder nicht ausreichend ist, werden ein Reinigungseinlauf verabfolgt und nach etwa einer Stunde das Kontrastmittel appliziert.

Kontrastmittel: Gut verträglich und kontrastreich sind trijodierte wasserlösliche Präparate. Dosierung nach Vorschrift (Urografin 76 %, Schering, 2 ml/kg KGW). Die Applikation der körperwarmen Lösung erfolgt intravenös in 2 – 3 Minuten.

Untersuchungstechnik:
a) 3 Minuten post injectionem 1. Aufnahme. Rückenlage, ventro-dorsaler Strahlengang.
   Die Aufnahme ist zur Beurteilung der Nierenfunktion und der Durchgängigkeit der Harnleiter zweckmäßig.
b) Kompression der Harnleiter durch Fixation einer Zellstoffrolle, eines Sandsacks oder ähnlichem mit Hilfe eines Bandkompressoriums.
c) 5 und 10 Minuten nach Anlegen der Ureterenkompression 2. und 3. Aufnahme, ventro-dorsaler Strahlengang.
   Die Aufnahmen sollen eine einwandfreie Nierenbeckendarstellung ergeben.
d) Nach Abnahme des Kompressoriums 4. Aufnahme, ventro-dorsaler Strahlengang.
   Da der gestaute kontrasthaltige Harn abfließt, müssen die Harnleiter zumindest in größeren Abschnitten dargestellt sein.
e) Umlagerung der Katze in rechte Seitenlage und 5. Aufnahme, latero-lateraler Strahlengang.
   Die Aufnahme ist zur Beurteilung der Harnblase und der Nieren erforderlich.

Das Kontrastmittel wird in etwa 30 Minuten ausgeschieden. Bei Störung der Nierenfunktion oder des Harntransports verzögert sich die Ausscheidung.

Excretory urography

Preparation of the patient: 12 hours prior to the examination the degree of filling of the intestinal tract should be checked (fluoroscopy, preliminary radiograph) and, if necessary a contact laxative or an evacuation enema administered; the cat should not be allowed food and liquids. Should preliminary preparation not be possible or sufficient, it is advisible to give an evacuation enema and to administer the contrast medium approximately after 1 hour.

Contrast media: Well tolerated and rich in contrast are tri-iodized, water soluble, organic compounds. Dosage according to prescription (Urografin 76 %, Schering, 2 ml/kg body weight). The medium should be injected intravenously at body temperature within 2–3 minutes.

Technique:
a) First radiograph to be taken 3 minutes after injection with ventrodorsal direction of the beam in dorsal recumbent position. This radiograph is suitable for the study of kidney function and the patency of the ureters.
b) Compression of the ureters by means of a compression block.
c) 5–10 minutes after applying the compression block, the second and third radiographs are taken with ventrodorsal direction of the beam. These radiographs should produce a diagnostic pyelogram.
d) After removal of the compression block, the fourth radiograph is taken with ventrodorsal direction of the beam.
   Since the retained urine containing contrast medium now flows freely, both ureters or at least the greater part of them should be outlined.
e) Placing of the patient into the right lateral position. Fifth radiograph is taken with laterolateral direction of the beam. This radiograph is necessary for evaluation of the urinary bladder and kidneys.

The contrast medium is normally excreted in approximately 30 minutes. In case of disturbed kidney function or urine flow the excretion of the contrast medium is slowed down.

Retrograde Zystographie

Die Untersuchung erfordert eine geeignete Ruhigstellung (Anästhesie). Der Katheter ist in die Harnblase einzuführen, die Harnblase zu entleeren.

Falls ein Pneumozystogramm gewünscht wird, sind 5 – 15 – 20 ml Luft in die Harnblase zu applizieren; der Katheter wird herausgezogen. Wenn ein flüssiges Harnblasenkontrastmittel verwendet wird, ist eine angemessene Menge zu applizieren, der Katheter wird herausgezogen.

Falls ein Doppelkontrast gewünscht wird, ist ein flüssiges Kontrastmittel zuerst zu applizieren; die Harnblase sollte leicht massiert werden, um das Kontrastmittel gleichmäßig zu verteilen (über Schleimhaut, Tumor, Konkremente); überschüssiges Kontrastmittel ist zu entfernen; die Luft wird insuffliert und der Katheter herausgezogen.

Retrograde cystography

Procedure requires a suitable sedation (anesthesia). The catheter is introduced into the urinary bladder and all urine removed.

If pneumocystogram is desired, 5 – 15 – 20 ml of air is insufflated into the bladder and the catheter withdrawn.

If liquid urologic contrast medium is used, an appropiate amount of it is instilled into the bladder and the catheter withdrawn. In case a double contrast is desired, a liquid contrast medium is instilled first; the bladder should be slightly massaged in order to spread the medium and make it adhere to the bladder (mucous membrane, tumors, concrements); after removing the excess of liquid, air is insufflated and catheter withdrawn.

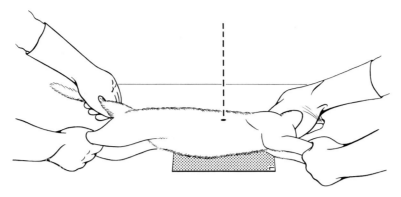

Abb. 344   Lagerung zur Aufnahme des Brustkorbs. Liegend. Latero-lateral.

Fig. 344   Positioning of thorax. Recumbent. Laterolateral.

Die Aufnahme sollte wegen der Lage des Herzens in linker Seitenlage angefertigt werden. Von dieser Regel sollte nur dann abgegangen werden, wenn eine krankhafte Veränderung in der rechten Hälfte des Brustkorbs (z.B. Tumor) oder an der rechten Brustwand (z.B. Rippenfraktur) lokalisiert wurde. Es ist darauf zu achten, daß die Medianebene parallel zur Kassette liegt. Bei fettleibigen Tieren läßt sich die korrekte Lagerung mit Hilfe von Schaumgummikeilen, die unter die Gliedmaßen zu legen sind, erleichtern.

Um feine Strukturen nachweisen zu können, sollte ein folienloser Film (Bleiunterlage nicht vergessen!) oder eine Kassette mit feinzeichnender Folie verwendet werden. Daraus resultiert eine längere Belichtungszeit und damit die Möglichkeit einer Unschärfe durch die Atembewegung. Zur Vermeidung der Bewegungsunschärfe sollte die Atmung möglichst am Ende einer Inspiration kurz unterbrochen werden.

Der Zentralstrahl sollte die 5. Rippe in halber Höhe treffen und im rechten Winkel auf den Film einfallen.

Due to the position of the heart radiographs should be taken in left lateral recumbency. The only exception to this rule should be in the event of pathology in the right half of the thorax (e.g. tumors) or in the right thoracic wall (e.g. rib fracture). It is essential that the midline is aligned parallel to the cassette. In obese animals the positioning can be facilitated by placing foam rubber pads under the legs.

To depict delicate structures a non-screen film (lead blocker not to be forgotten!) or a cassette with high definition screens should be used. This necessitates a longer exposure time and associated with it the possibility of blurring due to respiratory movements. To avoid the latter, breathing should be interrupted shortly at the end of an inspiratory phase.

The central beam should strike the 5th rib at the middle of its length and fall at the right angle on the film.

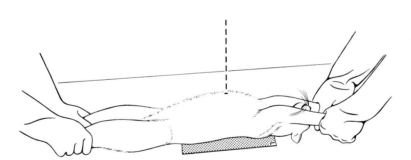

Abb. 345   Lagerung zur Aufnahme des Brustkorbs. Ventro-dorsal.

Fig. 345   Positioning of thorax. Ventrodorsal.

Die Katze ist in Rückenlage so zu lagern, daß die Medianebene senkrecht zum Röntgenfilm steht, also das Brustbein senkrecht über den Brustwirbeln liegt. Bei der Lagerung des narkotisierten Tieres kann es notwendig sein, Kopf-Hals im Gabelgriff so lange zu fixieren, bis die Katze korrekt gelagert ist und der Kopf durch die angedrückten Schultergliedmaßen gehalten werden kann.

Um feine Strukturen nachweisen zu können, sollten eine Kassette mit feinzeichnender Folie und zur Reduzierung der Streustrahlung eine Bucky-Blende verwendet werden.

Der Zentralstrahl sollte in Höhe des 7. Brustwirbels im rechten Winkel auf den Film einfallen.

The cat should be placed in the dorsal recumbent position with the median plane at right angles to the film, i.e., the sternum should lie perpendicularly above the thoracic vertebrae. In positioning a non-sedated animal it may be necessary to secure the head and neck in a fork grip until such time as the animal is aligned properly and the head can be secured by means of the front limbs being pressed against it.

To depict fine structures high definition screens should be used. To reduce scattered radiation the radiograph should be taken with a Bucky diaphragm.

The central beam should strike the film at the right angle on a level of the 7th thoracic vertebra.

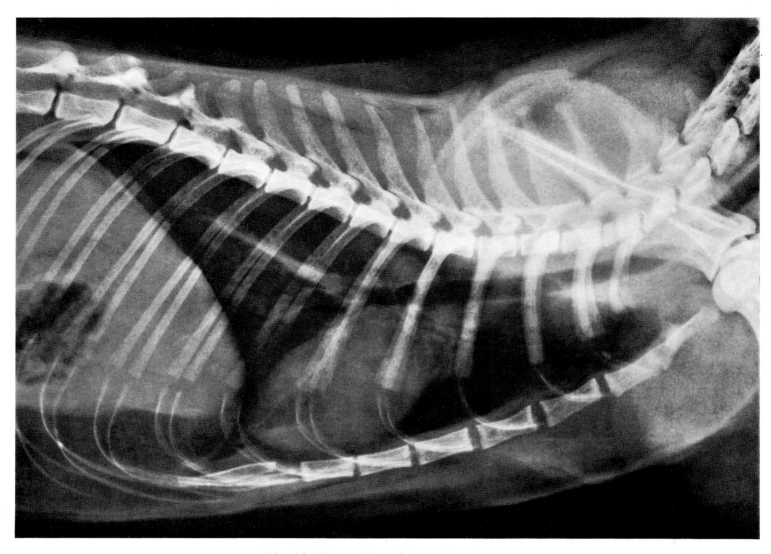

Abb. 346   Thorax. Liegend. Latero-lateral. Katze.
Bucky-Blende – Feinzeichnende Folie – FFA 120 cm – 65 kV – 8 mAs
Originalgröße (Ausschnitt aus 24 × 30 cm)
Lagerung Abb. 344

Fig. 346   Thorax. Recumbent. Laterolateral. Cat.
Bucky diaphragm – High definition screens – FFD 120 cm – 65 kV – 8 mAs
Original size (section 24 × 30 cm)
Positioning fig. 344

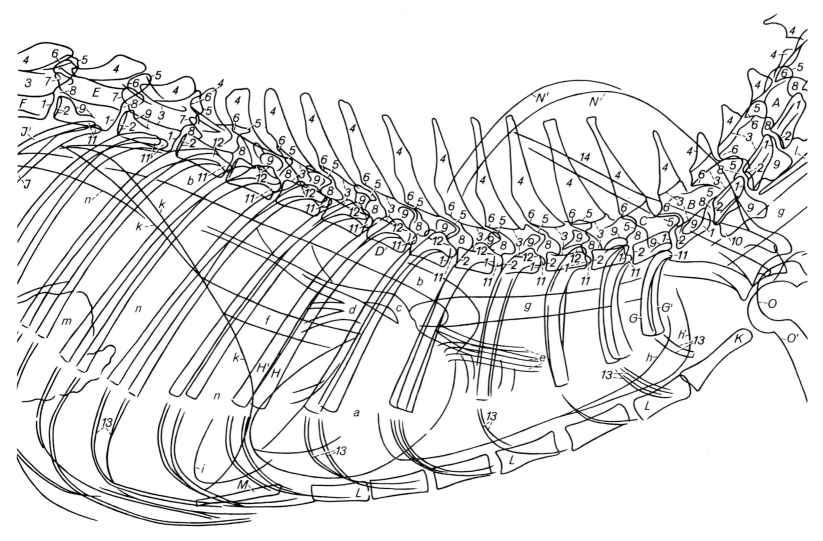

Abb. 347*   Röntgenskizze zu Abb. 346   Fig. 347*   X-ray sketch to fig. 346

A	4. Vertebra cervicalis;	
B	7. Vertebra cervicalis;	
C	1. Vertebra thoracica;	
D	7. Vertebra thoracica;	
E	13. Vertebra thoracica;	
F	1. Vertebra lumbalis;	
G	1. Os costale;	
H	7. Os costale;	
J	13. Os costale;	
K	Manubrium sterni;	
L	Sternebrae;	
M	Processus xiphoideus;	
N	Scapula;	
O	Humerus;	

a   Cor;
b   Aorta thoracica;
c   A. pulmonalis;
d, e   Aa. et Vv. pulmonales;
f   V. cava caudalis;
g   Trachea;
h   Cupula pleurae;
i   Lobus medius pulmonis dextri;
k   Diaphragma;
l   Oesophagus;
m   Ventriculus;
n   Hepar;

An den Wirbeln – On the vertebrae:

1   Facies terminalis cranialis;
2   Facies terminalis caudalis;
3   Foramen vertebrale;
4   Processus spinosus;
5   Processus articularis cranialis; an den kaudalen Brustwirbeln und an den Lendenwirbeln: Processus mamilloarticularis – Processus articularis cranialis; on the caudal thoracic vertebrae and on the lumbar vertebrae: mamilloarticular process;
6   Processus articularis caudalis;
7   Processus accessorius;
8   Foramina intervertebralia, durch die Incisurae vertebrae caudalis et cranialis zweier benachbarter Wirbel begrenzt – Foramina intervertebralia, limited by the caudal and cranial vertebral notches of two successive vertebrae;
9   Processus transversus;
10   Lamina ventralis vertebrae cervicalis VI;

An den Rippen – On the ribs:

11   Caput costae;
12   Tuberculum costae;
13   Cartilago costalis.

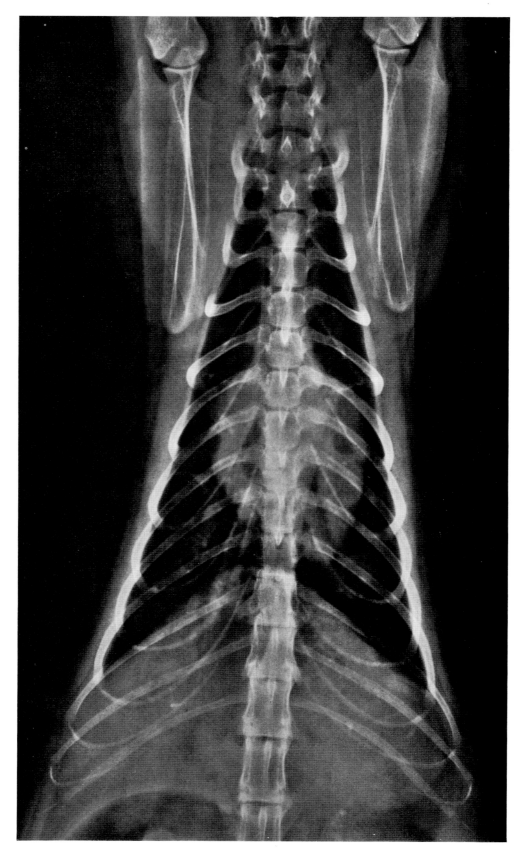

Abb. 348 Thorax. Ventro-dorsal. Katze.
Bucky-Blende – Feinzeichnende Folie – FFA 115 cm – 63 kV – 8 mAs
Originalgröße (Ausschnitt aus 24 × 30 cm)
Lagerung Abb. 345

Fig. 348 Thorax. Ventrodorsal. Cat.
Bucky diaphragm – High definition screens – FFD 115 cm – 63 kV – 8 mAs
Original size (section of 24 × 30 cm)
Positioning fig. 345

A   4. Vertebra cervicalis;
B   7. Vertebra cervicalis;
C   1. Vertebra thoracica;
D   7. Vertebra thoracica;
E   13. Vertebra thoracica;
F   1. Vertebra lumbalis;
G   1. Os costale;
H   7. Os costale;
J   13. Os costale;
K   Manubrium sterni;
L   Sternebrae;
N   Scapula;
O   Humerus;

a   Cor;
b   Aorta thoracica;
c   V. cava caudalis;
d   Cavum pleurae, laterale Wand – Cavum pleurae, lateral wall;
e   Diaphragma;
f   Hepar;
g   Darmschlinge – Intestinal loop;
h   Ren dexter;
i   Ren sinister;
k   Lien;

An den Wirbeln – On the vertebrae:

1   Facies terminalis cranialis;
2   Facies terminalis caudalis;
3   Pediculus arcus vertebrae, zugleich seitliche Begrenzung des Foramen
    vertebrale – Pediculus arcus vertebrae forming the lateral margin of
    the vertebral foramen;
4   Processus spinosus;
5   Processus articularis cranialis; an den kaudalen Brustwirbeln und an
    den Lendenwirbeln: Processus mamilloarticularis – Processus artic-
    ularis cranialis; on the caudal thoracic vertebrae and on the lumbar
    vertebrae: mamilloarticular process;
6   Processus articularis caudalis;
7   Processus accessorius;
8   Spatium interarcuale;
9   Processus transversus;

An den Rippen – On the ribs:

10   Caput costae;
11   Cartilago costalis.

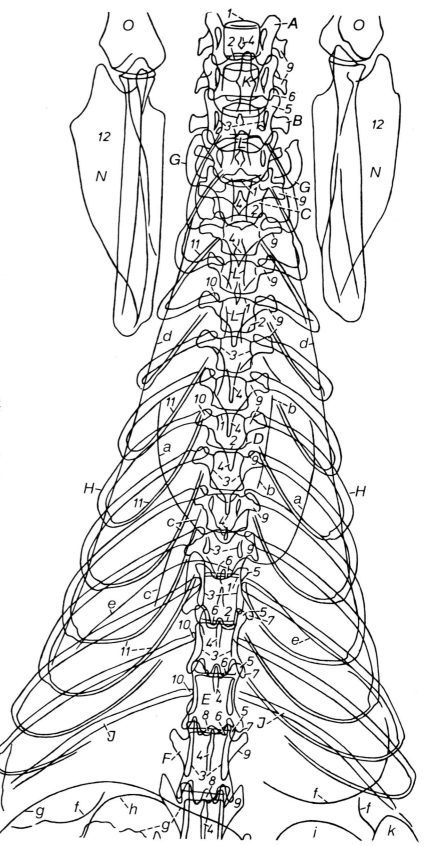

Abb. 349   Röntgenskizze zu Abb. 348
Fig. 349   X-ray sketch to fig. 348

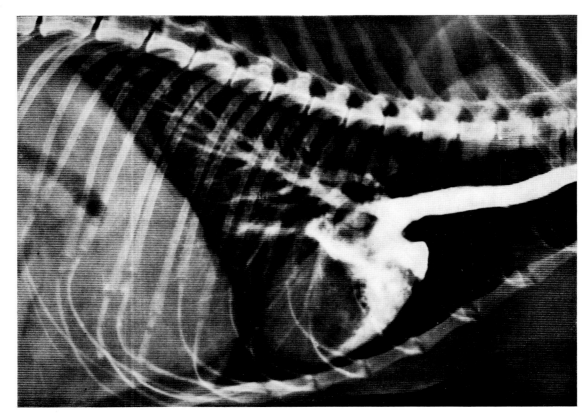

Abb. 350   Angiokardiographie (Urografin 76 %, Schering). Aufnahme aus Bildserie, aufgenommen
mit Angio-Kardio-Seriograph nach BUCHTALA. Injektion in die V. jugularis externa. Venöse Seite,
Endphase der Systole. Liegend. Latero-lateral. Katze.
Mittelverstärkende Folie – FFA 120 cm – 65 kV – 9 mAs
Originalgröße (Ausschnitt aus 24 × 30 cm)
Lagerung Abb. 344

Fig. 350   Angiocardiography (Urografin 76 %, Schering). Radiograph out of series, taken by
angio-cardio-seriograph after BUCHTALA. Injection into external jugular vein. Venous side, end phase of systole.
Recumbent. Laterolateral. Cat.
Standard screens – FFD 120 cm – 65 kV – 9 mAs
Original size (section of 24 × 30 cm)
Positioning fig. 344

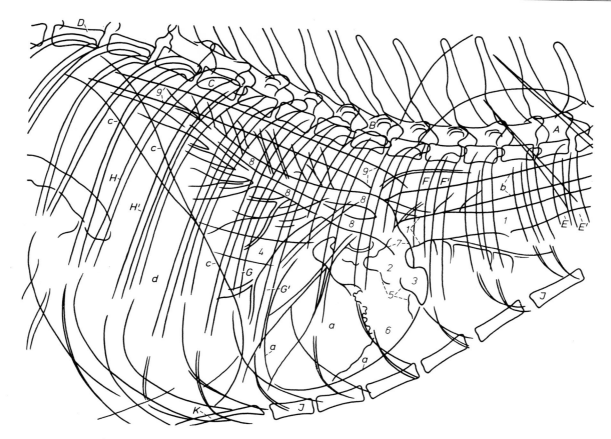

Abb. 351*   Röntgenskizze zu Abb. 350   Fig. 351*   X-ray sketch to fig. 350

A   1. Vertebra thoracica;
B   6. Vertebra thoracica;
C   10. Vertebra thoracica;
D   13. Vertebra thoracica;
E   1. Os costale;
F   4. Os costale;
G   7. Os costale;
H   10. Os costale;
J   Sternebrae;
K   Processus xiphoideus;

a   Cor;
b   Trachea;

c   Diaphragma;
d   Hepar;

1   V. cava cranialis;
2   Atrium dextrum;
3   Auricula dextra;
4   V. cava caudalis;
5   Ostium atrioventriculare dextrum;
6   Ventriculus dexter;
7   Truncus pulmonalis;
8   Aa. pulmonales;
9   Aorta thoracica.

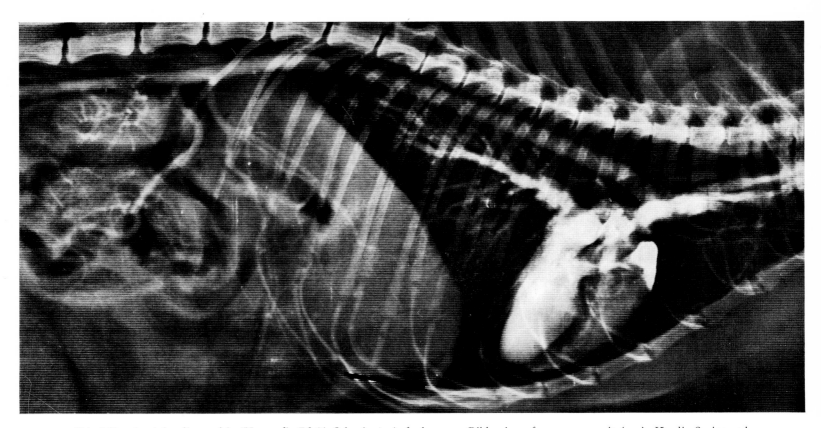

Abb. 352   Angiokardiographie (Urografin 76 %, Schering). Aufnahme aus Bildserie, aufgenommen mit Angio-Kardio-Seriograph
nach BUCHTALA. Injection in die V. jugularis externa. Arterielle Seite, Diastole. Liegend. Latero-lateral. Katze.
Mittelverstärkende Folie – FFA 120 cm – 65 kV – 9 mAs
Originalgröße (Ausschnitt aus 24 × 30 cm)
Lagerung Abb. 344

Fig. 352   Angiocardiography (Urografin 76 %, Schering). Radiograph out of series, taken by angio-cardio-seriograph
after BUCHTALA. Injection into external jugular vein. Arterial side, diastole. Recumbent. Laterolateral. Cat.
Standard screens – FFD 120 cm – 65 kV – 9 mAs
Original size (section of 24 × 30 cm)
Positioning fig. 344

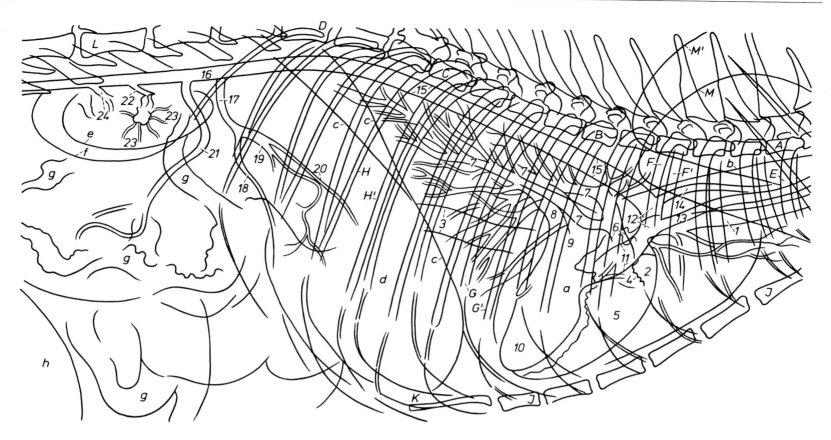

Abb. 353* Röntgenskizze zu Abb. 352   Fig. 353* X-ray sketch to fig. 352

A   1. Vertebra thoracica;
B   6. Vertebra thoracica;
C   10. Vertebra thoracica;
D   13. Vertebra thoracica;
E   1. Os costale;
F   4. Os costale;
G   7. Os costale;
H   10. Os costale;
J   Sternebrae;
K   Processus xiphoideus;
L   3. Vertebra lumbalis;

a   Cor;
b   Trachea;
c   Diaphragma;
d   Hepar;
e   Ren dexter;
f   Ren sinister;
g   Darmschlingen, zum Teil gashaltig – Intestinal loops partly filled with gas;
h   Gravider Uterus – Pregnant uterus;

1   V. cava cranialis;
2   Auricula dextra;
3   V. cava caudalis;
4   Ostium atrioventriculare dextrum;
5   Ventriculus dexter;
6   Truncus pulmonalis;
7   Aa. pulmonales;
8   Vv. pulmonales;
9   Atrium sinistrum;
10 Ventriculus sinister;
11 Bulbus aortae;
12 Arcus aortae;
13 Truncus brachiocephalicus;
14 A. subclavia sinistra;
15 Aorta thoracica;
16 Aorta abdominalis;
17 A. coeliaca;
18 A. lienalis;
19 A. gastrica sinistra;
20 A. hepatica;
21 A. mesenterica cranialis;
22 A. renalis dextra;
23 Aa. interlobares;
24 A. renalis sinistra.

Angiokardiographie

Allgemeines: Für die Herz- und Gefäßdiagnostik sind Aufnahmen in schneller Bildfolge erforderlich. Die Voraussetzung dafür ist eine besondere apparative Ausrüstung.

Die Passage des Kontrastmittels kann reproduzierbar festgehalten werden, wenn eine Bildverstärkeranlage mit einer kinematographischen Einrichtung oder mit einem Bildbandspeichergerät vorhanden ist. Eine Serie von Röntgenaufnahmen, also einzelne Phasen der Kontrastmittelpassage, können mit Hilfe eines Seriographen aufgenommen werden.

Die Abbildungen 350 und 352 stammen aus einer mit dem Angio-Kardio-Seriographen nach BUCHTALA aufgenommenen Bildserie. Das Gerät arbeitet mit 2 Filmtransportgeschwindigkeiten (0,4 und 0,8 Sekunden pro Aufnahme einschließlich Filmtransport). In 6 bzw. 12 Sekunden können 16 Röntgenaufnahmen im Format 24 × 30 cm angefertigt werden. Da das Gerät außerdem einen exakt einstellbaren verlängerten Ablauf der Bildgeschwindigkeit hat, können Spätfüllungen oder ein verzögerter Kontrastmitteldurchlauf erfaßt werden.

Um die Injektion des Kontrastmittels rasch durchführen zu können, wurden Apparate konstruiert, die das Kontrastmittel unter hohem Druck zu injizieren gestatten.

Bei Verwendung eines Angio-Seriographen ist eine Kontaktkupplung zwischen Injektionsspritze und dem Gerät zweckmäßig. Die erste Aufnahme wird dadurch nach der Applikation einer bestimmten Kontrastmittelmenge automatisch ausgelöst.

Kontrastmittel: Trijodierte organische Verbindungen, je nach Größe des Tieres 3–10 ml, intravenöse Applikation.

Untersuchungstechnik: Um Bewegungen des Tieres bei dem durch den Filmtransport verursachten Geräusch und um einer Überempfindlichkeitsreaktion gegenüber dem jodhaltigen Kontrastmittel vorzubeugen, ist es zweckmäßig, die Untersuchung am narkotisierten Patienten vorzunehmen.

Der Patient ist in linker Seitenlage auf dem Seriographen so zu lagern und zu fixieren, daß sich die Medianebene des Brustkorbs parallel zur Oberfläche des Geräts befindet (s. Abb. 344).

Angiocardiography

General consideration: For diagnostic radiography of the heart and the blood vessels, X-ray pictures taken in rapid succession are neccesary. This requires special equipment.

The passage of the contrast medium can be depicted with a machine equipped with image intensifier and movie camera or tape recorder and with a rapid cassette changer. Such a machine can produce multiple exposures, each less than 1/2 second apart.

The radiographs in figs. 350 and 352 are part of a series of pictures taken with an angio-cardio-seriograph after BUCHTALA. This machine features two speeds for film transportation (0.4 and 0.8 seconds per picture, including the film transportation). In 6 or 12 seconds, respectively, 16 radiographs, size 24 × 30 cm, can be taken. Since this apparatus also has an exact setting for slow-down of the film-passage, late filling or delayed passage of the contrast medium can be depicted.

There is special equipment on the market allowing rapid injection of the contrast medium under high pressure.

Should an angio-seriograph be employed the syringe (with contrast medium) should be connected to the machine. In this way, the first radiograph can be taken automatically as soon as a given amount of the contrast medium has been injected.

Contrast media: Tri-iodized organic compounds, 3–10 ml depending on the size of the cat, intravenous injection.

Technique: The patient should be anesthetized and placed in the left lateral position upon the seriograph. The median plane of the thorax should be parallel to the top of the table (fig. 344).

Bibliographie — Bibliography

ALKSNIS, A. (1938): Röntgenanatomische Studien über normale Gelenke beim Pferd, beim Hund und bei der Katze. Acta Univ. Latv. 2, 1

BARBER, D. L. (1975): Renal angiography in veterinary medicine. J. Am. Vet. Radiol. Soc. 16, 187

BECKER, P. (1958): Über den Strahlengang in der veterinär-medizinischen Röntgendiagnostik und Versuche seiner kinematographischen Darstellung. Berlin, FU, Diss. med. vet.

BELKIN, P. V. (1967): A device for positioning small animals. J. Am. vet. med. Ass. 150, 1367

BISHOP, E. J., W. MEDEVAG, and J. ARCHIBALD (1955): Radiological methods of investigating the thorax of small animals, including a technic for bronchography. North Am. Vet. 36, 477

BÖHME, G. (1959): Röntgenanatomische Untersuchungen am Katzenschädel. Berlin, FU, Diss. med. vet.

BUCHTALA, V., und H. P. JENSEN (1955): Die Probleme der zerebralen Angiographie. Fortschr. Röntgenstr. 82, 76

BULLOCK, L. P., and B. C. ZOOK (1967): Myelography in dogs, using water-soluble contrast mediums. J. Am. vet. med. Ass. 151, 321

CARLSON, W. D. (1961): Veterinary radiology. Lea and Febiger, Philadelphia

CHRISTOPH, H.-J. (1962): Abriß der Klinik der Hundekrankheiten. 2. Aufl. VEB Fischer, Jena

CHRISTOPH, H.-J., P. SCHWILLE und W. SCHNITZLEIN (1959): Pneumoperitoneum und Pneumoretroperitoneum bei Hund und Katze. Kleintier-Prax. 4, 73

DOUGLAS, S. W., and L. W. HALL (1959): Bronchography in the dog. Vet. Rec. 72, 901

DOUGLAS, S. W., and H. D. WILLIAMSON (1963): Principles of veterinary radiography. Ballière, Tindall and Cox, London

DRURY, F., K. M. DYCE, and R. H. A. MERLEIN (1953): The radiography of the normal urinary tract of the dog. Vet. Rec. 65, 647

DYCE, K. M. (1956): An experimental study of the biliary tract of the dog. Zbl. Veterinärmed. 3, 717

FICUS, H. J. (1965): Das Pyelogramm. Technik und Auswertung unter praxismäßigen Bedingungen. Kleintier-Prax. 10, 65

FICUS, H. J. (1966): Die Cholezystographie als Hilfsmittel in der Leberdiagnostik. Kleintier-Prax. 11, 105

FINCO, D. R., N. S. STILES, S. K. KNELLER, R. E. LEWIS, and R. B. BARRETT (1971): Radiologic estimation of kidney size of the dog. J. Am. vet. med. Ass. 159, 995

FRICK, W. (1960): Eigenschaften und Indikationsbereich neuer Gallenkontrastmittel. Dtsch. med. Wschr. 85, 1764

FROST, R. CH. (1956): Cholecystography and rupture of the diaphragm in small animals. Vet. Rec. 68, 1014

GAY, W. (1957): Jodinated organic compounds as radiographic contrast mediums in canine practice. J. Am. vet. med. Ass. 131, 149

GRAEGER, K. (1958): Die Nasenhöhle und die Nasennebenhöhlen beim Hund unter besonderer Berücksichtigung der Siebbeinmuscheln. Dtsch. tierärztl. Wschr. 65, 425; 468

GRAEME, S., and R. T. DIXON (1975): The choice of contrast media and optimum dose rates. J. Am. Vet. Radiol. Soc. 16, 98

GRANDAGE, J. B. (1975): Some effects of posture on the radiographic appearance of the kidneys of the dog. J. Am. vet. med. Ass. 166, 165

GUNSSER, J. (1977): Zur Untersuchung der Gallenblase der Katze. Berl. Münch. tierärztl. Wschr. 90, 184

GUTBROD, F. (1982): Die Doppelkontrastdarstellung der Blase bei Hund und Katze. Berl. Münch. tierärztl. Wschr. 95, 229

HABEL, R. E., R. B. BARRETT, C. D. DIESEM, and W. J. ROENIGK (1963): Nomenclature for radiographic anatomy. J. Am. vet. med. Ass. 142, 38

HABERMEHL, K.-H. (1975): Die Altersbestimmung bei Haus- und Labortieren. 2. Aufl. Parey, Berlin–Hamburg

HAMLIN, R. L. (1959): Angiocardiography for the clinical diagnosis of congenital heart disease in small animals. J. Am. vet. med. Ass. 135, 112

HAMLIN, R. L. (1960): Radiographic anatomy of heart and great vessels in healthy living dogs. J. Am. vet. med. Ass. 136, 265

HARTUNG, K. (1972): Zur Kontrastmittelwahl für die Ausscheidungsurographie beim Hund. Kleintier-Prax. 17, 185

HARTUNG, K., H.-M. BLAUROCK und W. CLAUSS (1968): Zur Technik der Lymphographie beim Hunde. Berl. Münch. tierärztl. Wschr. 81, 254

HERSMAN, R., L. J. KLEINE, and C. E. GILMORE (1972): A clinical evaluation of propyliodone bronchography. J. Am. Vet. Radiol. Soc. 13, 27

HORVATH, A. (1983): Röntgenanatomische Untersuchungen zur postnatalen Entwicklung des Hintergliedmaßenskeletts der Hauskatze (Felis catus). München, Diss. med. vet.

HORVATH, I. (1983): Röntgenanatomische Untersuchungen zur postnatalen Entwicklung des Vordergliedmaßenskeletts der Hauskatze (Felis catus). München, Diss. med. vet.

JENSEN, H.-P. (1954): Die zerebrale Seriographie mit dem Gerät nach Buchtala. Ärztl. Wschr. 9, 468

LORD, P. F., and ST.-E. OLSSON (1976): Myelography with metrizamide in the dog: A clinical study on its use for the demonstration of spinal cord lesions other than those caused by intervertebral disc protrusions. J. Am. Vet. Radiol. Soc. 17, 42

LORD, P. F., R. C. SCOTT, and K. F. CHAN (1974): Intravenous urography for evaluation of renal diseases in small animals. J. Am. Anim. Hosp. Ass. 10, 139

NICKEL, R., A. SCHUMMER und E. SEIFERLE (1982): Lehrbuch der Anatomie der Haustiere. II. Bd. Eingeweide. 5. Aufl. Parey, Berlin-Hamburg

NICKEL, R., A. SCHUMMER und E. SEIFERLE (1984): Lehrbuch der Anatomie der Haustiere. III. Bd. Kreislaufsystem, Haut und Hautorgane. 2. Aufl. Parey, Berlin-Hamburg

NICKEL, R., A. SCHUMMER und E. SEIFERLE (1984): Lehrbuch der Anatomie der Haustiere. I. Bd. Bewegungsapparat. 5. Aufl. Parey, Berlin-Hamburg

NOMINA ANATOMICA VETERINARIA – NOMINA HISTOLOGICA (1983): Publ. by the Int. Committee on Vet. Gross Anat. Nomenclature of the World Ass. of Vet. Anat., 3rd ed. Ithaca, N. Y.

O'BRIEN, J. A., L. E. HARVEY, and J. A. TUCKER (1969): The larynx of the dog. Its normal radiographic anatomy. J. Am. Vet. Radiol. Soc. 10, 38

OSBORNE, C. A., and C. R. JESSEN (1971): Double-contrast cystography in the dog. J. Am. vet. med. Ass. 159, 1400

PATTERSON, S. A., and W. D. CARLSON (1956): Radiographic diagnostic methods. J. Am. vet. med. Ass. 128, 246

RHODES, W. H., D. F. PATTERSON, and D. K. DETWEILER (1960): Radiographic anatomy of the canine heart, part I. J. Am. vet. med. Ass. 137, 283

RHODES, W. H., D. F. PATTERSON, and D. K. DETWEILER (1963): Radiographic anatomy of the canine heart, part II. J. Am. vet. med. Ass. 143, 137

RISER, W. H. (1973): The dog as a model for hip dysplasia. Some aspects of growth, form, and development of the normal and dysplastic hip joint. Bern, Diss. med. vet.

RISER, W. H. (1975): Growth and development of the normal canine pelvis, hip joints and femurs from birth to maturity: A radiographic study. Vet. Pathol. 12, 264

ROOS, H., H. SCHEBITZ und B. VOLLMERHAUS (1979): Zur postnatalen Entwicklung der kurzen Röhrenknochen des Hundes. Berl. Münch. tierärztl. Wschr. 92, 329

ROOS, H., B. VOLLMERHAUS, H. SCHEBITZ, L. BRUNNBERG und H. WAIBL (1981): Zur Anatomie der Fugenknorpel langer Röhrenknochen des Hundes. 2. Mitteilung: Fugenknorpel des Radius und der Ulna. Kleintier-Prax. 26, 81

SALEH, M. (1964): Beitrag zum röntgenologischen Zahnstatus sowie Röntgendiagnose der Zahnkrankheiten des Hundes. München, Diss. med. vet.

SCHALLER, O. (1955): Anatomische Grundlagen der Röntgendarstellung des Hundeherzens. Acta anat. 17, 273

SCHEBITZ, H. (1963): Zur Harnblasen-Harnröhrenverletzung beim Hund. Berl. Münch. tierärztl. Wschr. 76, 21

SCHINZ, H. R., W. E. BAENSCH, E. FRIEDL und E. UEHLINGER (1952): Lehrbuch der Röntgendiagnostik, Bd. I: Skelett, Teil I; Bd. II: Skelett, Teil II; Bd. III: Innere Organe, Teil I; Bd. IV: Innere Organe, Teil II. Thieme, Stuttgart

SCHLAAFF, S. (1964): Die Ausscheidungsurographie beim Hund. Kleintier-Prax. 9, 70

SCHLAAFF, S. (1965): Zur Bronchographie beim Kleintier. Mhefte Vet-Med. 20, 63

SCHNELLE, G. B. (1950): Radiology in small animal practice. North Am. Vet., Inc., Evanston, Ill.

SCHROEDER, M. (1978): Beitrag zur Entwicklung des Skeletts der Vordergliedmaße beim Deutschen Schäferhund. München, Diss. med. vet.

SCHWILLE, P. (1958): Diagnostisches Pneumoperitoneum. Leipzig, Diss. med. vet.

SHIVELEY, M. J., and D. C. van SICKLE (1982): Developing coxal joint of the dog: Gross morphometric and pathologic observations. Am. J. vet. Res. 43, 185

SILVERMAN, S., and P. F. SUTER (1975): Influence of inspiration and expiration on canine thoracic radiographs. J. Am. vet. med. Ass. 166, 502

SISSON, S., and J. D. GROSSMAN (1953): The anatomy of the domestic animals. 5th ed. Vol. 2. W. B. Saunders Comp., Philadelphia – London – Toronto

SUTER, P. F., J. P. MORGAN, T. A. HOLLIDAY, and T. R. O'BRIEN (1971): Myelography in the dog: Diagnosis of tumors of the spinal cord and vertebrae. J. Am. Vet. Radiol. Soc. 12, 29

STUTZ, E., und H. VIETEN (1955): Die Bronchographie. Thieme, Stuttgart

THRALL, D. E., R. E. LEWIS, M. A. WALKER, S. K. KNELLER, and J. M. LOSONSKY (1975): The basis for dosing water soluble myelographic medium for lumbar administration: Body weight or crown rump length. J. Am. Vet. Radiol. Soc. 16, 130

TICER, J. W. (1975): Radiographic technique in small animal practice. W. B. Saunders Comp., Philadelphia–London–Toronto

VOLLMERHAUS, B., H. SCHEBITZ und H. ROOS (1981): Über die Entwicklung der Insertio ligamenti patellae beim wachsenden Hund. Berl. Münch. tierärztl. Wschr. 94, 255

VOLLMERHAUS, B., H. SCHEBITZ, H. ROOS, L. BRUNNBERG, J. KLAWITTER-POMMER und H. WAIBL (1981): Zur Anatomie der Fugenknorpel langer Röhrenknochen des Hundes. 1. Mitteilung: Einleitung und Fugenknorpel des Humerus. Kleintier-Prax. 26, 75

VOLLMERHAUS, B., H. SCHEBITZ, H. WAIBL, R. KÖSTLIN, J. KLAWITTER-POMMER und H. ROOS (1981): Zur Anatomie der Fugenknorpel langer Röhrenknochen des Hundes. 4. Mitteilung: Fugenknorpel der Tibia und der Fibula und Schlußdiskussion. Kleintier-Prax. 26, 95

WACK, P. (1958): Untersuchungen zur Technik der Bronchographie bei Hunden. München, Diss. med. vet.

WAIBL, H., B. VOLLMERHAUS, H. SCHEBITZ, U. MATIS und H. ROOS (1981): Zur Anatomie der Fugenknorpel langer Röhrenknochen des Hundes. 3. Mitteilung: Fugenknorpel des Os femoris. Kleintier-Prax. 26, 89

WALTER, P. (1956): Die Röntgenstereoskopie nach der Methode von Prof. Dr. A. Hasselwander in der Veterinäranatomie mit Untersuchungen an Herz und Zwerchfell des Hundes. Anat. Anz. 103, 38

WIDMER, W. (1978): Beitrag zur Entwicklung des Skeletts der Hintergliedmaße beim Deutschen Schäferhund. München, Diss. med. vet.

ZONTINE, W. J. (1975): Canine dental radiology: Radiographic technique development, and anatomy of the teeth. J. Am. Vet. Radiol. Soc. 16, 75